AF710151

ASCE Manuals and Reports on Engineering Practice No. 69

Sulfide in Wastewater Collection and Treatment Systems

Prepared by the
Sulfide Task Group of the
Water Pollution Management Committee
of the Environmental Engineering Division
of the American Society of Civil Engineers

Published by the
American Society of Civil Engineers
345 East 47th Street
New York, New York 10017-2398

ABSTRACT

ASCE Manual 69, Manual of Practice on Sulfide in Wastewater Collection and Treatment Systems, is intended to help planners, designers, and operators of wastewater facilities understand sulfide problems and methods of quantifying and controlling them. It provides a compilation of theoretical and practical information based on current scientific understanding and state-of-the-art engineering practice. It also includes a review of the physical and chemical characteristics and facility design parameters to sulfide generation and release. The odorous and corrosive effects of sulfide and the safety considerations for hydrogen sulfide are explained. Practical methods of evaluating and resolving sulfide problems in existing collection systems are presented, as are methods of designing new collection systems to minimize such problems. Example problems and case histories have been included to help engineers apply this information to field problems. Key sulfide-related problems discussed in this manual include: (1) Corrosion of facilities; (2) odor nuisance; (3) health hazards; and (4) treatment plant upsets. The design procedures to correct or prevent these problems are described in Chaps. 9 through 12.

Library of Congress Cataloging-in-Publication Data

Sulfide in wastewater collection and treatment systems/prepared by the Sulfide Task Group of the Water Pollution Management Committee of the Environmental Division of American Society of Civil Engineers.

p. cm.—(ASCE manuals and reports on engineering practice; no. 69)

Includes bibliographies and index.

ISBN 0-87262-681-4

1. Sewerage—Corrosion—Handbooks, manuals, etc. 2. Sewage disposal plants—Corrosion—Handbooks, manuals, etc. 3. Sewerage—Odor control—Handbooks, manuals, etc. 4. Sewage disposal plants—Odor control—Handbooks, manuals, etc. 5. Hydrogen sulphide—Handbooks, manuals, etc. I: American Society of Civil Engineers. Committee on Water Pollution Management. Sulfide Task Group. II. Series.

TD691.S84 1989 89-177
626.3—dc19 CIP

The material presented in this publication has been prepared in accordance with generally recognized engineering principles and practices, and is for general information only. This information should not be used without first securing competent advice with respect to its suitability for any general or specific application.

The contents of this publication are not intended to be and should not be construed to be a standard of the American Society of Civil Engineers (ASCE) and are not intended for use as a reference in purchase specifications, contracts, regulations, statutes, or any other legal document.

No reference made in this publication to any specific method, product, process, or service constitutes or implies an endorsement, recommendation, or warranty thereof by ASCE.

ASCE makes no representation or warranty of any kind, whether express or implied, concerning the accuracy, completeness, suitability or utility of any information, apparatus, product, or process discussed in this publication, and assumes no liability therefor.

Anyone utilizing this information assumes all liability arising from such use, including but not limited to infringement of any patent or patents.

Authorization to photocopy material for internal or personal use under circumstances not falling within the fair use provisions of the Copyright Act is granted by ASCE to libraries and other users registered with the Copyright Clearance Center (CCC) Transactional Reporting Service, provided that the base fee of $1.00 per article plus $.15 per page is paid directly to CCC, 27 Congress Street, Salem, MA 01970. The identification for ASCE Books is 0-87262/88. $1 + .15. Requests for special permission or bulk copying should be addressed to Reprints/Permissions Department.

Copyright © 1989 by the American Society of Civil Engineers, All rights reserved.
Library of Congress Catalog Card No. 89-177
ISBN 0-87262-681-4
Manufactured in the United States of America

MANUALS AND REPORTS ON ENGINEERING PRACTICE

(As developed by the ASCE Technical Procedures Committee, July 1930, and revised March 1935, February 1962, and April 1982.)

A manual or report in this series consists of an orderly presentation of facts on a particular subject, supplemented by an analysis of limitations and applications of these facts. It contains information useful to the average engineer in his everyday work, rather than the findings that may be useful only occasionally or rarely. It is not in any sense a "standard," however; nor is it so elementary or so conclusive as to provide a "rule of thumb" for nonengineers.

Furthermore, material in this series, in distinction from a paper (which expresses only one person's observations or opinions), is the work of a committee or group selected to assemble and express information on a specific topic. As often as practicable the committee is under the general direction of one or more of the Technical Divisions and Councils, and the product evolved has been subjected to review by the Executive Committee of that Division or Council. As a step in the process of this review, proposed manuscripts are often brought before the members of the Technical Divisions and Councils for comment, which may serve as the basis for improvement. When published, each work shows the names of the committee by which it was compiled and indicates clearly the several processes through which it has passed in review, in order that its merit may be definitely understood.

In February 1962 (and revised in April 1982) the Board of Direction voted to establish:

> A series entitled 'Manuals and Reports on Engineering Practice,' to include the Manuals published and authorized to date, future Manuals of Professional Practice, and Reports on Engineering Practice. All such Manual or Report material of the Society would have been refereed in a manner approved by the Board Committee on Publications and would be bound, with applicable discussion, in books similar to past Manuals. Numbering would be consecutive and would be a continuation of present Manual numbers. In some cases of reports of joint committees, bypassing of Journal publication may be authorized.

AVAILABLE* MANUALS AND REPORTS OF ENGINEERING PRACTICE

Number

10 Technical Procedure for City Surveys
13 Filtering Materials for Sewage Treatment Plants
14 Accommodation of Utility Plant Within the Rights-of-Way of Urban Streets and Highways
31 Design of Cylindrical Concrete Shell Roofs
33 Cost Control and Accounting for Civil Engineers
34 Definitions of Surveying and Associated Terms
35 A List of Translations of Foreign Literature on Hydraulics
36 Wastewater Treatment Plant Design
37 Design and Construction of Sanitary and Storm Sewers
40 Ground Water Management
41 Plastic Design in Steel—A Guide and Commentary
42 Design of Structures to Resist Nuclear Weapons Effects
44 Report on Highway and Bridge Surveys
45 Consulting Engineering—A Guide for the Engagement of Engineering Services
46 Report on Pipeline Location
47 Selected Abstracts on Structural Applications of Plastics
49 Urban Planning Guide
50 Report on Small Craft Harbors
51 Survey of Current Structural Research
52 Guide for the Design of Steel Transmission Towers
53 Criteria for Maintenance of Multilane Highways
54 Sedimentation Engineering
55 Guide to Employment Conditions for Civil Engineers
56 Subsurface Investigation for Design and Construction of Foundations of Buildings
57 Operation and Maintenance of Irrigation and Drainage Systems
58 Structural Analysis and Design of Nuclear Plant Facilities
59 Computer Pricing Practices
60 Gravity Sanitary Sewer Design and Construction
61 Introductory Manual on Computer Services
62 Existing Sewer Evaluation and Rehabilitation
63 Structural Plastics Design Manual
64 Manual on Engineering Surveying
65 Construction Cost Control
66 Structural Plastics Selection Manual
67 Wind Tunnel Model Studies of Buildings and Structures
68 Aeration—A Wastewater Treatment Process
69 Sulfide in Wastewater Collection and Treatment Systems

*Numbers 1, 2, 3, 4, 5, 6, 7, 8, 9, 11, 12, 15, 16, 17, 18, 19, 20, 21, 22, 23, 24, 25, 26, 27, 28, 29, 30, 32, 38, 39, 43, and 48 are out of print.

Task Group Members:

* Kenneth K. Kienow, P.E., Chairman
  Gary L. Amy
* Curtis Bryant, P.E.
* Dan Glasgow, P.E.
  Howard Hoffman, P.E.
  Jack Keller, P.E.
  William Knopf, P.E.
  Richard Lanyon, P.E.
  J. A. Oleszkiewicz, P. Eng.
  Richard D. Pomeroy, P.E.
  Manuel Ponte, P.E.
  Clifford W. Randall
* John Redner, P.E.
  Thomas Rolen, P.E.
  Ernest Rogers, P.E.
* Perry Schafer, P.E.
  Sterling Schultz, P.E.
* George Tchobanoglous, P.E.

* Blue Ribbon Review Panel Member

MANUAL OF PRACTICE ON SULFIDE IN WASTEWATER COLLECTION AND TREATMENT SYSTEMS

TABLE OF CONTENTS

Chapter 5: Effects Of Sulfide In Wastewater Systems

Chapter 6: Predicting Sulfide And Corrosion

CHAPTER 1

INTRODUCTION

A. BACKGROUND

Sulfide (defined in this manual as inorganic divalent sulfur) has been recognized as a major contributor to objectionable odors and corrosion at wastewater collection and treatment facilities. The most prevalent form of sulfide associated with problems of odors and corrosion in wastewater facilities is hydrogen sulfide, which also poses a safety hazard to plant personnel. Sulfide has been the cause of countless odor complaints, has cost industry and utility companies millions of dollars in corrosion damages, and has caused serious illness or death to many wastewater facility personnel.

The odor, corrosion, and safety problems associated with sulfide in wastewater have been known for decades, and have in recent years become a major concern to the wastewater industry. Long-term exposure to sulfide-laden air has resulted in extensive deterioration of collection and treatment facilities, and has necessitated major capital expenditures for remedial measures. Expanding residential development, accompanied by increased public sensitivity, has sharpened the attention focused on wastewater odors. The serious safety hazard associated with hydrogen sulfide is further emphasized by the number of fatalities among wastewater system employees each year. Hydrogen sulfide gas is one of the leading causes of death among workers in sanitary sewer systems.

B. PURPOSE

This Manual of Practice is intended to help planners, designers, and operators of wastewater facilities understand sulfide problems and the methods of quantifying and controlling them. It provides a compilation of theoretical and practical information based on current scientific understanding and state-of-the-art engineering practice. It also includes a review of the physical and chemical characteristics of sulfide, and a discussion of the relationship of wastewater characteristics and facility design parameters to sulfide generation and release. The odorous and corrosive effects of sulfide and the safety considerations for hydrogen sulfide are explained. Practical methods of evaluating and resolving sulfide problems in existing collection systems are presented, as are methods of designing new collection systems to minimize such problems. Example problems and case histories have been included to help engineers apply this information to field problems.

Key sulfide-related problems discussed in this manual include: (1) Corrosion of facilities; (2) odor nuisance; (3) health hazards; and (4) treatment plant upsets. Table 1-1 shows the relationship of various sections of this manual to particular sulfide problems. The design procedures to correct or prevent these problems are described in Chaps. 9 through 12.

TABLE 1-1 Overview of Sulfide Problems Considered in This Manual

Report section (1)	Corrosion (2)	Odor (3)	Safety (4)	Process problems (5)
2 Chemistry of Sulfides B. Sulfur Compounds C. Hydrogen Sulfide: Properties and Characteristics D. Biological Transformations Involving Sulfur	Corrosive properties subsection Oxidation subsection	Organic sulfur compounds subsection Odor subsection	Toxicity sub-section	
3 Wastewater Characteristics E. Wastewater Composition		Table 3-1	Table 3-1 (Toxicity)	
4 Occurrence of Sulfide in Wastewater B. Sulfide Generation in Sewers	Sewer slope subsection Grit and debris subsection Sewer size subsection Surcharging subsection	Sewer slope subsection Sewer size subsection		Solids deposits in sewers, Fig. 4-3 Highly variable influence sulfide, Fig. 4-3
D. Wastewater Pump Stations		Influent sulfide subsection		
E. Wastewater Treatment Facilities	Throughout	Throughout		
F. Industrial Wastes		High sulfide wastes subsection		
5 Effects on Wastewater Systems B. Interceptor Systems	Corrosion of cementitious materials subsection Corrosion of appurtenances subsection	Odor subsection	Toxic H_2S levels subsection	
C Treatment Plants	Corrosion of cementitious materials subsection Corrosion of equipment subsection	Odor subsection		Sludge bulking discussion
6 Predicting Sulfide and Corrosion	Sec. B General Sec. C Corrosion evaluation Sec. D8 Corrosion calculations Sec. D9 Peak corrosion rates	Sec. D13 Units of measure used in odor control		

TABLE 1-1 Overview of Sulfide Problems Considered in This Manual

Report section (1)	Corrosion (2)	Odor (3)	Safety (4)	Process problems (5)
7 Reliability of Models	Sec. 7B Sec. C			
8 Evaluating Existing Systems	Sec. C2 Physical inspection of manholes		Sec. C2 Physical inspection of manholes	
9 Designing Sewerage Systems			Sec. E2 Wetwell safety	
10 Design of Corrosion Control	Entire chapter		Sec. 10H2 Manhole safety	Imposed flow variability, Sec. 10G
11 Odor Control		Entire chapter		
12 Design of Sulfide Control	Sec. C Corrosion resistant materials	Sec. D Odor control of gas systems		
13 Safety Aspects			Entire chapter	

C. REFERENCES

Engineering and scientific literature contains many books and articles concerned with sulfide. They cover the full range from theoretical discussions to sulfide control in the field. A computerized literature search of the Engineering Index revealed over 8,000 references to the key word "sulfide" for the years 1970 through 1981. An attempt to narrow the Engineering Index list was made by using a combination of the key words sulfide and wastewater; this produced over 7,000 references. These examples make two points. First, despite all the information available, there are still large unknown areas, such as unknown reactions, occurrence of unknown intermediates, and uncertainty as to when sulfide will be a problem. The choice of the most positive and cost-effective method for control is sometimes difficult. A great need remains for more research and field studies on sulfide. The second point is that it will be impossible to list all of the significant publications in this manual. An attempt has been made to list in each chapter bibliography most of the major works dealing with sulfide in sewers.

Users of this manual should be aware that interpretation of sulfide data is often difficult because several different data bases are used in reporting. In the literature, sulfide concentrations are reported in units of milligrams per liter (mg/L), in parts per million (ppm) on a weight basis, and in ppm on a volume basis. Often ppm is used without specifying whether it is based on weight or volume. Whenever ppm is used, the report must specify "ppm (w/w)" when the data are expressed as a weight ratio, and "ppm (v/v)" when a volume ratio is used. Convention will allow "ppm" to be a weight ratio and "ppmv" to be a volume ratio. Also, environmental engineering publications sometimes convert data in terms of mg/L to ppm on the assumption that the terms are equivalent. This assumption may result in extremely erroneous values. For further information on units of measurement and conversion factors, see Appendix C.

CHAPTER 2

SULFIDE: PHYSICAL, BIOLOGICAL, AND CHEMICAL CHARACTERISTICS

A. INTRODUCTION

The literature associated with sulfur, its various chemical forms, and the many biological and chemical transformations involved in the sulfur cycle is vast. What follows is a brief description of: (1) Various sulfur compounds; (2) important properties and characteristics of hydrogen sulfide; (3) biological transformations involving sulfur; (4) chemical oxidation of sulfide; and (5) chemical precipitation of hydrogen sulfide. An extensive reference list has been included so that individual topics may be pursued in greater depth.

B. SULFUR COMPOUNDS

Domestic and industrial wastewater normally contains one or several species of sulfur compounds. There are at least 30 possible ionic and molecular inorganic species; however, only six are thermodynamically stable at room temperature: bisulfate (HSO_4^-), sulfate (SO_4^{2-}), sulfur (S^0), hydrogen sulfide (H_2S), hydrosulfide (HS^-), and sulfide (S^{2-}) (23). Thiosulfates, polysulfides, and polythionates are found in nature; however, they are relatively unstable and are generally not found in significant concentrations. Sulfites (SO_3^{2-}) are found in some industrial wastewaters. Inorganic sulfur is incorporated into organic cellular material and is described as a part of the sulfur cycle (10, 11).

Many of the organic sulfur compounds, such as dimethyl sulfide, are volatile and are characterized by unpleasant odors. The organic sulfur compounds are made up of three major groups: (1) The thiols or mercaptans, which contain an R-SH group or radical; (2) the thioethers, which have two organic radicals joined to one sulfur atom R_1-S-R_2; and (3) the disulfides, which have two connecting sulfur atoms joined with organic radicals on each end, R_2-S-S-R_2 (22). The sulfur cycle is highly dependent on bacterial action, which produces a mixture of oxidized and reduced sulfur species as well as various organic sulfides in wastewater.

C. HYDROGEN SULFIDE: PROPERTIES AND CHARACTERISTICS

Most odor, toxicity, and corrosion problems in sewerage systems are caused by gaseous hydrogen sulfide. Hydrogen sulfide is a colorless, flammable, poisonous gas with the odor of rotten eggs. The ability of humans to detect hydrogen sulfide in air is affected by both the concentration and the length of exposure. Under controlled conditions, H_2S can be detected by humans in concentrations as low as 0.21 ppb (parts per billion) on a volume basis (v/v) (13). Hydrogen sulfide gas will burn in air with a pale blue flame, and has a lower explosive limit of 4.3% by volume and an upper limit of 46% by volume. It is slightly heavier than air at the same temperature and pressure, and has a specific gravity of 1.192 (16, 19).

TABLE 2-1. Henry's Law Constants for Hydrogen Sulfide[a]

Temperature, in °C	$K_{H_2S} \times 10^5$ atm^{-1}
0	373.1
10	272.5
15	236.5
20	207.0
25	183.4
30	164.2
40	134.2
50	113.1
60	97.1

[a] Adapted from Ref. 20.

In the following paragraphs, solubility, toxicity, odor, and corrosive properties of H_2S are reviewed.

1. Solubility

To discuss the solubility of hydrogen sulfide gas, it is useful to consider the dissociation reactions for H_2S in water, which are, for the first dissociation

$$H_2S \rightarrow HS^- + H^+ \tag{2-1}$$

and for the second dissociation

$$HS^- \rightarrow S^{-2} + H^+ \tag{2-2}$$

The corresponding equilibrium expressions are

$$\frac{[HS^-][H^+]}{[H_2S]} = K_{H_2S} = 1.12 \times 10^{-7}\ (25^\circ C) \tag{2-3}$$

$$\frac{[S^{-2}][H^+]}{[HS^-]} = K_{HS^-} = 1.0 \times 10^{-15}\ (25^\circ C) \tag{2-4}$$

The products of the equilibrium expressions are given in terms of moles per liter, and low concentrations are assumed with activity coefficient f equal to unity. This assumption is warranted because of the known low solubility of hydrogen sulfide gas. Although, as shown, a second stage of dissociation can occur, essentially no sulfide (S^{2-}) is formed below a pH of 12, which is well above the normal pH of water and wastewaters.

TABLE 2-2. Solubility of Hydrogen Sulfide[a]

Temperature, in °C	H_2S, in milligrams per liter
0	7,100
5	6,040
10	5,160
15	4,475
20	3,925
25	3,470
30	3,090
40	2,520
60	1,810

[a]Computed using the Henry's Law constants given in Table 2-1.

TABLE 2-3. Percentages of H_2S and HS^- and Solubility of Hydrogen Sulfide as a Function of pH

pH (1)	Percentage of H_2S (2)	Percentage of HS^- (3)	Solubility,[a] mg/L (4)
4	99.9	0.1	3,470
5	98.9	1.1	3,510
6	90.1	9.9	3,840
7	47.7	52.3	7,270
7.5	22.5	77.5	15,400
8	8.3	91.7	41,800[a]
8.5	2.80	97.20	124,000[a]
9	0.89	99.11	390,000[a]
10	0.09	99.91	

[a] Solubility, neglecting ionic strength correction, at 25°C.

Because hydrogen sulfide is a relatively insoluble gas, its solubility can be described by Henry's Law, which usually is given as

$$x_g = K_H P_g \tag{2-5}$$

in which x_g = mole fraction of gas at equilibrium in the liquid phase; K_H = Henry's Law constant, in atm^{-1}; and, P_g = partial pressure of gas in atmosphere, in atm. Values of Henry's Law constant for hydrogen sulfide are given in Table 2-1.

The relationship between temperature and the solubility of hydrogen sulfide (as H_2S) in pure water in contact with pure hydrogen sulfide gas at a pressure of 1 atmosphere is reported in Table 2-2 (3).

For pH values in the range from 4 to 10, the relative percentage of H_2S and HS^- present in a solution and the solubility of hydrogen sulfide gas in contact with 1 atmosphere of pure hydrogen sulfide are shown in Table 2-3 (3).

The percentage distribution values given in Table 2-3 can be found by noting that the percentage of H_2S is given by

$$H_2S, \% = \frac{[H_2S]\ 100}{[H_2S] + [H_2S^-]} \tag{2-6}$$

Solving for HS^- in Eq. 2-3, and substituting the result in Eq. 2-6, yields the following expression which can be solved in terms of the hydrogen ion concentration:

$$H_2S, \% = \frac{100}{1 + K_{H_2S}/[H^+]} \tag{2-7}$$

As shown in Table 2-3, the molar fraction of hydrogen sulfide in solution is dependent on the pH of the solution, assuming there are no other reactions, such as

TABLE 2-4. Effects of Hydrogen Sulfide Gas on Humans[a]

Concentration, in parts per million by volume	Effect
<0.00021	olfactory detection threshold
0.00047	olfactory recognition threshold
0.5 to 30	strong odor
10 to 50	headache, nausea, and eye, nose, and throat irritation
50 to 300	eye and respiratory injury
300 to 500	life threatening (pulmonary edema)
700 or more	immediate death

[a] Adapted from Refs. 12, 13, 18, 29

precipitation with a metal ion. Further, the percentage of hydrogen sulfide that is ionized increases with pH. At pH 4, the percentage of HS^- formed is only 0.1, with 99.9% of the sulfide present as molecular hydrogen sulfide.

Because air contains essentially no hydrogen sulfide, the gas can be removed from water by stripping at low pH levels. Although raw wastewater normally does not have a pH of 8, it is possible to reduce or eliminate odors by raising the pH (see Table 2-3). This rise in pH shifts the equilibrium to the HS^- ion and decreases the hydrogen sulfide concentration available for stripping.

2. Toxicity

Hydrogen sulfide is an extremely toxic substance, and workers must avoid exposure to it. Collapse, coma, and death from respiratory failure may occur within a few seconds (15, 18). The presence of hydrogen sulfide and an oxygen deficiency have been reported to be the two leading atmospheric causes of death for people who work in sewers and associated structures (9). However, it has been pointed out that death due to hydrogen sulfide may be attributed to lack of oxygen, and that absence of oxygen rarely occurs in sewers (21). The effects of hydrogen sulfide gas on humans at different concentrations are reported in Table 2-4.

Concentrations of 20 to 150 ppm (v/v) are reported to cause irritation of the eyes, and slightly higher concentrations cause irritation of the upper respiratory tract (29). Similar problems have been reported at concentrations above 10 ppm (v/v) (21). Higher concentrations of hydrogen sulfide affect the nervous system, with 30-minute exposure to 500 ppm (v/v) resulting in headache, dizziness, excitement, staggering gait, diarrhea, and dysuria (25). The reported concentrations at which headache, dizziness, and other neurological manifestations occur vary significantly. For example, unconsciousness has occurred at hydrogen sulfide concentrations ranging from 230 to 1,000 ppm (v/v) with a 25-minute exposure or less (21). If a person overcome by hydrogen sulfide is not promptly hospitalized, death will be almost certain. At lower concentrations, the gas tends to cause depression; at higher concentrations, it acts as a stimulant. At very high concentrations, the hydrogen sulfide causes paralysis of the respiratory center.

It is also reported that hydrogen sulfide does not combine with the hemoglobin in the blood, but that its toxic action is entirely due to paralysis of the respiratory system (21). Exposure to concentrations of 800 to 1,000 ppm (v/v) may be fatal in 30 minutes or less; higher concentrations are instantly fatal (21, 25). Thus, the concentration of hydrogen sulfide which will cause death is not well defined, and lower concentrations will cause death if unconsciousness occurs and the victim is not removed from the contaminated area. Rescuers removing an unconscious victim from a hydrogen sulfide atmosphere must use proper protective equipment to avoid being overcome by the gas themselves.

3. Odor

Hydrogen sulfide is an insidious gas because it causes the sense of smell to fail after initial exposure. This happens quickly at the more dangerous concentrations, above about 50 ppm (v/v). Therefore, headache, dizziness, nausea, and lassitude may occur without the victim's realizing that he or she is being exposed to dangerous concentrations of hydrogen sulfide.

The olfactory threshold has been measured as low as the concentrations shown in Table 2-4. Other researchers have measured the olfactory threshold as much as 1 to 2 orders of magnitude higher. The reasons for such large variability include the different techniques and degree of controls used in testing. Also, some researchers do not indicate whether the threshold measured was "recognition," i.e., being able to recognize the odorant as hydrogen sulfide, or "detection," i.e., the lowest concentration at which the respondent detects the slightest perception of an odor. At the detection threshold, the specific odorant cannot normally be recognized or identified. Also, variability is associated with different olfactory responses in people.

4. Corrosive Properties

Hydrogen sulfide gas is corrosive to iron, steel, and copper. Therefore, selection of materials used for equipment that will be in contact with hydrogen sulfide must include consideration of its corrosive nature. It is also a major cause of corrosion in collection system and treatment plant structures, and in concrete, asbestos cement, and cast or ductile iron sewer lines. Hydrogen sulfide is oxidized aerobically to sulfate by the genus *Thiobacillus*, which forms sulfuric acid. The sulfuric acid attacks concrete, asbestos cement, or mortar linings by dissolving the constituents of Portland cement and forming calcium sulfate. The chemical and bacteriological process of concrete corrosion caused by hydrogen sulfide is considered in detail in the latter chapters of this manual. Representative literature citations include Refs. 12, 17, 19, and 24.

D. BIOLOGICAL TRANSFORMATIONS INVOLVING SULFUR

The purpose of this section is to introduce the reader to the sulfur cycle and to some of the many biological transformations that are involved with sulfur and compounds containing sulfur. Specific emphasis is given to the reactions involving hydrogen sulfide.

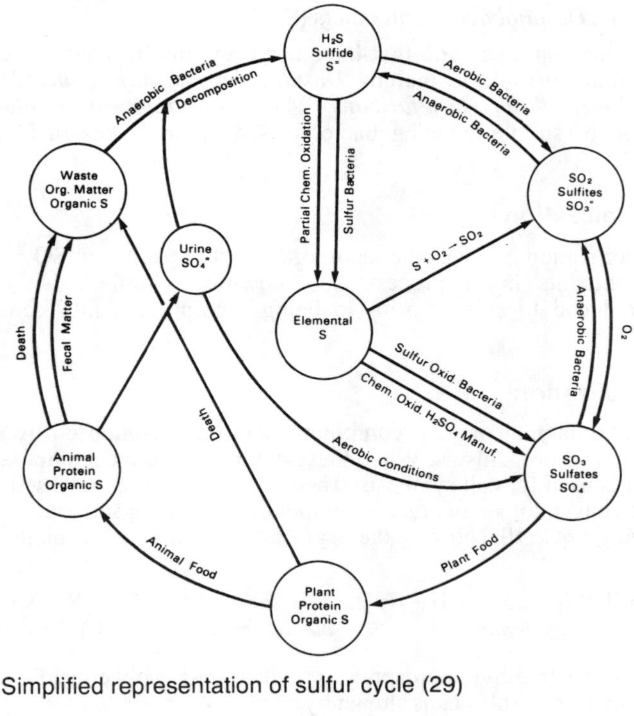

Figure 2-1. Simplified representation of sulfur cycle (29)

1. The Sulfur Cycle

A simplified form of the sulfur cycle is shown in Fig. 2-1. The principal biological transformations that are involved, starting with SO_4^{2-} are: reduction, immobilization, mineralization, and oxidation (2, 10, 11, 14, 27, 28). Each of these transformations is considered in the following subsections. Oxidation in sewers is also described.

2. Bacterial Reduction

The bacterial reduction of SO_4^{2-} to H_2S is known as dissimilatory or respiratory SO_4^{2-} reduction. This reduction occurs when conditions favor the growth of anaerobic bacteria.

In dissimilatory reduction, the sulfate acts as the terminal electron acceptor. Electron donors include organic acids, fatty acids, and alcohols. Some anaerobic organisms are also able to use H_2 as an electron donor. Typical reactions illustrating the biological reduction of SO_4^{2-} to sulfide using different electron donors are as follows (10,14):

$$2CH_3\,CHOH + SO_4^{2-} \rightarrow 2CH_3\,COOH + S^{2-} + 2H_2O + CO_2 \qquad (2\text{-}8)$$

$$4H_2 + SO_4^{2-} \rightarrow S^{2-} + 4H_2O \qquad (2\text{-}9)$$

The principal group of organisms involved in the transformation represented by Eq. 2-8 include *Desulfovibrio*, *Desulfomonas*, and *Desulfotomaculum* (2, 27). Organisms involved in the transformation represented by Eq. 2-9 include *Desulfobulbus*, *Desulfobacter*, and others (2, 27).

Other anaerobic bacteria that have been reported to have the capability of reducing sulfate to sulfide include *Desulfovibrio desulfuricans, Desulfovibrio salexigens, Vibrio thermodesulfuricans* (24), and *Spirovibrio desulfuricans* (23). Identification of sulfate-reducing bacteria as a group is described in *Standard Methods* (26).

3. Immobilization

Immobilization involves the incorporation of the sulfur in SO_4^{2-} into cellular organic constituents. In this process, also known as assimilatory SO_4^{2-} reduction, sulfate is reduced to sulfide prior to being incorporated into cellular organic compounds (10).

4. Mineralization

Sulfur-containing organic compounds can be decomposed by a variety of heterotrophic microorganisms. When these compounds are decomposed, a portion of the sulfur is used for cell synthesis. The remainder is often released as hydrogen sulfide. The conversion of an organic compound containing sulfur can be illustrated by considering Eq. 2-10, showing the degradation (reduction) of methionine (10).

$$\underset{\textit{methionine}}{CH_3\,SCH_2\,CHNH_2\,COOH} + H_2\,O \rightarrow \underset{\textit{methyl mercaptan}}{CH_3SH} + NH_3 + \underset{\textit{ketobutyrate}}{CH_3CH_2\,COCOOH} \qquad (2\text{-}10)$$

In turn, the methyl mercaptan is hydrolyzed biochemically into methyl alcohol and hydrogen sulfide, as shown by

$$\underset{\textit{methyl mercaptan}}{CH_3SH} + H_2O \rightarrow \underset{\textit{methyl alcohol}}{CH_4OH} + \underset{\textit{hydrogen sulfide}}{H_2S} \qquad (2\text{-}11)$$

5. Oxidation

Chemical oxidation of sulfide to sulfate or elemental sulfur, or both, is possible. However, the chemical reaction is dependent on a number of factors described in Sec. E of this chapter, and it is much slower than biological oxidation of sulfide (22). Further, only dissolved hydrogen sulfide is reported to be chemically oxidized with oxygen; the hydrosulfide ion, HS^-, is not oxidized (3). Therefore, bacteriological oxidation reactions are important to the environmental engineer. These reactions are especially significant in sewers, where corrosion occurs at the crown as a direct result of sulfuric acid produced by bacterial oxidation of hydrogen sulfide.

Sulfur-oxidizing bacteria can use H_2S, elemental sulfur, or partially reduced oxides of sulfur as energy sources. The sulfur energy sources are converted to sulfate. Carbon dioxide is used as a carbon source for the production of new cellular protoplasm. The genus *Thiobacillus*, a strictly aerobic microorganism, is the primary organism involved in sulfide oxidation. *Thiobacillus thioparus*, *Thiobacillus denitrificans*, *Thiodoxidans*, and *Thiobacillus concretivorus* are reported capable of oxidizing sulfur, sulfide, and thiosulfates (6, 17, 24). Members of the genus *Thiobacillus* are autotrophic bacteria, and are capable of growth without the metabolism of any organic material.

6. Oxidation in Sewers

The actual reactions involved in the oxidation or conversion of sulfide to sulfate in sewer lines are complex and not completely understood. Parker has postulated that chemical reactions "fix hydrogen sulfide as thiosulfuric and polythionic acids" followed by biological oxidation of thiosulfate and polythionate to sulfur and sulfuric acid (19). Others have suggested that the overall bacterial reaction results in oxidation of reduced sulfur compounds to thiosulfate, with little explanation for the final reactions to sulfate (22).

From a study of sulfide oxidation in activated sludge, Aulenbach and Heukelekian reasoned that a combination of chemical and biological oxidation reactions is involved in the complete oxidation of sulfide to sulfate (1). They concluded that elemental sulfur may be produced from chemical oxidation reactions; also there may be direct biological oxidation of sulfide to thiosulfate. Finally, the reactions are completed by oxidation of thiosulfate to sulfate through one or several intermediates, tetrathionate possibly being one of them.

E. CHEMICAL OXIDATION OF SULFIDE

The chemical oxidation of sulfide can be accomplished with a variety of oxidants. Oxidation with oxygen, chlorine, and hydrogen peroxide are considered in the following subsections.

1. Oxidation with Oxygen

Based on experimental studies, it is known that the oxidation of sulfide with oxygen is a chain reaction that can occur over a relatively long period (5). It has been shown that the oxidation of sulfide with oxygen (and possibly with other oxidants) is affected by certain elements and organics that can either inhibit or act as catalysts for the oxidation. Some of the metals that can act as catalysts include the divalent ions of manganese, nickel, iron, cobalt, and copper (5). Glycerol, carbitol, the disinfectant Roccal, and EDTA have been reported to inhibit the oxidation of hydrogen sulfide.

Thus, although dissolved hydrogen sulfide gas may not be found in streams containing oxygen, it is possible for conditions to exist where inhibitors are present, and oxygen and sulfide may coexist for long periods when no bacterial activity occurs. Also, as a result of the different effects of catalysts and inhibitors, the rate of chemical oxidation of sulfide can be expected to vary widely between different waste streams.

There are two reasons why hydrogen sulfide gas may be limited when oxygen is present: (1) Sulfur-reducing bacteria cannot grow in the presence of oxygen, because they are strict anaerobes; and (2) hydrogen sulfide can be oxidized chemically by oxygen to either free sulfur or sulfate, as illustrated by

$$2H_2S + O_2 \rightarrow 2H_2O + 2S \tag{2-12}$$

$$H_2S + 2O_2 \rightarrow 2H^+ + SO_4^{2-} \tag{2-13}$$

Only the hydrogen sulfide gas in solution is oxidized, because oxygen will not oxidize hydrogen sulfide in the gas phase except at higher temperatures (3). The hydrosulfide ion is not oxidized by oxygen.

TABLE 2-5. Solubility Data for Selected Metal Sulfides

Metal sulfide (1)	Solubility product[a] (18°C to 25°C) (2)	Sulfide concentration, in moles per liter (3)
Manganous sulfide	1.4×10^{-15}	3.7×10^{-8}
Ferrous sulfide	3.7×10^{-19}	6.1×10^{-10}
Zinc sulfide	1.2×10^{-23}	3.5×10^{-12}
Nickel sulfide	1.4×10^{-24}	1.2×10^{-12}
Stannous sulfide	1.0×10^{-25}	3.2×10^{-13}
Cobalt sulfide	3.0×10^{-26}	1.7×10^{-13}
Lead sulfide	3.4×10^{-28}	1.8×10^{-14}
Cadmium sulfide	3.6×10^{-29}	6.0×10^{-15}
Silver sulfide	1.6×10^{-49}	3.4×10^{-17}
Copper sulfide	8.5×10^{-45}	9.2×10^{-23}
Mercuric sulfide	2.0×10^{-49}	4.5×10^{-25}

[a] From Ref. 4.

To determine the rate of oxidation of sulfide, each specific wastewater must be evaluated individually. In most cases, it is advisable not to depend on chemical oxidation of sulfide with oxygen as the sole method for elimination of sulfides in sewer lines.

2. Oxidation with Chlorine

Sulfide in solution is oxidized by chlorine as shown by

$$Cl_2 + H_2S \rightarrow 2H^+ + 2Cl^- + S^0 \qquad (2\text{-}14)$$

$$4Cl_2 + H_2S + 4H_2O \rightarrow 10\,H^+ + 18Cl^- + SO_4^{2-} \qquad (2\text{-}15)$$

Experimental studies have shown that more free sulfur is produced at higher pH values (3). Conditions for free sulfur formation include slow chlorine addition and intensive mixing (22). However, excess chlorine is normally used, and the wastewater is frequently at or slightly below neutral pH. Both of these conditions tend to favor the oxidation of the sulfide to sulfate. Other oxidation products, such as thiosulfate, trithionate, and sulfite, are likely to result from using intermediate amounts of chlorine or inadequate mixing, or both (22).

3. Oxidation with Hydrogen Peroxide

Hydrogen peroxide is an effective and powerful oxidant for destroying hydrogen sulfide. Ozone is the only commercially available oxidant that is stronger. Hydrogen peroxide reactions with sulfide are similar to those of oxygen in that they are pH-dependent. However, hydrogen peroxide is capable of oxidizing the hydrosulfide ion, which oxygen is incapable of doing. The reaction time for oxidizing the hydrosulfide ion is two to three times longer than that required to oxidize dissolved hydrogen sulfide gas. At a pH of 7 or less, sulfides are oxidized to elemental sulfur. Sulfate is the end product at pH values above 8 (8). These chemical reactions are described by

$$H_2S + H_2O_2 \rightarrow 2H_2O + S \quad (pH = 7) \qquad (2\text{-}16)$$

$$HS^- + H_2O_2 + H^+ \rightarrow 2H_2O + S \quad (pH = 7) \qquad (2\text{-}17)$$

$$H_2S + 4\,H_2O_2 \rightarrow 4H_2O + SO_4^{2-} + 2H^+ \qquad (2\text{-}18)$$

Hydrogen peroxide has an advantage over other chemical oxidants in that it decomposes to oxygen and water. In a sewer line, the oxygen remaining after oxidation of the sulfides will prevent formation of additional sulfide.

F. CHEMICAL PRECIPITATION OF SULFIDE

Sulfide removal may occur by precipitation with metals present in wastewater, such as iron, chromium, copper, zinc, nickel, or cadmium (7). The solubility of some of the different sulfides and their solubility products are shown in Table 2-5 (7).

These are given to show that wastewaters containing metals will precipitate sulfide as it is formed. Sulfide precipitation is especially important in industrial wastewaters, and in anaerobic sludge digesters where metals have been concentrated in the sludge. Of the metals shown in Table 2-5, only iron has been used as a chemical additive to intentionally remove sulfide; ferrous ions will precipitate sulfide as ferrous sulfide. Because the solubility product of ferrous sulfide is 3.7 x 10^{-19} at 18°C, the reaction

$$Fe^{2+} + S^{2-} \rightarrow FeS \quad (2\text{-}19)$$

is very effective in precipitating sulfides. Ferric ions will also precipitate sulfide through reduction of the iron to the ferrous ion and the sulfide is oxidized to sulfur. The ferrous ion is then available for direct precipitation with other sulfides as shown in

$$2Fe^{3+} + S^{2-} \rightarrow 2Fe^{2+} + S^0 \quad (2\text{-}20)$$

Several other iron sulfides may also form, including pyrrhotite, ferric sulfide (Fe_2S_3), smythite (Fe_3S_4), and FeS_2 (8).

Improved sulfide removal can be achieved through the addition of both ferric and ferrous salts. Experience in the field has shown that dissolved sulfide can be reduced to 0.2 mg/L in this manner (22). The theoretical reaction to precipitate sulfide with ferrous and ferric ions is assumed to take the form shown by

$$Fe^{2+} + 2\,Fe^{3+} + 4\,HS^- \rightarrow Fe_3S_4 + 4\,H^+ \quad (2\text{-}21)$$

With the solubility concentrations reported in Table 2-5, it should be possible to reduce the dissolved sulfide to much less than 0.2 mg/L. However, it appears that it is not possible to reduce the hydrosulfide ion in the complex solutions found in wastewater to the concentration level shown in Table 2-5.

REFERENCES

1. Aulenbach, D. B., and Heukelekian, H., "Transformation and Effects of Reduced Sulfur Compounds in Activated Sludge Treatment," *Sewage and Industrial Wastes*, Vol. 27, No. 4, April 1955, pp. 434-448.
2. Brock, T. D., Smith, D. W., and Madigan, M.T., *Biology of Microorganisms*, 4th ed., Prentice-Hall, Inc., Englewood Cliffs, NJ, 1984.
3. Camp, T. R., and Meserve, R. L., "Composition of Natural and Polluted Waters and of Liquid Wastes," *Water and Its Impurities*, 2nd ed., Dowden, Hutchinson, and Ross, Inc., Stroudsburg, PA, 1964, pp. 64-75.
4. *Chemistry and Physics Handbook*, 62nd ed., CRC Press, Inc., Boca Raton, FL, 1982.
5. Chen, K. Y., and Morris, J. C., "Oxidation of Sulfide by O_2: Catalysis and Inhibition," *Journal of the Sanitary Engineering Division,* ASCE, Vol. 98, No. SA1, Proc. Paper 8710, Feb. 1972, pp. 215-227.
6. Click, C. N., and Reed, J. C., "Atmospheric Release of Hydrogen Sulfide and Ammonia from Wet Sludges and Wastewaters," *Proceedings of the Second National Conference on Complete Water Reuse*, American Institute of Chemical Engineers and the Environmental Protection Agency Technology Transfer, May 1975, pp. 426-437.
7. *Control and Treatment Technology for the Metal Finishing Industry - Sulfide Precipitation*, EPA 625/8-30-003, Industrial Environmental Research Laboratory, Cincinnati, OH, April 1980.

8. Dohnalek, D. A., and FitzPatrick, J. A., "The Chemistry of Reduced Sulfur Species and Their Removal from Groundwater Supplies," *Journal of the American Water Works Association*, Vol. 75, No. 6, June 1983, pp. 298-308.
9. Gerardi, M. H., "A Review of Dangerous Gases in Sanitary Sewers," *Public Works*, Vol. 113, No. 10, Oct. 1982, pp. 34- 36.
10. Higgins, I. J., and Burns, R.G., *The Chemistry and Micro-biology of Pollution,* Academic Press, New York, NY, 1975.
11. Jocelyn, P. C., *Biochemistry of the SH Group*, Academic Press, New York, NY, 1972.
12. Keating, E. J., "Regular Testing Can Control Hydrogen Sulfide," *Water & Sewage Works*, Vol. 125, No. 7, July 1978, pp. 68-70.
13. Manufacturing Chemists Association, "Research On Chemical Odors, Part 1 - Odor Thresholds For 53 Commercial Chemicals," 1968.
14. Mara, D. D., *Bacteriology For Sanitary Engineers*, Churchill Livingstone, London, England, 1974.
15. McKinney, R. E., *Microbiology for Sanitary Engineers*, McGraw-Hill Book Co., Inc., New York, NY, 1962, pp. 171-172.
16. *The Merck Index*, Windholz, M., ed., 10th ed., Merck & Co., Inc., Rahway, NJ, 1983; Hydrogen Sulfide, pp. 697-698.
17. Meyer, A. H., and Ledbetter, W. B., "Sulfuric Acid Attack on Concrete Sewer Pipe," *Journal of the Sanitary Engineering Division*, ASCE, Vol. 96, No. SA5, Proc. Paper 7624, Oct. 1970, pp. 1167-1182.
18. National Institute for Occupational Safety and Health. *Health Evaluation Report - Papillion Creek Wastewater Treatment Plant, Omaha, Nebraska*, Publication No. HETA 83- 440-1537,1984.
19. Parker, C. D., "Mechanics of Corrosion of Concrete Sewers by Hydrogen Sulfide," *Sewage and Industrial Wastes*, Vol. 23, No. 12, Dec. 1951, pp. 1477-1485.
20. Perry, R. H., and Green, D., *Perry's Chemical Engineering Handbook,* 6th ed., McGraw-Hill Book Co., New York, NY, 1984.
21. Pomeroy, R. D., "Lethal Atmospheres in Sewers," *Water Pollution Control Federation Deeds and Data*, Vol. 17, No. 10, Oct. 1980, pp. 9-15.
22. *Process Design Manual for Sulfide Control in Sanitary Sewerage System*, U.S. EPA, 1974.
23. Rubin, A. J., ed., "Chemistry of Sulfur Species and Their Removal From Water Supply," *Chemistry of Water Supply, Treatment, and Distribution*, Ann Arbor Science Publishers, Inc., Ann Arbor, MI, 1974, pp. 109-135.
24. Santry, I. W., Jr., "Hydrogen Sulfide in Sewers," *Journal of the Water Pollution Control Federation,* Vol. 35, No. 12, Dec. 1963, pp. 1580-1588.
25. Sax, N. I., *Dangerous Properties of Industrial Materials*, 6th ed., Van Nostrand Reinhold Co., New York, NY, 1984, p. 1552.
26. *Standard Methods for the Examination of Water and Wastewater*, 16th ed., American Public Health Association, Washington, DC, 1985.
27. Stanier, R. Y., Ingraham, J. L., Wheels, M. L., and Painter, P. R., *The Microbial World,* 5th ed., Prentice-Hall, Englewood Cliffs, NJ, 1986.
28. Stevenson, F. J., *Cycles of Soil: Carbon, Nitrogen, Phosphorus, Sulfur, Micronutriants*, Wiley-Interscience Publication, John Wiley & Sons, New York, NY, 1986.
29. U. S. Environmental Protection Agency, *Design Manual - Odor and Corrosion Control in Sanitary Sewerage Systems and Treatment Plants*, Publication No. EPA 1600/1-85/018, 1985.

CHAPTER 3

WASTEWATER CHARACTERISTICS AND THEIR EFFECT ON SULFIDE

A. INTRODUCTION

To assess the development of hydrogen sulfide in sewerage works, it is important to understand many aspects of wastewater, such as wastewater components, flow variations, characterization, and composition. The purpose of this chapter is to describe these aspects in general, with an emphasis on wastewater characteristics of importance to the generation of hydrogen sulfide. Additional details may be found in Refs. 4, 6, 11, and 12.

B. WASTEWATER COMPONENTS

The components of flow that comprise the wastewater from a community will depend on the type of collection system, and may include:

1. Domestic wastewater (also called sanitary sewage): wastewater discharged from residences and from commercial, industrial, and similar facilities.

2. Commercial (nonsanitary) wastewater: wastewater from commercial facilities, excluding domestic wastewater.

3. Industrial (nonsanitary) wastewater: wastewater flows arising from industrial activities, excluding domestic wastewater.

4. Infiltration: the water entering a sewer system, including sewer service connections from the ground, through such means as (but not limited to) defective pipes, pipe joints, connections, or manhole walls. Infiltration does not include, and is distinguished from, inflow.

5. Inflow: the water discharged into a sewer system, including service connections, from such sources as (but not limited to) roof leaders; cellar, yard, and area drains; foundation drains; cooling-water discharges; drains from springs and swampy areas; manhole covers; and cross connections from storm sewers and combined sewers. Inflow in combined sewers (see discussion below) will include flow from catch basins, storm waters, surface runoff, street wash waters, or drainage. Inflow does not include, and is distinguished from, infiltration.

6. Storm water: water resulting from precipitation runoff.

Three types of sewer systems are used for the removal of wastewater and storm water: sanitary sewer systems, storm water sewers, and combined sewers (4). Where separate sewers are used for the collection of wastewater (sanitary sewers) and storm water (storm sewers), the wastewater will be made up of the first five components identified above. Where a single combined sewer is used, the wastewater will consist of all six of the listed components.

The particular mix of flow will depend on the type of development in the tributary service area (i.e., mix of residential, commercial, industrial), on the age and condition of the wastewater collection system, and on the time of year. In older systems that have not been maintained, infiltration and inflow may account for more than 50% of the wastewater flow rate, and even more in areas of persistent high groundwater.

C. VARIATIONS IN WASTEWATER FLOWS

Wastewater flow rates are subject to wide variation. They vary hourly, daily, weekly, yearly, and seasonally. For the purpose of this manual, it will be appropriate to consider short-term (one week or less), seasonal, and industrial variations.

1. Short-term Variations

Daily wastewater flow rates tend to follow a somewhat diurnal pattern, such as is shown in Fig. 3-1.

The two curves presented in Fig. 3-1 are representative of what would be expected for a small community and a large one. As shown, the variation between the minimum and peak flows is reduced as the total flow increases. In both cases, the lowest flows occur in the early morning hours, when water consumption is lowest. In some systems, the early morning low flows will consist primarily of leakage, infiltration and inflow, and small amounts of domestic wastewater. Two peaks will generally be observed. The first peak occurs around noon, and a second peak occurs in the early evening, usually between 7 p.m. and 9 p.m. In new systems,

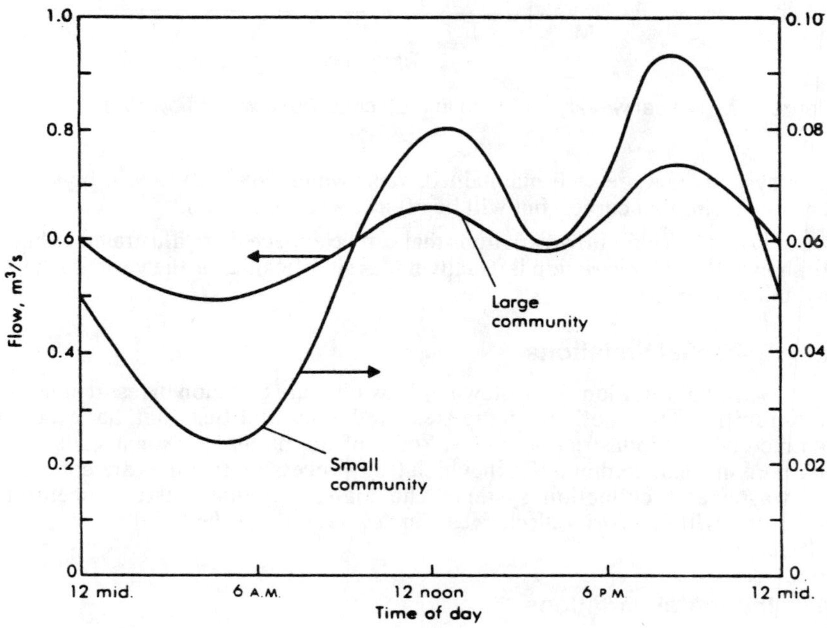

Figure 3-1. Daily variation in wastewater flow for small community and large community

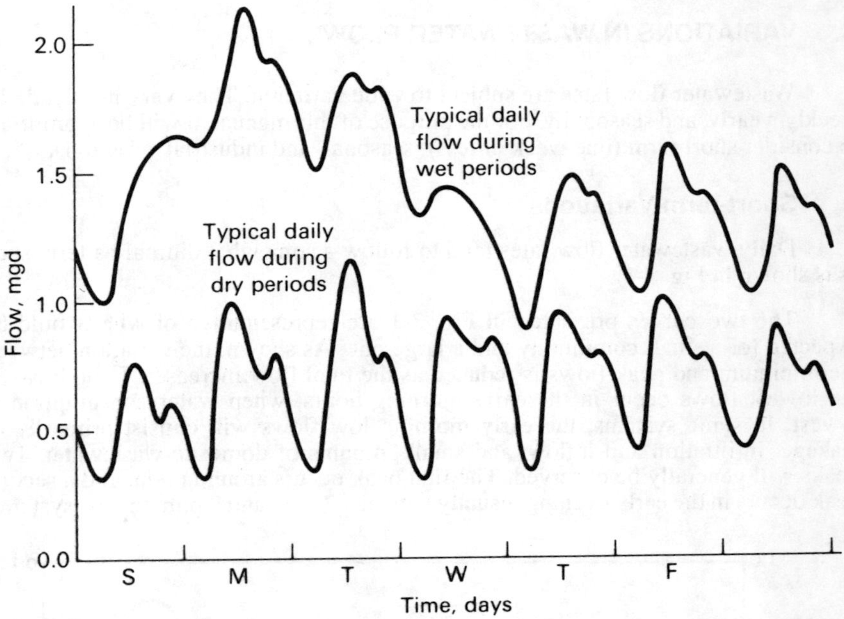

Figure 3-2. Typical weekly variation in municipal wastewater flow rates

or in systems that are well maintained, wastewater flow curves will be similar to water consumption curves, but will be offset by several hours.

Typical weekly flow variations that can be expected are illustrated in Fig. 3-2. As shown, the daily variation is usually not as significant as is the variation between dry and wet periods.

2. Seasonal Variations

Seasonal variations in wastewater flow rates are common in resort areas, small communities with college campuses, and communities that have seasonal commercial and industrial activities. Some of the highest seasonal variations are observed in small communities in which fruit-processing facilities are connected to the wastewater collection system. The higher organic waste concentrations associated with industrial discharges can severely affect the build-up of hydrogen sulfide.

3. Industrial Variations

Because many conventional industrial activities are undergoing significant change with respect to water usage, it is difficult to quantify industrial wastewater flow rates. The impact of industrial discharges is usually greatest in small communities. The discharge of wastewater at elevated temperatures is of special concern if there is a tendency for hydrogen sulfide to be generated in the collection system, since sulfide production rate increases as wastewater temperatures rise.

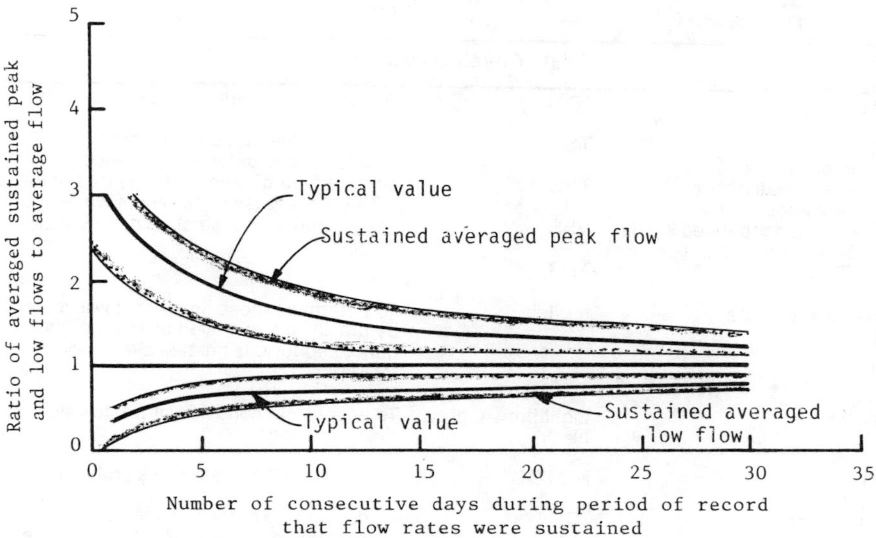

Figure 3-3. Curve for estimation of averaged sustained peak wastewater flows

4. Sustained Flows

With respect to the production of hydrogen sulfide in sewers, it is important to know how long sustained peak flows can be expected. Sustained peak flows can be estimated using the curve shown in Fig. 3.3.

D. WASTEWATER CHARACTERIZATION

The measures used to characterize wastewater or to assess wastewater quality are usually categorized as physical, chemical, and biological. The measures most commonly used are reported in Table 3-1.

As shown, the measures can be divided into five major categories: physical properties, inorganic constituents, organic constituents, gases, and biological characteristics. Wastewater quality parameters routinely measured and reported are mostly in terms of gross pollutional parameters, such as biochemical oxygen demand (BOD) and suspended solids (SS). They are of interest in water pollution control activities, and are described in more detail later in this chapter.

Perhaps one of the most difficult problems in assessing the characteristics of wastewater is sample collection and analysis. In general, three questions must be asked:

1. Where should the sample be taken?
2. How can a representative sample be taken?

TABLE 3-1 Common Tests Used to Assess Gross and Specific Characteristics of Wastewater

Test (1)	Abbreviation or Definition (2)	Use (3)
(a) Physical Properties		
Turbidity	NTU	To assess the clarity of water.
Solids: total solids total volatile solids suspended solids volatile suspended solids Total dissolved solids (TS-SS)	TS TVS SS VSS TDS	To assess the reuse potential of a wastewater and to determine the most suitable type of process for its treatment. TDS tests assess suitability of water sources for public, industrial, and agricultural uses.
Settleable solids	mL/L	To determine those solids that will settle by gravity in a specified time period; test data used for design of sedimentation facilities.
Color	Light brown, gray, black	To assess the condition of wastewater (fresh or septic).
Odor	MDTOC[b]	To determine if odors will be a problem.
Temperature	°C	To design and carry on biological and other treatment processes; to determine the saturation concentration of various gases.
(b) Organic Constituents		
Five day carbonaceous biochemical oxygen demand	BOD ($CBOD_5$)	To measure amount of oxygen required to stabilize a waste water biologically.
Ultimate carbonaceous biochemical oxygen demand	BOD_u ($CBOD_u$)	
Nitrogenous oxygen demand	NOD	To measure the amount of oxygen required to oxidize biologically the nitrogen in the wastewater to nitrate.
Chemical oxygen	COD	To measue the amount of oxygen required to stabilize the waste completely. For simple organic wastes, the COD corresponds to BOD_u.
Total organic carbon	TOC	Often used as a substitute for the BOD_t test.
Specific organic compounds		To determine presence of pesticides, phenols, surfactants, grease, and other organic materials.
(c) Gases		
Oxygen	O_2	To assess the performance of aeration facilities.
Hydrogen sulfide	H_2S	To assess the production of hydrogen sulfide and the odor potential of a wastewater.

TABLE 3-1 Common Tests Used to Assess Gross and Specific Characteristics of Wastewater (cont'd)

Test (1)	Abbreviation or Definition (2)	Use (3)
(d) Inorganic constituents		
Calcium, magnesium, sodium	Ca^{2+},Mg^{2+},Na^{+}	To determine the sodium absorption ratio where agricultural reuse is being considered.
Chloride	Cl^{-}	To assess the suitability of wastewater for agricultural reuse.
Nitrogen:		
organic nitrogen	Org N	To measure the nutrients present and the degree of decomposition in the wastewater; oxidized forms can be taken as a measure of the degree of oxidation.
free ammonia	NH_3	
ammonium	NH_4-	
nitrites	NO_2	
nitrates	NO_3	
total Kjeldahl nitrogen[c]	TKN	
Phosphorus:		
total phosphorus	TP	
organic phosphorus	Org P	
inorganic phosphorus PO_4	Inorg P	
Sulfate	$SO^2{}_4$	To assess the treatability of the waste sludge.
Sulfide:		
dissolved	H_2S, HS^- , $S^=$	
insoluble	FeS, etc.	
Alkalinity	A	To measure the buffering capacity of the wastewater.
pH	pH = log 1/(H^+)	To measure the acidity or basicity of an aqueous solution.
(e) Biological characteristics		
Heavy metals		To assess the suitability of the wastewater for reuse and for toxicity effects in treatment and on receiving waters.
Trace elements		May be important in biological treatment and for agricultural use.
Total coliform bacteria	MPN	
Fecal coliform bacteria	MPN	
Viruses	PFU[d]	

a Detailed procedure for the tests listed in this table are presented in Standard Methods (9).
b MDTOC = minimum detectable threshold odor concentration.
c Total Kjeldahl nitrogen = organic nitrogen + ammonia.
d PFU = plaque-forming unit.
Source: Adapted from Ref. 10.

Figure 3-4. Field determination of hydrogen sulfide using portable field test kit

3. What preservation measures are required to assure that reliable and accurate data are obtained? Wastewater sampling is considered in greater detail in Appendix B.

For the most part, the constituents listed in Table 3-1 are determined in the laboratory using the procedures detailed in Standard Methods (9). An important exception is hydrogen sulfide, which is often determined in the field (see Fig. 3-4).

The laboratory methods and techniques used to determine the various forms of sulfide, and the test kits available for the field measurement of hydrogen sulfide, are described in Appendix D.

E. WASTEWATER COMPOSITION

Composition refers to the actual amounts of physical, chemical, and biological constituents present in wastewater. In general, the composition of untreated wastewater depends, as noted previously, upon the composition of municipal water supply, the nature of the residential community, the number and type of commercial and industrial establishments, and the characteristics of the wastewater collection system. Thus, as expected, there is a wide variation in the composition of wastewater from community to community.

Typical data on the composition of untreated domestic wastewaters in the United States, based on 24-hour composite samples, are presented in Table 3-2.

TABLE 3-2. Typical Composition of Untreated Municipal Wastewater

Constituent (1)	Concentration Range[a] Strong (2)	Medium (3)	Weak (4)	U.S. Average[a] (5)
Solids, total:	1,200	720	350	
dissolved, total[b]	850	500	250	
fixed	525	300	145	
volatile	325	200	105	
Settleable solids, mL/L	20	10	5	
Biochemical oxygen demand, 5-day				
20°C (BOD_5 20°C)	400	220	110	181
Total organic carbon (TOC)	290	160	80	102
Chemical oxygen demand (COD)	1,000	500	250	417
Nitrogen (total as N):	85	40	20	34
Org-N	35	15	8	13
NH_3-N	50	25	12	20
NO_2-N	0	0	0	
NO_3-N	0	0	0	0.6
Phosphorous (total as P):	15	8	4	9.4
organic	5	3	1	2.6
inorganic	10	5	3	6.8
Chlorides[b]	100	50	30	
Alkalinity (as $CaCO_3$)[b]	200	100	50	211
Grease	150	100	50	
Total coliform bacteria MPN/100mL				22×10^6
Fecal coliform bacteria MPN/100mL				8×10^6
Viruses, PFU/100mL				3.6

[a] All values are expressed in mg/L, except as noted.
[b] Values should be increased by amount in domestic water supply.
Source: From Ref. 4.

As with wastewater flows, the concentration of the individual constituents found in wastewater will vary depending on the time of day, day of the month, and month of the year. Typical data on the variation of BOD and SS concentrations with the time of day are presented in Fig. 3-5.

The variation in the concentration of BOD with time must be considered when assessing the generation of hydrogen sulfide in wastewater collection systems.

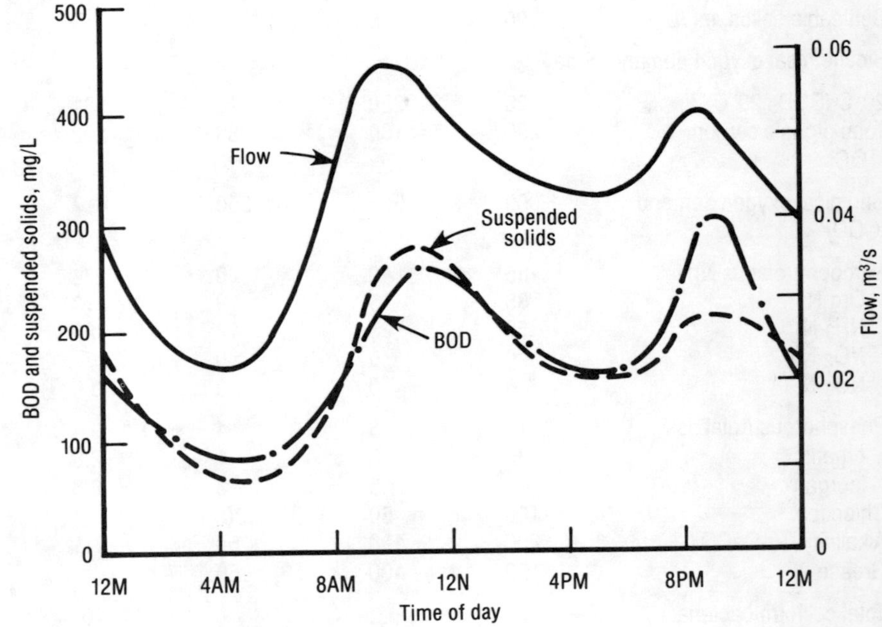

Figure 3-5. Typical variation in hourly wastewater flow, BOD and SS values for small community

Not reported in Table 3-2, the gases found in wastewater include nitrogen (N_2), oxygen (O_2), carbon dioxide (CO_2), ammonia (NH_3), hydrogen sulfide (H_2S), and methane (CH_4). The first three are the common gases of the atmosphere, and will be found in wastewater exposed to the atmosphere. Ammonia is produced as a result of bacterial metabolism under both aerobic and anaerobic conditions. Ammonia in wastewater exists almost entirely as the ammonium ion NH^+_4. Under reducing (anaerobic) conditions in wastewater, H_2S and CH_4 are produced in association with CO_2 and NH_3. As conditions become more anaerobic, hydrogen sulfide and methane will be produced sequentially and, in many cases, simultaneously.

F. WASTEWATER CHARACTERISTICS OF IMPORTANCE TO GENERATION OF HYDROGEN SULFIDE

Although all characteristics of wastewater are important, seven are of particular interest with respect to the generation of hydrogen sulfide: sulfur compounds, temperature, suspended solids particle size distribution, biochemical oxygen demand (BOD), effective biochemical oxygen demand (EBOD), pH, and dissolved oxygen (DO). Each of these characteristics is considered separately in the following subsections.

1. Sulfur Compounds

Sulfur compounds of concern in the generation and release of hydrogen sulfide include: (1) Sulfur in the oxidized form, principally present as sulfate (SO_4^{-2}); (2) sulfur in organic compounds, including proteins and amino acids; and (3) sulfur already present in the wastewater in the form of inorganic ionic sulfides (primarily HS^-, with lesser amounts of S^{-2}), and sulfide compounds, such as hydrogen sulfide (H_2S). The sulfates in wastewater are: (1) The sulfates in the water supply; (2) those added by domestic usage; (3) those added by industrial discharges; and (4) the sulfates in groundwater that may infiltrate into the collection system. Organic sulfur in wastewater is derived primarily from domestic discharges from grinder units used to dispose of kitchen wastes. Sulfide in wastewater is derived principally from the reduction of sulfate. The chemistry of the sulfur compounds is considered in greater detail in Chap. 2.

2. Temperature

The temperature of wastewater will generally be warmer than that of the water supply. Depending on the geographic location, the average temperature of wastewater will vary from 10°C to 22°C, with a representative average being 15.6°C. In some of the southwestern states, where the temperature of the water supply exceeds 26°C in the summer, wastewater temperatures as high as 30°C are common. Relatively high wastewater temperatures also are common in sewers used to collect wastes from commercial and industrial activities. For example, laundromats cause a significant increase in the local temperature of wastewater.

With respect to the presence of hydrogen sulfide in sewerage facilities, temperature is important because it affects the solubility of hydrogen sulfide, the equilibrium constants for the dissociation of hydrogen sulfide, and the rate of biochemical reaction under both aerobic and anaerobic conditions.

Temperature effects on gas solubility.—The importance of temperature on gas solubility can be assessed from a cursory review of the values given in Table 2-2. As shown in the table, there is about a 40% difference in the saturation concentration for hydrogen sulfide between 10°C and 30°C.

Temperature effects on H_2S dissociation.—Temperature will also affect dissociation of hydrogen sulfide. For example, the logarithm of the first dissociation constant for hydrogen sulfide will vary from 7.16 at 10°C to 6.88 at 30°C (12).

Temperature effects on rates of reaction.—Rates of reaction are affected significantly by temperature changes, although the stoichiometry is relatively unaffected (10). In biological reactions, the effect of temperature is accounted for by adjusting the value of the rate constant k using the following approximate equation derived from the van't Hoff-Arrhenius temperature relationship (3):

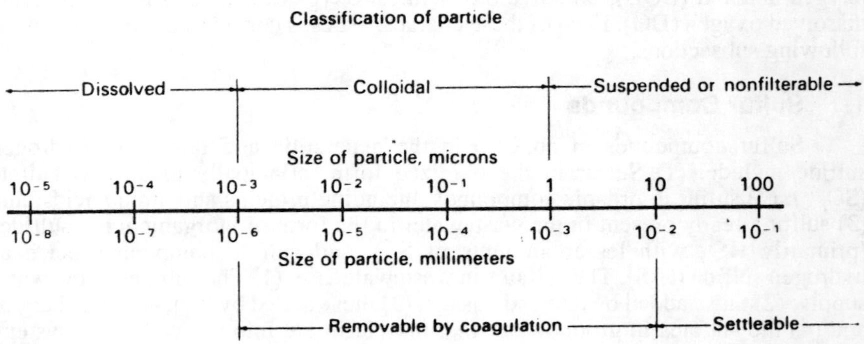

Figure 3-6. Size distribution of solids found in wastewater

$$K_{T_1} = K_{T_2}\,\theta^{(T_2 - T_1)} \tag{3-1}$$

in which K_T = rate constant at temperature T, θ = temperature coefficient, and T = temperature, °C.

3. Suspended Solids Particle Size Distribution

The solid and semisolid material in wastewater can be classified by size and state as settleable, suspended, colloidal, and dissolved (see Fig. 3-6). The size range for the various categories is somewhat arbitrary and is often related to the techniques used for their measurement. For example, settleable solids are determined using an Imhoff cone and noting the quantity of solids that have settled after one hour. Settleable solids are reported as mL/L. The difference between settleable and suspended solids is a practical one. Suspended solids do not settle out in an Imhoff cone and would not normally be removed by sedimentation. Colloidal solids are arbitrarily defined as being in the range from 0.001 μ to 1. 0 μ in size.

In practice, the solids in wastewater are divided into two broad categories: dissolved (including colloidal and small suspended particles), and suspended (including settleable). The distinction is made using a glass fiber membrane filter with a nominal pore size of about 1.2 μ. Particles passing the filter are arbitrarily classed as dissolved, even though it is known that a large fraction of the solids passing the 1.2 μ filter may be colloidal. More recently, various methods have been

TABLE 3-3. Typical Data on Distribution of Filterable Solids in Wastewater

Sample (date)[a]	Concentration[b] in mg/L	Percent of Mass Retained in Size Range in Microns[c] >0.1 <1.0	> 1.0 < 3.0	> 3.0 < 5.0	> 5.0 < 8.0	> 8.0 < 12.0	> 12.0
UCD (7/14/82)	62.2	12.5	12.9	5.8	3.8	6.1	58.8
UCD (7/21/82)	129.9	16.1	25.1	0	0	0	58.8
Las Vegas (8/3/83)	146.1	14.2	32.4	6.9	0	6.5	40.0
Las Vegas	284.0	1.8	32.6	11.5	11.1	1.8	41.2
Los Banos	268.0	20.5	18.7	6.7	3.0	10.1	41.0

[a] Samples collected at University of California, Davis, Las Vegas, Nevada, and Los Banos, California.

[b] Mass of solids retained on a filter with a pore size of 0.1 μ.

[c] $1\ \mu = 10^{-3}$ mm.

Source: Adapted from Ref. 10.

used to characterize the solids in wastewater, including serial filtration using membrane filters, particle size counters, and gel chromatography (2).

Typical data on the distribution of solids in wastewater determined using polycarbonate filters with a lower limit of 0.1 μ are reported in Table 3-3.

As shown, a significant portion of the solids that would be classified as dissolved (below 1 μ) with the conventional solids test are, in fact, colloidal and suspended in size (greater than 0.1 μ but less than 1.0 μ). Particles in the size range from 0.1 μ to 1.0 μ are important because of the amount of organic matter that may be associated with them. If, for example, all of the dissolved BOD (see following subsection) in a wastewater were associated with the solids in this range, the expected rate of BOD exertion would be quite different from that of a wastewater where the BOD was present in a dissolved form (say less than 0.00001 μ). The relationship of BOD to particle size is important, because it has been observed that the rate of hydrogen sulfide generation is greater (other things being equal) in wastewater with a high fraction of soluble BOD or chemical oxygen demand (COD).

4. Biochemical Oxygen Demand (BOD)

BOD is the amount of oxygen used in the biological metabolism of biodegradable organic compounds. In municipal wastewater, BOD values are a measure of the oxygen required at a specified time and temperature for the carbonaceous oxidation of a nonspecific mixture of soluble and nonsoluble organic compounds. Because the reported BOD values are only for the biodegradation of the carbonaceous organic material, the federal government is encouraging the use of the term carbonaceous biochemical oxygen demand (CBOD).

The principal end products of the biological oxidation of the carbonaceous organic matter in wastewater are cell tissue, carbon dioxide (CO_2), ammonia (NH_3) and water (H_2O). In equation form, the BOD reaction can be represented as follows:

$$\underset{\substack{\text{organic}\\ \text{matter}}}{\text{COHN}} + \text{nutrients} + \underset{\text{oxygen}}{O_2} \xrightarrow{\text{bacteria}} CO_2 + NH_3 + H_2O \tag{3-2}$$

The oxygen required for completion of the reaction in Eq. 3-2 represents the ultimate carbonaceous BOD.

The ammonia produced during the carbonaceous oxidation of the organic matter can be oxidized further to nitrate (NO_3^-). Nitrification is the term applied to the biological oxidation of ammonia to nitrate, and the oxygen required is termed the nitrogenous biochemical oxygen demand (NBOD). The overall nitrification reaction can be represented as

$$NH_4^+ + 2O_2 \rightarrow NO_3^- + 2H^+ \; H_2O \tag{3-3}$$

Because the exertion of NBOD is slow, typically, it is not noted in the BOD test at times less than 5 to 8 days. When nitrification does occur in a BOD test, determination of the carbonaceous BOD is generally not possible. Therefore, it is now common practice to add a chemical to the BOD test to suppress the nitrification reaction (9).

BOD kinetics.—The rate at which BOD is exerted is modeled, for practical purposes, as a first-order decay reaction. It should be noted, however, that a number of other rate models can be and have been used to model the BOD reaction. In the first-order model, it is assumed that the oxygen uptake (BOD exertion) is only a function of the BOD remaining. For the reaction occurring in a BOD bottle it can be shown that

$$\frac{d\,(BOD_r)}{dt} = -k_1(BOD_r) \tag{3-4}$$

in which BOD_r = BOD remaining at time t in mg/L, and k = reaction rate constant, d^{-1}.

The integrated form of Eq. 3-4 is

$$BOD_r = BOD_u \, e^{-k_1 t} \tag{3-5}$$

in which BOD_u = ultimate carbonaceous BOD, in mg/L.

The value of BOD_t (BOD exerted at time t) is determined using

$$BOD_t = BOD_u \, (1 - e^{-k_1 t}) \tag{3-6}$$

The effect of temperature is accounted for by correcting the rate constant using Eq. 3-1. Because the reference temperature for the BOD test is 20°C, the k value at 20°C is used as the reference value. The most commonly used value θ for the BOD bottle test, when the temperatures are near 20°C, is 1.047 (3, 10).

The Five-day, 20°C BOD.—Because of the long time required to reach the ultimate BOD_u, the five-day, 20°C BOD value (BOD_5) has been adopted as the reference standard. The BOD_5 value is also used to assess the effectiveness of wastewater treatment facilities. A serious shortcoming of the BOD_5 reference value is that it has little or no relationship to the stoichiometry of the biological reactions occurring in the BOD bottle.

5. Effective Biochemical Oxygen Demand (EBOD)

The term EBOD was introduced by Pomeroy and Bowlus in 1946 (5), in a paper in which they presented an empirical velocity criterion to avoid the build-up of sulfide. The velocity criterion was related to the EBOD, a combined BOD-temperature function defined as

$$EBOD = BOD_5 \times 1.07^{(T-20)} \tag{3-7}$$

in which BOD_5 = five-day, 20°C BOD, in mg/L; 1.07 = empirical temperature coefficient; and T = wastewater temperature, °C. In adopting the term EBOD, it was assumed that the wastewater was municipal, temperatures were moderate, and the soluble fraction of the total BOD was not unusually high (5).

Recognizing the importance that the proportion of soluble BOD might have in the generation of hydrogen sulfide, it has been suggested that filtered BOD or COD values might be used in place of the total BOD value now used (1).

6. Acidity and Alkalinity

The pH of water is defined as

$$pH = -\log [H^+] = \log \frac{1}{[H^+]} \tag{3-8}$$

in which H^+ = hydrogen ion concentration, in mol/L.

Although pH data are always interpreted in terms of the hydrogen ion concentration, it should be remembered that pH is a measure of the hydrogen ion activity and not concentration (7, 8). The pH scale ranges from 0 to 14; a pH value of 7 is assumed to represent neutrality. Solutions with pH values in the ranges from 0 to 7 and 7 to 14 are termed acidic and basic, respectively. When contaminants having ionizable H^+ or OH^- groups dissolve in water and wastewater, the equilibrium between H_2O, H^+, and OH^- shifts, and the pH value decreases (becomes more acidic) or increases (becomes more basic).

The hydrogen ion concentration is important because it affects chemical reactions and chemical equilibria. The equilibrium expressions for hydrogen sulfide considered previously in Chap. 2 are a prime example. In wastewater, the pH is largely controlled by the alkalinity. In water, the alkalinity is defined as (7, 8, 10):

$$A = [HCO_3^-] + 2[CO_3^{-2}] + [OH^-] + [H^+] \tag{3-9}$$

in which A = alkalinity, in meq/L; HCO_3^- = concentration of bicarbonate ion, in mol/L; CO_3^{-2} = concentration of carbonate ion, in mol/L; OH^- = concentration of hydroxide ion, in mol/L; and, H^+ = concentration of hydrogen ion.

In wastewater, the alkalinity is also affected by the presence of compounds such as acetate, borate, HPO_4^{2-}, etc. For this reason, the alkalinity in wastewater is measured in terms of the amount of acid needed to lower the pH to a specified level, most commonly the methyl orange end point. In general, because most of the alkalinity of wastewater is in the form of bicarbonate, the pH range for most wastewaters is in the range from 6.5 to 8.0.

7. Dissolved Oxygen (DO)

The presence of dissolved oxygen (DO) in wastewater is significant because oxygen will react with hydrogen sulfide, principally forming thiosulfate, according to the following reaction (7):

$$2HS^- + 2O_2 \rightarrow S_2O_3^{2-} + H_2O \tag{3-10}$$

Sulfide may also be oxidized by bacteria of the genera *Thiothrix* and *Beggiotoa*. The sulfide, which is oxidized to sulfur, is stored within the cells. The intracellular reaction mechanisms are unknown. The overall change can be expressed by

$$2HS^- + O_2 \rightarrow \underset{\substack{\text{insoluble}\\\text{sulfur}}}{2S} + 2OH^- \tag{3-11}$$

In wastewater collection systems, the dissolved oxygen that may be present in wastewater is usually derived from one or more of the following sources: (1) Oxygen that may be present in wastewater discharged to the collection system; (2) oxygen absorbed from the gas in the sewer atmosphere due to turbulence, splashing, and other flow disturbances which cause gases to be entrained in the wastewater; (3) oxygen absorbed from the sewer and manhole atmosphere; (4) oxygen absorbed from air entrained in fittings; and (5) oxygen that may be present in local inflow and groundwater infiltration.

REFERENCES

1. Hvitred-Jacobsen, T., and Halkjaer Nielsen, P., "The Impact of Wet Weather Flow on the Hydrogen Sulfide Formation in Combined Sewers," TNO Committee on Hydrological Research, *Proceedings and Information*, No. 36, 1986.
2. Levine, A.D., Tchobanoglous, G., and Asano, T., "Characterization of the Size Distribution of Contaminants in Wastewater: Treatment and Reuse Implications," *Journal WPCF*, Vol. 57, No. 7, July 1985, pp. 805-816.
3. Metcalf & Eddy, Inc., *Wastewater Engineering: Treatment, Disposal, Reuse*, written and edited by G. Tchobanoglous, McGraw-Hill Book Co. New York, NY, 1979.
4. Metcalf & Eddy, Inc., *Wastewater Engineering: Collection and Pumping of Wastewater*, written and edited by G. Tchobanoglous, McGraw-Hill Book Co., New York, NY, 1981.
5. Pomeroy, R., and Bowlus, F. D., "Progress Report on Sulfide Control Research," *Sewage Works Journal*, Vol.18, No. 4, 1946, pp. 597-640.
6. *Process Design Manual for Sulfide Control in Sanitary Sewerage Systems*, U.S. Environmental Protection Agency, Technology Transfer, Washington, DC, 1974.
7. Sawyer, C.N., and McCarty, P.L., *Chemistry For Environmental Engineering*, 3d ed., McGraw-Hill Book Co., New York, NY, 1978.
8. Snoeyink, V.L., and Jenkins, D., *Water Chemistry*, John Wiley and Sons, New York, NY, 1980.
9. *Standard Methods for the Examination of Water and Wastewater*, 16th ed., American Public Health Association, New York, NY, 1985.
10. Tchobanoglous, G., and Schroeder, E.D., *Water Quality: Characteristics, Modeling, Modification*, Addison-Wesley, Reading MA, 1985.
11. Thistlethwayte, D.K.B., ed. *Control of Sulfides in Sewerage Systems*, Ann Arbor Science Publishers, Inc., Ann Arbor, MI, 1972.

12. U. S. Environmental Protection Agency, *Design Manual —Odor and Corrosion Control in Sanitary Sewerage Systems and Treatment Plants*, EPA625/1-85-018, Center for Environmental Research Information, Cincinnati, OH, 1985.

CHAPTER 4

OCCURRENCE OF SULFIDE IN WASTEWATER

A. INTRODUCTION

Sulfide is found in wastewater as the result of either the anaerobic decomposition of sulfur compounds or the direct input of sulfide-bearing wastes. The objectives of this chapter are to discuss those factors influencing the generation of sulfide, those wastewaters likely to contain sulfide, and those areas in wastewater facilities where sulfide can occur including sewers, force mains, inverted siphons, pump stations, and treatment processes. Control of sulfide generation, odor emission, and corrosion are discussed in Chaps. 9, 10, 11, and 12.

B. SULFIDE GENERATION IN SEWERS

To understand the conditions occurring in sewers which exacerbate sulfide generation, it is necessary to understand the basic mechanism of the process and contributing factors, such as velocity effects, sewer slope, grit and debris, poor construction, turbulence, sewer size, and surcharging. Each of these factors is considered in the following subsections.

1. The Mechanism of Sulfide Generation

For sulfate to be reduced to sulfide requires a medium completely devoid of free oxygen or any other active oxidizing agent. The stream of wastewater in a partly-full sewer is not completely anaerobic because it is exposed to the sewer atmosphere. Oxygen absorbed at the surface of the stream generally reacts quite rapidly, and in large sewers its concentration is typically held to a very low level. Even so, enough oxygen is usually present to prevent sulfate reduction in the stream.

The place where strictly anaerobic conditions most often develop is in the slime layer that forms on the submerged pipe wall. This layer is a matrix of filamentous microbes and gelatinous material (zoogleae) embedding various smaller bacteria. A typical slime layer may be considered to be 0.04 in. (1 mm) thick, but if the velocity of the stream is high, it may be no more than 0.01 in. (0.25 mm) thick, and if abrasive material is carried by the water, the wall may be scoured clean. At low velocities the slime may be as much as 1/8 in. (3 mm) thick, or even more.

If oxygen is present in the stream, it diffuses into the slime layer, but the aerobic bacteria found there use it so rapidly that it advances only a very short distance. Except where oxygen concentrations are high, the aerobic zone is less than 0.01 in. (0.25 mm) thick. Beneath that, the slime layer is anaerobic, and it is there that sulfide generation occurs.

Calculations based upon diffusion coefficients and observed rates of sulfide production show that sulfate or the organic nutrients available to the sulfate-1reducing bacteria, or both, are used up in a very short distance, and that the thickness of the sulfide-producing zone is generally on the order of 0.01 in. (0.25 mm). At deeper levels, the slime layer is anaerobic but largely inactive, because of the lack of a nutrient supply.

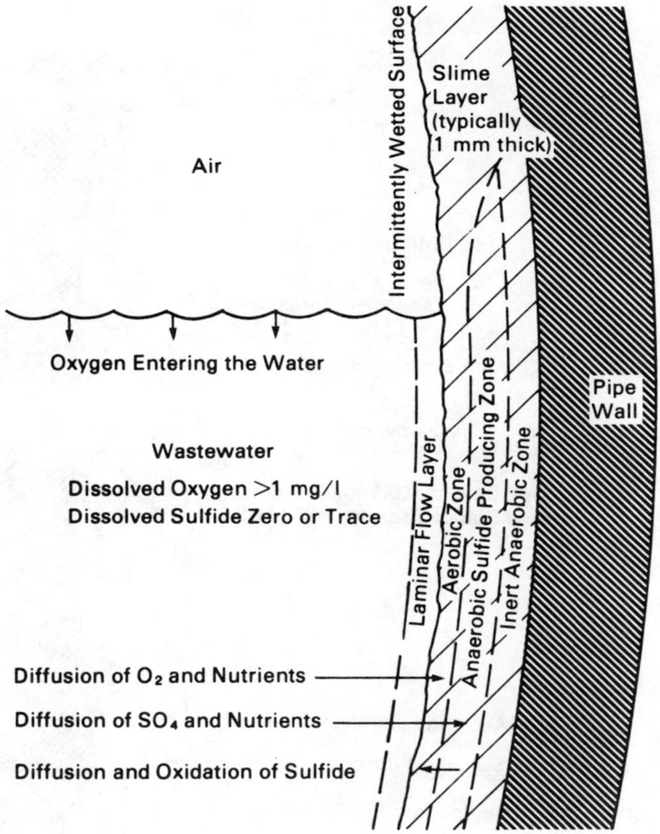

Figure 4-1. Processes occurring in sewers with sufficient oxygen to prevent sulfide from entering stream

As long as the surface of the slime layer is aerobic, sulfide diffusing out of the anaerobic zone will be oxidized there. No sulfide will be found in the stream unless it is from some extraneous or upstream source (6, 15). Figure 4-1 illustrates the processes just described, with the slime layer depicted on a greatly magnified scale.

It is the oxygen supply, rather than the kinds of bacteria present, that determines where the aerobic zone ends and the anaerobic zone begins. The oxygen concentration in the stream, as well as temperature and the concentration of organic food materials, determine how deep the oxygen will penetrate. If the oxygen supply increases, the aerobic zone will become thicker. Aerobes are present in the anaerobic zone, ready to oxidize the organic nutrients, but they remain dormant as long as oxygen is absent. The anaerobes, on the other hand, cannot carry on their processes as long as oxygen is present, but they are ready to resume their activity when oxygen disappears. Since these determining factors are not constant, the boundaries between

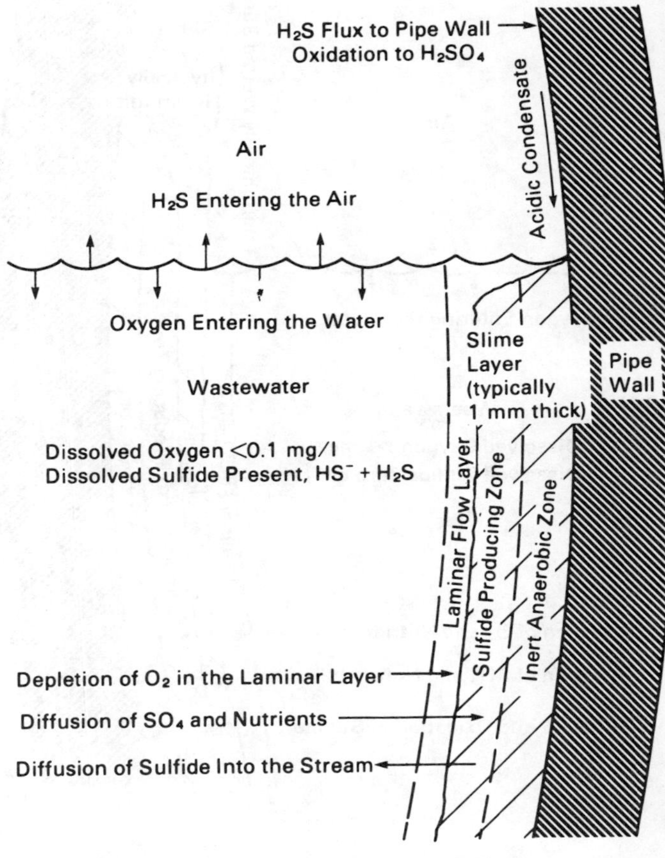

Figure 4-2. Processes occurring in sewers under sulfide build-up conditions

the zones are not fixed. Furthermore, bits of the slime are continually sloughing off into the stream, adding to the variability of the zone boundaries.

If the oxygen concentration in the stream drops to a low level, generally a few tenths of a milligram per liter, not enough oxygen will reach the surface of the slime layer to oxidize all of the sulfide that is produced. Sulfide can then escape into the stream. The process is illustrated in Fig. 4-2.

The oxygen concentration that is critical for preventing sulfide access to the stream is generally in the range of 0.1 to 1. 0 mg/L. It depends, among other things, on temperature and on the turbulence of the stream (9). If the water is stationary or moving slowly, oxygen becomes depleted near the pipe wall, and sulfide may escape from the slime layer even when the main bulk of the wastewater contains several milligrams per liter of oxygen. If there are organic solids slowly rolling along the bottom of the pipe, they will release sulfide even when the stream has a high oxygen content. When the oxygen concentration is low enough that sulfide leaks into the stream, only a minor part of it may nevertheless get through. Completely anaerobic

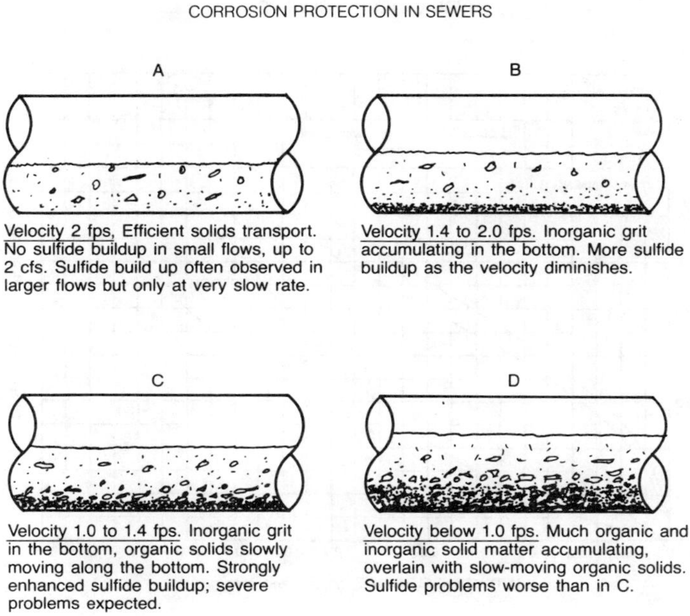

Figure 4-3. Solids accumulation at various flow velocities

conditions must be approached before all of the sulfide produced can pass into the stream.

The environment of a sewer is dynamic, tending to result in variations in the concentration and amount of sulfide produced. However, by understanding how various aspects of a sewer affect its sulfide production, it is possible to determine which sewers are likely to have the greatest levels of sulfide production.

2. Velocity Effects

In a sewer flowing less than full, the higher the velocity of the flow, the lower will be the generation of sulfides. This is best understood from the previous description of the mechanism of sulfide generation. At high relative velocities, the following will happen.

1. Solids will move along with the flow.
2. The slime layer will remain thin.
3. The aerobic zone will remain thick.
4. More oxygen will diffuse into the wastewater as oxygen is consumed by the aerobic decomposition of organics.

Less sulfide is generated because:

1. The oxygen-consuming solids keep moving downstream (presumably to a treatment facility).
2. The thin slime layer produces little sulfide.

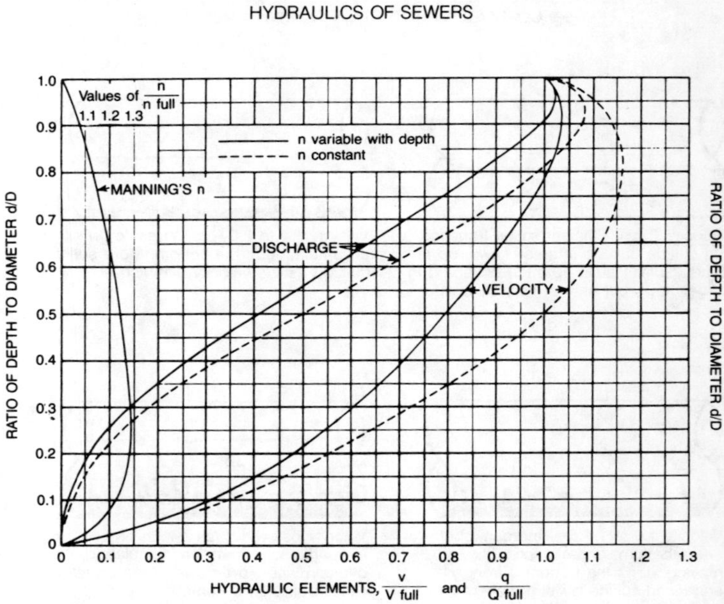

Figure 4-4. Proportional velocity and discharge

3. The thick aerobic zone oxidizes the small amounts of sulfide produced.

4. The oxygen-laden wastewater supplies oxygen to the aerobic zone of the slime layer.

The effect of velocity on solids in sewers is shown in Fig. 4-3. If sulfide is already present in a wastewater stream, then higher velocities will tend to cause the H_2S to be released into the air (provided that the sewer is not flowing full). Once it reaches the air space above the wastewater, gaseous hydrogen sulfide is available for oxidation to sulfuric acid, which actually causes the attack on concrete structures, steel grating, manhole rungs, electrical switchgear, or concrete, asbestos cement, and ferrous pipes. However, if sulfide build-up in a sewer is great enough, then it will not be necessary to have higher velocities to cause the sulfide to be released to the air space. At concentrations of 2 or 3 mg/L, sulfides will diffuse out of quiescent flowing wastewater and be available for oxidation to sulfuric acid (8).

The rate of release of H_2S into the stream is proportional to the concentration of un-ionized H_2S, and also is proportional to the 3/8 power of the rate of energy dissipation, which is the product of slope times velocity.

3. Sewer Slope

When considering sewer slope (or any other aspect of sulfide generation), it is important to remember that the generation of sulfide is distinct from the corrosion or odor problems caused by the presence of sulfide. Conditions leading to sulfide generation need not be the same as conditions leading to corrosion or odor problems.

TABLE 4-1. Typical Minimum Sewer Slopes (14)

Pipe size, in. (1)	Corresponding slope, ft/ft (2)	Capacity flow, mgd (3)
6	0.0050	0.25
8	0.0035	0.46
10	0.0025	0.71
12	0.0020	1. 03
15	0.0015	1. 62
18	0.0012	2. 34

Note: 1 in. = 25.4 mm; 1 mgd = 3,785,000 L/day

Sewer slope affects sulfide production in several ways. However, all other things being equal (flow, diameter, wastewater characteristics), the less the slope in a sewer, the greater its production of sulfide. This effect is, in turn, due to the effect of sewer slope on the velocity of the wastewater flow. Particularly in older sewer systems, sewer slopes were determined by ground slope. Sizing of lines, if done scientifically, was done in order to convey a design flow at the available ground slope. Once the relationship among flow, pipe diameter, pipe friction, and pipe slope for sewers was understood, it became apparent that inadequate slope resulted in low velocities and a tendency for solids to deposit in the sewer.

Most sewering agencies now have standards for minimum slope for each size of sewer line. Unfortunately, this alone does not provide assurance of adequate slope. For example, Table 4-1 indicates typical minimum slopes for sewers. Flowing at capacity, which is at a wastewater depth of 80% of the sewer depth, and with a Manning's roughness coefficient of 0.013, the velocity is approximately 2 ft/s (0.6 m/s). In a typical case, for a peak flow of 0.60 mgd, a 10-in. (25-cm) diameter sewer would be adequate. The minimum slope (from Table 4-1) is 0.0025. In some cases, a designer will find that the minimum diameter pipe at minimum slope over a considerable distance will result in a deep line. The greater depth may result in the need for the installation of more expensive pipe installed by more expensive equipment, and with more expensive backfill requirements. In some cases, a lift station would be necessary. Faced with this, a designer might select an 18-in. (45-cm) pipe laid at a minimum slope of 0.0012.

Over a distance of 2 miles (3. 2 km), such a larger, flatter pipe would save 13. 73 ft. (4. 18 m) of pipe invert drop, possibly eliminating the need for and the cost of a lift station. Figure 4-4 shows the relationships between velocity and flow at varying depth for circular sewers. The graphs are plotted as a function of the percentage depth of flow so that they apply to varying pipe diameters.

In the example, a flow of 0.60 mgd in an 18-in. sewer with a capacity of 2. 34 mgd yields a ratio of q/Q_{full} = 0.60/2. 34 = 0.26. The flow ratio of q/Q_{full} = 0.26 corresponds to a depth percentage of 35%. The corresponding velocity ratio, v/V_{full} = 0.83. Since V_{full} is 2 ft/s, an 18-in. sewer at peak flow of 0.60 mgd results in a peak velocity of 0.83 x 2 ft/s = 1. 66 ft/s. This is less than desirable. At average flow conditions of, say, 0.30 mgd, the flow ratio would be 0.13, the depth percentage

TABLE 4-2. Comparison of Flows (14)

Factors affecting flow (1)	10-in. (25-cm) sewer (2)	18-in. (45-cm) sewer (3)
Minimum slope	0.0025	0.0012
Capacity flow, Q_{full} mgd	0.71	2. 34
Flow ratio, q/Q_{full} at 0.60 mgd	0.85	0.26
Percentage depth of flow at 0.60 mgd	70%	35%
Velocity ratio, v/V_{full}, at 0.60 mgd	1.14	0.83
Velocity at 0.60 mgd (ft/s)	2.28	1.66
Flow ratio, q/Q_{full}, at 0.30 mgd	0.42	0.13
Percentage depth of flow at 0.30 mgd	45%	25%
Velocity ratio, v/V_{full}, at 0.30 mgd	0.95	0.65
Velocity at 0.30 mgd (ft/s)	1. 90	1. 30

Note:1 mgd = 3,785,000 L/day; 1 ft/s = 0.305 m/s.

25%, and the velocity ratio 0.65. Thus the average velocity would be 1. 30 ft/s. A comparison between a 10-in. and an 18-in. sewer under these flows is shown in Table 4-2.

Clearly, the flatter, larger sewer results in inadequate velocities, which in turn are likely to result in relatively high sulfide production rates. The actual rates will depend upon other key factors related to the wastewater, such as strength (BOD), composition, and temperature. Figure 4-5 compares profiles of two alternative sewers, one properly sized and one oversized, based on the example described.

While the flatter sewer may produce considerable sulfides, it may not be subject to sulfide corrosion (13, 14). In the first place, the pipe could be of vitrified clay or plastic. Secondly, the low wastewater velocity will tend to keep sulfides in solution until some turbulence is encountered. As an example, Fig. 4-6 demonstrates what would happen if the relatively flat, oversized line entered a steeper sewer. The likely result would be sulfide release to the sewer atmosphere all along the steeper sewer. Usually, this will result in sulfide corrosion. The most severe corrosion would probably occur just downstream of the junction manhole.

In some older sewer systems constructed of corrodible pipe, sulfide corrosion has not been a problem because the lines run in a surcharged condition most of the time. A continuously surcharged pipe will not corrode, because there is no sewer air space where hydrogen sulfide can be oxidized to sulfuric acid. While this will produce relatively high concentrations of sulfide, concrete or asbestos cement pipe will not corrode if it is surcharged daily. Surcharging tends to wash sulfuric acid from the pipe walls before it has had a chance to build up to the point when corrosion begins. Additional discussion of surcharging is included in a later subsection devoted to the topic.

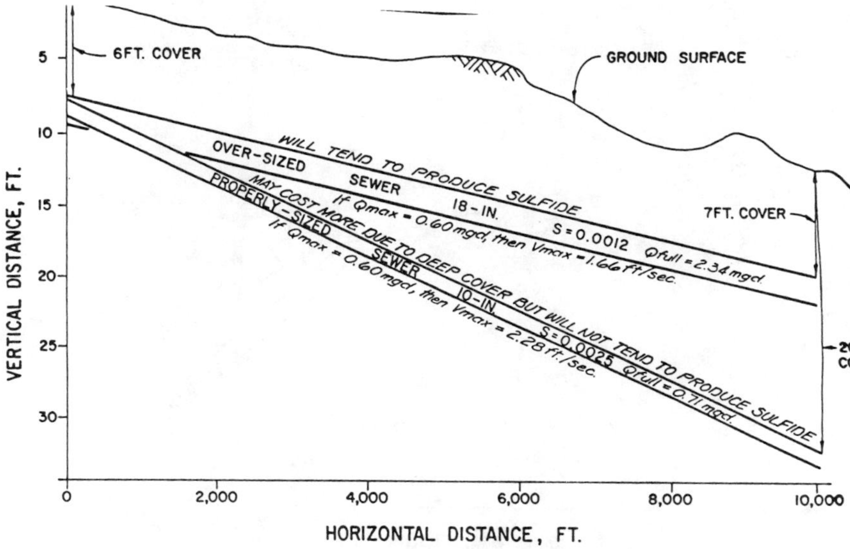

Figure 4-5. Profile comparison of 10-in. and 18-in. sewers

4. Grit and Debris

A sewer which would otherwise tend to be free of sulfide can produce sulfide if the invert is clogged with grit and debris. Being heavier than organic solids, grit is less likely to be moved along in wastewater flow with generally adequate velocities. Accumulations of grit can produce shoals which increase friction, in turn slowing the wastewater flow. The slower the flow, the more grit accumulates, and organic solids will begin to accumulate. Similarly, any other settleable debris or roots can impede the smooth flow in a sewer and tend to trap organic solids.

Various sewering agencies have developed sewer cleaning programs aimed specifically at reducing accumulations of grit and debris in order to minimize sulfide generation. Although proper design and construction of sewers is the preferred approach to avoiding sulfide generation, a good maintenance program can go a long way toward minimizing the problem in many existing sewers.

5. Poor Construction

Anything which tends to retard the flow of solids in a sewer is likely to increase the production of sulfide in the lines. Improperly installed pipe may not have a continuous slope; thus, an otherwise adequate slope for a line may include localized sections where the slope is inadequate. At these points, flow velocity will fall, and grit and organic solids may tend to accumulate, allowing the conditions for sulfide generation to develop.

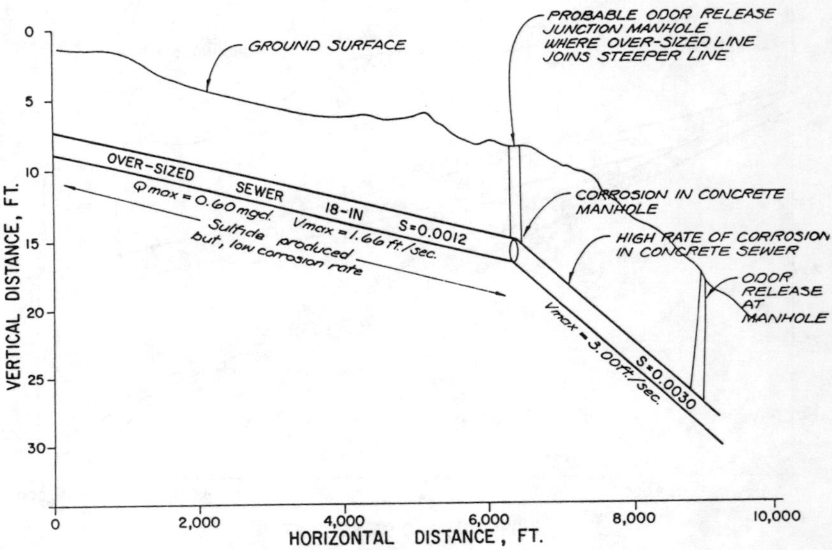

Figure 4-6. Profile of oversized sewer joining steeper sloped sewer

If gasketed pipe joints are improperly made up, the rubber gaskets can end up inside the pipe. This can impede the flow and cause the collection of rags and other debris which normally would flow downstream. The resulting slowdown of the flow may result in sulfide generation.

Poorly made-up joints or split or cracked pipe may also permit silt and sand to leak into the sewer in large quantities. This may lead to even more serious direct consequences, such as street cave-ins. In the meantime, the silt and sand (grit) tend to result in sulfide generation as previously described.

Conversely, in some cases, poorly made-up joints or split or cracked pipe can actually reduce sulfide build-up and its effects. In areas with high groundwater, large quantities of water can infiltrate the sewers. The increased flow will tend to have a greater velocity, and the more dilute wastewater will tend to produce sulfide at a slower rate. However, the other disadvantages of infiltration preclude consideration of poorly constructed joints as a method of sulfide control.

6. Turbulence

As is often the case when considering sulfide in sanitary sewer systems, turbulence can be both beneficial and detrimental. When turbulence improves the oxygenation of wastewater flowing through a sewer system, sulfide is not generally produced. In cases where oxygenated flows enter anaerobic flows, oxygen from the oxygenated flow can oxidize sulfide to dilute sulfate, rendering it harmless within the flow. Also, such inputs of oxygen can prevent or reduce the formation of additional sulfide.

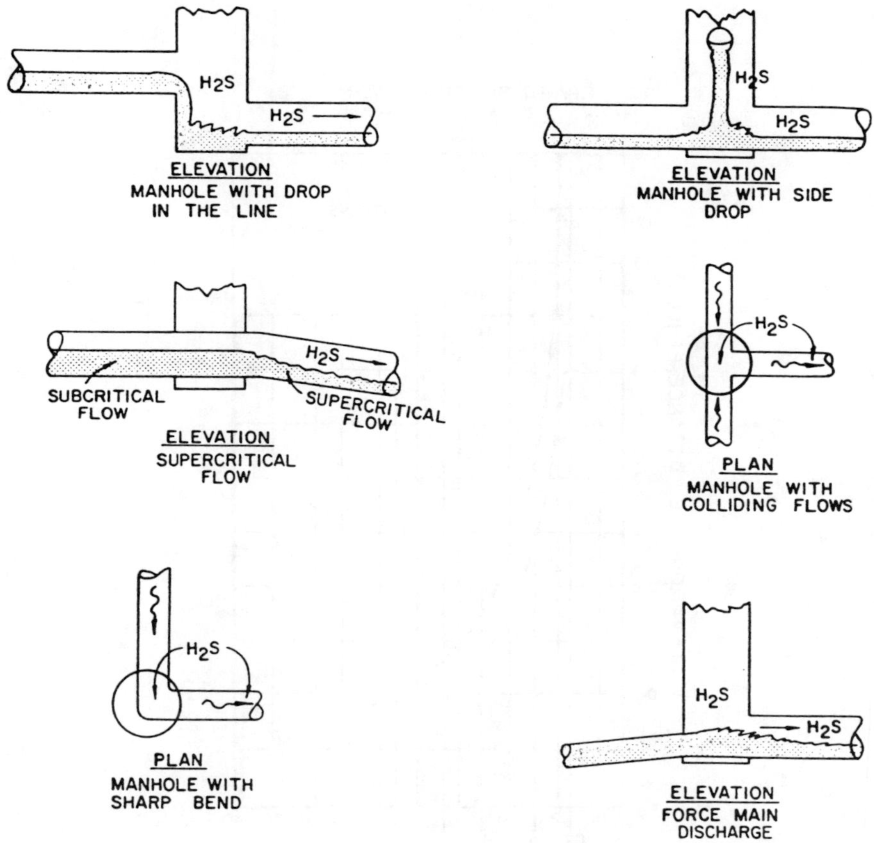

Figure 4-7. Sources of turbulence in sewer systems

Turbulence can also be utilized to cause sulfide to be released in places where it will be harmless. For example, a plastic or plastic-lined section of a primarily concrete trunk sewer can be installed at a turbulent spot. Sulfide released to the air space may be oxidized by aerobic bacteria on the inside pipe crown. The resulting sulfuric acid will not attack the plastic. Therefore, it will eventually drip down into the flow, where it will be diluted and rendered harmless to unlined concrete pipe downstream.

On the other hand, turbulence in a sewer system can result in severe localized sulfide release and corrosion of unlined concrete or asbestos cement pipe. This occurs in flows where significant concentrations of sulfide are already present in the wastewater. For example, an area with relatively flat sewers or long force mains will tend to produce sulfide. If the trunk sewer serving this area takes a drop or has a sudden increase in slope, the area immediately downstream (and possibly upstream) will be the site of considerable sulfide release. If there is an air space for oxidation

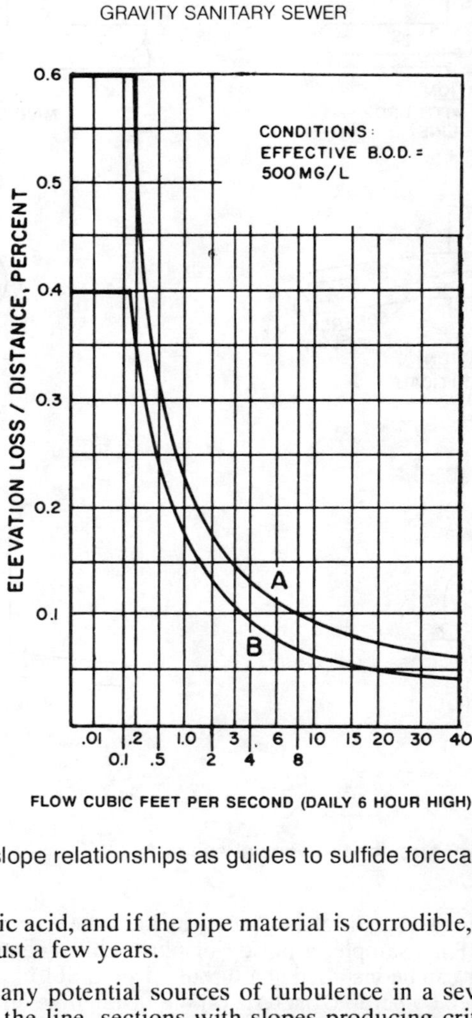

Figure 4-8. Flow-slope relationships as guides to sulfide forecasting

of sulfide to sulfuric acid, and if the pipe material is corrodible, the sewer pipe may be eaten away in just a few years.

There are many potential sources of turbulence in a sewer system. Among these are drops in the line, sections with slopes producing critical or near-critical flow, manholes with flows dropping in from the side, manholes with flows colliding, manholes with sharp bends, and discharges from force mains. Examples of sources of turbulence are shown in Fig. 4-7.

Critical flow can be reached at surprisingly low slopes in larger sewers. For example, in a 60-in. (152-cm) diameter line (with Manning's $n = 0.013$) carrying a flow of 50 mgd, critical flow is reached at a slope of approximately 0.0020 [or 2 ft per 1,000 ft (2 mm per 1,000 mm)]. This corresponds to a velocity of approximately 5.5 ft/s (1.68 m/s). Thus, in a 60-in. diam sewer with a slope of 0.0020, flows approaching or exceeding 50 mgd would tend to become quite turbulent without any apparent sources of the turbulence. Near the point where a transition to subcritical slope occurs, a hydraulic jump will form. This can be a most energetic point of sulfide release.

7. Sewer Size

The conditions existing in small collecting sewers are highly erratic. Because of the shallow depth and the low oxygen reacting rate in the fresh wastewater, and the effects of service laterals along the way, the stream may remain free of sulfide even though the velocity is so slow that solids are deposited.

Little is known with certainty about flows in sewers serving a few homes. The flushing of a single toilet will produce a flow of about 0.35 ft^3/s (0.1 m^3/s), which is more than 100 times the daily average flow from a residence. The flow from a single flush would soon level out to a slower rate in a sewer, but more than one fixture may be in use at the same time, so occasional flows of 0.035 ft^3/s (0.001 m^3/s) in the sewer are not rare. A reasonable estimate is that an occasional peak flow from two houses might be 0.05 ft^3/s (0.014 m^3/s). Considering the observed hydraulic conditions in small sewers, the peak flows from one or two residences in a sewer laid at a slope of 0.006 will not produce velocities much greater than 1 ft/s (0.3 m/s). For a part of the time, the flow will be practically zero, even from several homes. At these times, the pipe surfaces and part of the deposited solids will become quite well oxidized if solids accumulations are not excessive. This condition, and the fact that the fresh wastewater contains dissolved oxygen, will preclude sulfide generation close to the wastewater sources unless solids are stranded in such amounts that they produce stagnant pools.

When the collected flow is great enough so that the bottom of the pipe remains essentially free of organic deposits, the determining factor in respect to the occurrence of sulfide is the oxygen balance in the stream. For flows up to one or two cubic feet per second (or more if the temperature is low), and at a velocity not less than 2 ft/s (0.6 m/s), surface aeration is generally sufficient to maintain an oxygen residual of a few tenths of a milligram per liter in the stream and to suppress sulfide build-up. In larger flows, a greater oxygen absorption rate at the surface of the stream would be necessary to prevent sulfide build-up because of the greater volume of water that must be supplied with oxygen; but the actual absorption rate is generally less in the large sewers, because of smaller slopes. However, in these larger flows, the effects of junctions and other points of turbulence become more important than surface aeration of the normally flowing stream.

Based upon an evaluation of available data, Fig. 4-8 has been drawn as an aid in forecasting, in a qualitative way, the possible sulfide conditions that may occur for various slope-flow combinations in sewerage systems. (As a matter of convenience, a cube-root scale is used.) The vertical scale is the overall slope, that is, the elevation loss averaged over distances equal to 1-hr flow times, except for the small collecting sewers where it means the actual pipe slope. All substantial tributary flows should be considered in the calculation of energy loss. Even a small tributary flow has an important effect if it drops several feet into the larger stream.

Interpretations of Fig. 4-8 will be restricted to sewers with flow depths not exceeding two-thirds of the pipe diameter. This means that there cannot be any pressure mains unless measures are taken to offset their oxygen-depleting and sulfide-producing potentials. Except in small sewers, the vertical scale can be considered to be H/L, in which H = elevation, and L = distance in the same units, calculated upstream from the point of reckoning over distances representing flow times of about an hour. It will be referred to as "effective slope." In the collecting sewers of minimum size, the actual pipe slope should be used.

The interpretation of Fig. 4-8 must be qualified as being related to wastewater streams of specified characteristics. The temperature of the wastewater in a sewer follows an annual cycle, BOD_5 and quantity of flow have diurnal cycles, and flows usually have long-time trends. It is useful to define a climactic condition as the combination of the average BOD_5 and flow for the highest 6-hr flow period of the day and the average temperature for the warmest three months of the year. Where diurnal BOD_5 curves have not been made, it may be assumed that the BOD_5 for the 6-hr high-flow period is 1. 25 times the BOD_5 of a flow-proportioned 24-hr composite. The climactic EBOD (effective biochemical oxygen demand) is defined by

$$[EBOD]_c = [BOD]_c \times 1.07^{(T_c - 20)} \qquad (4\text{-}1)$$

in which $[EBOD]_c$ = climactic EBOD, mg/L; $[BOD]_c$ = climactic BOD, mg/L; and T_c = climactic temperature, °C.

Let it be assumed that the climactic EBOD for a sewer is 500 mg/L. Then the interpretation of Fig. 4-8 is as follows:

Curve A. While the condition prevails, a system functioning with slope-flow relationships as shown by Curve A may be expected to produce very little sulfide, rarely more than 0.1 or 0.2 mg/L of dissolved sulfide. The annual average dissolved sulfide concentration is expected to be only a few hundredths of a milligram per liter.

Curve B. While the condition prevails, a system functioning with slope-flow relationships as shown by Curve B may produce dissolved sulfide at concentrations of several tenths of a milligram per liter.

It must be emphasized that Fig. 4-8 cannot be used to predict sulfide conditions accurately. The purpose is to indicate an apparent trend in the effect of flow-slope relationship on sulfide build-up. Additional data are necessary to refine and fully substantiate the indicated relationships.

If the climactic EBOD is higher or lower than 500, the positions of the curves will be altered, except that the cut-off effective slope of 0.6% for Curve A at small flows will remain the same, since the determining factors are hydraulic. The relationship between EBOD and required slope is complex. The nutrients available for the sulfide-generating bacteria in the slime layer are presumed to be roughly proportional to the BOD_5, but the oxygen requirement to prevent sulfide build-up may not increase in the same ratio. If slope, and hence velocity, is increased, the greater turbulence not only increases oxygen absorption, but also provides more effective transfer of oxygen to the slime surface. As a result, a lower oxygen concentration suffices to prevent the escape of sulfide from the slime layer into the stream. Equally important, the length of time that the wastewater spends in transit is reduced.

Considering all factors, it is judged that similar sulfide conditions will result if the effective slopes are increased or decreased in proportion to the square root of the effective BOD. Thus, Fig. 4-8 shows that for the Curve A condition, a flow of 2.0 ft^3/s (0.6 m^3/s) requires an effective slope of 0.175%. If the EBOD were 600 mg/L instead of 500 mg/L, the effective slope for the Curve A condition at 2. 0 ft^3/s would be $0.175\% \times \sqrt{600/500}$.

For a system in which effective slopes are substantially higher than that required by Curve A, sulfide concentrations will be negligible for practical purposes. For conditions substantially below Curve B, higher sulfide concentrations must be anticipated, generally requiring odor and corrosion control measures, or continuous treatment of the wastewater to prevent sulfide build-up.

8. Surcharging

Accepted engineering practice for sewer design is usually based upon the assumption of free surface flow even under peak flow conditions; i. e., sewers are designed not to flow full. A sewer flowing full, or surcharged, has no air space. Without the availability of oxygen from the air above, the wastewater in the sewer will usually become anaerobic. Of course, other factors, such as biochemical oxygen demand, temperature, detention time, and a lack of bactericidal agents (e. g., chlorine), also enter into the determination of the state of wastewater in a sewer.

Sewers that are not adequate for the flows being carried will tend to surcharge. During periods of surcharge, sewer lines which might otherwise not produce sulfide will usually produce sulfide. However, the surcharged lines will not be subject to sulfide corrosion. If a particular line remains surcharged most of the time, then there will be no oxidation of sulfide gas to sulfuric acid. Hence, there will be no corrosion. If a particular line produces sulfide but is subject to daily surcharging, then the sulfuric acid produced will be washed off daily and tend to produce only minimal corrosion.

The problems associated with surcharging tend to occur downstream. In some systems this will occur at a pump station wet well, or at the headworks of a wastewater treatment plant. In other systems, the problems occur in adequately sized, free-flowing sewers downstream of surcharged sewers.

For example, at one time, the City of Modesto, California, had trunk sewers that were surcharged much of the time. Concentrations of sulfide of up to 3 or 4 mg/L would arrive at the treatment plant, causing odors and corrosion at the headworks. Then, in the late 1960s, Modesto undertook a major sewer system expansion. Some of the new relief sewers drew sufficient flows away from the old trunk sewers to permit them to flow in an unsurcharged condition. Since much of the older service area had sewers with marginal slopes, sulfide entering and produced in the old trunk sewers was now available for attacking the concrete pipe walls. Pipe failure and street cave-ins occurred at several locations.

Thus, in general, surcharging, or other lines flowing full, such as long force mains and inverted siphons, are to be avoided if possible.

C. FORCE MAINS AND INVERTED SIPHONS

Any pipe that flows full in a sewer system will tend to produce sulfide unless control measures of some sort are employed. As previously described, standard design practice precludes most collection system pipes from flowing full, even during peak flow periods. Force mains and inverted siphons are exceptions. These pipes are intended to flow full under all conditions.

Although force mains receive pumped flow and inverted siphons receive gravity flow, the production of sulfide is essentially the same in both cases. There are some special considerations regarding sulfide and inverted siphons. However, the following discussion applies equally to both types of pipes.

1. Sulfide Production in Pipes Flowing Full

For any pipe flowing full, including surcharged gravity sewers as well as force mains and inverted siphons, a general equation has been developed for the prediction of sulfide build- up. It is presumed that there is no reaeration, and that any oxygen in a pipe full of raw wastewater will rapidly consume whatever oxygen may have been present originally. As long as there is sulfate or other reducible sulfur compounds available, then the sulfide production will principally depend upon time and wastewater strength. The general equation takes the form

$$\frac{d[\mathrm{S}]}{dt} = 3.28\, M\, \frac{(1{+}0.48R)}{R}\, [\mathrm{BOD}]\, (1.07)^{T-20} \qquad (4\text{-}2)$$

in which $d[S]/dt$ = rate of change of sulfide concentration, mg/L-hr; 3.28 = conversion factor from meters to feet; M = specific sulfide flux coefficient, m/hr; R = hydraulic radius, ft (1/4 the diameter in a filled pipe);$(1+0.48R)$ = empirical factor for sulfide production in the stream; BOD = biochemical oxygen demand concentration, 5-day, mg/L; and T = wastewater temperature, °C.

Simplifying Eq. 4-2 for circular pipes, and assuming an experimentally determined typical M value of 0.001 m/hr, yields

$$\frac{d[S]}{dt} = 0.0131\ \frac{(1+0.12D)}{D}\ [BOD]\ (1.07)^{T-20} \tag{4-3}$$

in which D = pipe diameter, ft.

For example, applying Eq. 4-3 for a BOD_5 of 200 mg/L, a wastewater temperature of 20°C, in a 12-in. force main with a length of 1,000 ft and a velocity of 3 ft/s, results in

$$\frac{d[S]}{dt} = 4.116\ \text{mg/L-hr}$$

$$t = \frac{\text{length}}{\text{velocity}} = \frac{1{,}000\ \text{ft}}{3\ \text{ft/s}} = 333\text{s} = 5.55\ \text{min} = 0.0925\ \text{hr}$$

and $(S) = 4.116\ \text{mg/L-hr} \times 0.0925\ \text{hr} = 0.381\ \text{mg/L}$

Thus, with an initial sulfide concentration of 0.0 mg/L, a concentration of 0.381 mg/L would be produced in a relatively short force main (1,000 ft).

In this example, if the force main were reduced to 10- in. diameter, then

$$\frac{d[S]}{dt} = 4.851\ \text{mg/L-hr},\quad t = 0.0642\ \text{hr, and}\ (S) = 0.312\ \text{mg/L}.$$

Therefore, for a smaller force main, the slightly higher sulfide production rate is more than offset by the shorter detention time, resulting in a decrease in the sulfide concentration produced. In fact, oversized force mains, especially long ones, can produce very high sulfide concentrations.

2. Ventilation

Another aspect of force mains and inverted siphons is that they prevent sewer gases from flowing downstream as they normally would. This prevents natural ventilation from diluting the hydrogen sulfide. Unless ventilation is provided at a sewage pump station wet well, or at the upstream manhole for an inverted siphon, hydrogen sulfide is likely to accumulate.

Mechanical positive ventilation is usually provided for pump station wet wells. Wet-well air must then be either scrubbed or discharged, along with the wastewater, to a downstream (preferably airtight) manhole. Gravity ventilation through an "air jumper" pipe is usually sufficient for an inverted siphon. Some authorities prefer to provide corrosion protection to wet wells, and not to bother with ventilation.

As a potential source of turbulence, the discharge end of a force main or an inverted siphon is frequently the site of sulfide release. Sections C8 and C9 of Chap. 6 discuss methods of design to account for increased corrosion due to turbulence.

D. WASTEWATER PUMP STATIONS

Wastewater pump stations are frequently a source of sulfide in sanitary sewer systems. Also, they serve as points of accumulation for sulfide produced elsewhere. Both influent sulfide and wet-well sulfide production are considered in the following subsections.

1. Influent Sulfide

Sulfide gas, usually traveling downstream along with the flow in a sewer, will tend to collect in sewage pump station wet wells unless adequate ventilation is provided. In some cases, the wet-well atmosphere is merely blown into the force main discharge manhole and allowed to continue downstream. In other cases, air is blown into the wet well. This will tend to push hydrogen sulfide fumes, along with the rest of the sewer atmosphere, out of whatever vents are available. If nearby upstream manhole covers are vented or do not seat tightly, then localized odor problems may develop.

In some cases, air from wet wells is sucked out and discharged to the atmosphere. Unless hydrogen sulfide and other offensive gases are very dilute, or unless they can be discharged in a place where the odors do not matter, the discharge air poses an odor problem (at a minimum, and possibly a health and corrosion problem as well). At some pump stations, wet-well exhaust is scrubbed or oxidized to reduce or eliminate sulfide problems.

It is important that the influent sewer to a pump station is not surcharged by either a high wet-well level or by any screens or comminutors upstream of the actual wet well. (The problems associated with surcharging were described in Sec. B8.) If relatively high wet-well levels or plugged influent screens cause the influent sewer to surcharge, then the surcharge may even occur at relatively low flows. Settling of oxygen-demanding solids will tend to be severe and sulfide will be produced.

2. Wet–well Sulfide Production

Some authorities feel that wet-well sulfide production is not nearly as great as has been thought in the past. Rather, wet-well sulfide problems have been the result of hydrogen sulfide release caused by turbulence. However, sulfide may be generated in pump station wet wells in much the same way as it is generated in large filled pipes. Essentially, sulfide is generated as a result of detention of wastewater in the relative absence of oxygen,and as a result of solids settling out for extended anaerobic decomposition. All other factors being equal (strength of wastewater, pH, temperature, presence of precipitating metal ions), a pump station wet well will tend to produce sulfide if: (1) Solids tend to settle out and are not removed during pumping cycles; or (2) long on/off pump cycles or simply an oversized wet-well result in a long detention time.

Unless oxygen is supplied to the wet-well, or some other chemical control is practiced, even a well-ventilated wet-well will not supply sufficient oxygen through the wastewater surface to maintain aerobic conditions. Therefore, the less time wastewater stays in the wet well, and the less time that solids spend in the wet well, the lower the production of sulfide.

Settling of organic solids is a function of time and of wet-well geometry. It is important that pumps remove all organic solids that tend to settle in wet wells. Usually this can be achieved by providing sufficient slope (typically 1 vertical: 2 horizontal) for wet-well bottoms, except for areas directly affected by pump suction (7). Solids may also tend to settle out in wet-well areas around pump suction pipes if the connected pump runs infrequently. This problem can be alleviated by alternating pumps, usually with automatic controls.

Detention time of wastewater in a wet well is a function of the wet-well size divided by the inflow. Since the inflow usually varies considerably over the course of a day (and usually more so over a week or a year), it is usually desirable to have multiple pumps or variable-speed pumps, or both. Pump station control systems can be designed to provide a maximum interval between pump starts. It is not advisable to use a very small wet well because this would result in excessively short start/stop cycles which will damage pump motors.

Pump station wet wells are frequently designed for large future flows and may be oversized initially. Thus, during early years, oversized wet wells (in a fashion similar to initially oversized trunk sewers) may tend to generate sulfide. One effective approach for dealing with this problem is to use a divided wet well. In the early years, only one-half or one-third (or some other fraction) of the available wet well is used. When flows increase, more of the available wet-well is placed into service. Also, start/stop elevations may be set relatively close together in the early years of a pump station's operation in order to shorten the pump cycle interval.

E. WASTEWATER TREATMENT FACILITIES

Odor and corrosion can be significant problems at wastewater treatment facilities. Most unit processes are potential sources of odor release, but usually the farther wastewater travels through the treatment facility the less likely it is to be a source of odors and hydrogen sulfide, particularly when aerobic processes are utilized. Note that odor emissions and hydrogen sulfide generation can be significant problems throughout the treatment facility as a result of improper operation and maintenance practices, and the recycling of odorous or sulfide-containing waste streams. The areas discussed in the following subsections are headworks, preliminary treatment, primary clarifiers, aerobic biological treatment, secondary clarifiers, solids processing, recycle sidestreams, septage handling, and plant operating techniques.

1. Headworks

The term headworks refers to the transition area between the collection system and the treatment facility. From this area, wastewater is distributed to the treatment processes. Headworks buildings at many wastewater facilities often house influent pumps and preliminary treatment equipment. The influent channel at a treatment facility, or the influent wet well, can often be a point of turbulence as the result of elevation drops or sharp directional changes from the influent collection system. This is often the point in the treatment facility where high organic strength and septic sidestreams from septage haulers or subsequent treatment processes are returned. Turbulence in these areas can be controlled by minimizing differences in elevation in design or through operational methods. Submerging recycled sidestreams can reduce turbulence and localized odor release. Regular cleaning and flushing can go a long way towards minimizing odor emission.

2. Preliminary Treatment

Bar screens, comminutors, and screening devices can be sources of odor generation if organic matter is allowed to accumulate as a result of infrequent cleaning and flushing. Grit chambers and grit conveyance systems can also be sources of odor, particularly if a grit is stored for any length of time before disposal. Improved housekeeping, with frequent cleaning and grit disposal, can significantly reduce the odors. Off gases from aerated grit chambers can be a significant source of odors. They can be isolated with covers, and controlled through odor- scrubbing equipment.

3. Primary Clarifiers

The primary point of odor release is at the effluent weirs as a result of the fall in wastewater. Designing submerged draw-off or back-flooding effluent channels, to reduce the fall over the weirs, will be beneficial. Sulfide concentrations in the primary effluent can be reduced by assuring adequate and complete removal of settled solids, scum, and grease. For those plants having sufficient primary clarifier capacity, the number of clarifiers in service can be reduced during warmer periods to reduce the detention time. Any sumps associated with primary treatment, such as grease collection sumps, need to be tightly covered during periods when they are accumulating grease.

4. Aerobic Biological Treatment

Properly designed, operated, and maintained aerobic process units, whether fixed film or activated sludge, are not normally significant sources of odor. Overloaded process units, poor distribution of wastewater, poor distribution of air, and improper maintenance can lead to the increased presence of anaerobic zones. The return sludge wet well can be a source of odor if overcapacity or methods of operation result in significant detention times.

The complex biology involved in oxidation ponds can frequently be a source of odor during seasonal temperature changes or climatic conditions that reduce the transfer of oxygen. Maintaining a low loading level to oxidation ponds is critical for the control of odor emissions.

5. Secondary Clarifiers

Secondary clarifiers should not be a source of odors if upstream aerobic treatment processes are properly designed and operated. Inadequate rates of sludge withdrawal and surface skimming are major causes of odor emissions in secondary clarifiers.

6. Solids Processing

Sludge thickening, conditioning, digestion, dewatering and storage, whether aerobic or anaerobic processes are used, are significant sources of odor emission and sulfide generation. The extent of odor generation will depend upon the type and characteristics of the sludge and the method of processing used. Considerable attention should be given to the design and operation of the odor-control equipment that must be part of these facilities. Sludge incineration or solids reduction process units, as well as gas utilization equipment, all require emission control equipment.

7. Recycle Sidestreams

Liquid sidestreams resulting from many of the processing operations in treatment facilities include supernatants, centrates, filtrates, elutriates, decant liquors, and filter backwash waters. These sidestreams all can carry high organic loads and significant ammonia and hydrogen sulfide concentrations.

8. Septage Handling

The receipt of septic tank, cesspool, chemical toilet, or other hauled wastewaters is often a source of significant organic and sulfide load and odor for treatment plants. Odors are often released during the transfer operation from the

TABLE 4-3. Raw Wastewater Characteristics

Characteristic (1)	Average (2)	Minimum (3)	Maximum (4)
Hide production, lb	74,000	12,000	106,000
Flow, mgd	0.44	0.20	0.71
BOD, mg/L	3,734	2,700	4,600
Suspended solids, mg/L	4,806	2,539	8,287
Sulfide, mg/L	300	100	700
Chrome, mg/L	82	20	183
Grease, mg/L	1,000		

Note: 1 lb = 0.37 kg; 1 mgd =3,785,000 L/day
Source: Ref. 4.

septage hauling vehicle to the disposal facility because turbulent conditions are generated or poor procedures used.

9. Plant Operation

Odor emissions in treatment facilities can be substantially reduced by basic operational practices. They include keeping process units free and clean from accumulations of grease, solids, and debris; minimizing the holding periods of waste streams; closely monitoring both the hydraulic and organic load on process units; maintaining gas piping in leak-free condition; and assuring proper operation of any odor control equipment.

F. INDUSTRIAL WASTES

A wide variety of industrial wastes contain sulfide (1, 2, 4, 5, 10, 12). In addition, there are industrial wastes which, by virtue of their low pH, high temperature, or high BOD, may greatly accelerate the production and liberation of sulfide in wastewater systems. Conversely, industrial wastes may contain significant amounts of metal ions which will tie up sulfide ions as insoluble salts. This will reduce the liberation of hydrogen sulfide gas from the wastewater stream (3). Specific wastes to be considered are high sulfide wastes, high sulfate wastes, high BOD wastes, high temperature wastes, and priority pollutants.

1. High Sulfide Wastes

Among the industries which discharge major quantities of soluble sulfide to wastewater collection and treatment systems, some of the most significant are tanneries, refineries, whey processors, paper mills, and slaughterhouses. When one of these industrial facilities has its own wastewater treatment and disposal system, the discharge requirements will usually limit the effluent sulfide (both as a concentration and on a mass emission basis). When an industrial facility discharges to a sewer system connected to a publicly-owned treatment works (POTW), U. S.

TABLE 4-4. Original Pretreatment Plant Results

Effluent characteristic (1)	Effluent concentration, mg/L (2)	Removal, percentage (3)
BOD	900	71
Suspended solids	375	92
Sulfides	7	98
Chrome	3	96
Grease	128	87

Environmental Protection Agency effluent guidelines and standards will usually restrict the quality of the discharge.

This may require pretreatment. However, in many cases, specific effluent limitations for sulfide have not been developed. On the other hand, most sewer use ordinances will address the problem of sulfide discharge directly (through industrial effluent limitations) or indirectly (through restrictions on discharge of substances deleterious to the wastewater collection or treatment systems, or both).

Some examples of the significance of sulfide from tanning and petroleum refining industrial discharges follow.

Tanning. —The processes that may be used in the cleaning, preparation, and tanning of animal hides vary considerably. Similarly, there is a wide variation in the characteristics of raw wastewater from tanneries. Table 4-3 lists the raw wastewater characteristics for one tannery in Michigan (4).

The impact of this raw wastewater on any sewer system would not be difficult to imagine. The tannery did have a pretreatment facility consisting of a grit chamber, two-stage clarification with coagulant and polymer addition, and effluent caustic addition to reduce the impact of the effluent on the sewer system. The original pretreatment facility produced an effluent with characteristics as shown in Table 4-4.

Even with 98% reduction of effluent sulfide, the effluent described in Table 4-4 caused "tremendous odor problems" in the long municipal sewer system lines. Even if the effluent did not contain 7 mg/L sulfide, the high BOD would tend to produce sulfide in the sewers and force mains.

Subsequent to pilot studies, a much more involved treatment process, including biological secondary treatment and systems for recovery of chrome, grease, and sulfide, was put into operation. The results of this improved pretreatment process are presented in Table 4-5, based on a flow of 0.395 mgd.

The raw wastewater strength at this particular tannery and the problems which it creates are not unusual (1, 5). U. S. Environmental Protection Agency pretreatment standards for leather tanning and finishing permit tanneries to discharge up to 24 mg/L of sulfide. Recognizing that this may be insufficiently stringent, the regulations specifically state that "publicly-owned treatment works receiving wastewater from a [tannery] may require more stringent pretreatment standards for sulfide than those established by this part without EPA approval."At the same time, the guidelines allow for a less stringent sulfide standard, provided that the municipal collection and treatment systems will not be subject to enhanced opportunities for release of hydrogen sulfide gas.

TABLE 4-5. Improved Pretreatment Plant Results

Wastewater characteristic (1)	Influent, mg/L (2)	Effluent, mg/L (3)	Removal, percentage (4)
BOD	4,681	437	91
Suspended solids	7,349	463	94
Sulfide	120	0	100
Chrome	108	2.6	98
Grease	1,300	100	92

Petroleum refining.—Sulfide and other sulfur ions and compounds are found in crude oil to one degree or another. Sulfur content is so important to refineries that crude oils are usually classified, in part, based upon the sulfur content. Thus, a "sweet" crude oil is one with a low sulfur content, and a "sour" crude oil is one with a high sulfur content.

Since sulfur in crude oil results in air pollution and fouling problems during the combustion of petroleum products (primarily fuel oil and gasoline), it is usually an important part of the refining process to minimize all forms of sulfur in the product. The result is sulfur in the waste streams, and a significant amount ends up as soluble sulfide in wastewater flows. Typical refinery wastewater may contain 1 or 2 mg/L sulfide, but certain units, such as sour water strippers, may produce sulfide concentrations as high as 150 mg/L (2).

In a related industrial waste, sour water from a pilot coal liquefaction plant was reported to have a sulfide concentration of 9,500 mg/L (12).

2. High Sulfate Wastes

In most sewers, oxidized sulfur is usually not limiting with respect to sulfide production. In particular, for domestic wastewaters, sulfate ions and organic sulfur compounds are usually available even after long sewer runs [possibly 20 miles (32 km) or more]. However, with strong organic industrial wastes, such as those from food processing, all of the available oxidized sulfur may be converted to sulfide under anaerobic conditions. Under these conditions, wastes containing high concentrations of sulfate ion or organic sulfur compounds may increase sulfide production in the wastewater collection and treatment systems.

For example, in vegetable oil processing, sulfuric acid is used to separate water from the product. Even with pretreatment, vegetable oil wastewaters tend to have very high sulfate concentrations. Other characteristics of these wastewaters typically include low pH and very high BOD; so, unless they are kept aerobic, or controlled by other measures, their discharge to sewer systems will often result in serious sulfide problems.

Because metal processing often involves using sulfuric acid, metal processing wastewaters usually contain metal ions in addition to sulfate ions. Since the heavier metal ions will usually form insoluble sulfide precipitates, such wastewaters rarely contribute to sulfide concentrations in wastewater systems. However, it is important that such wastes are not strongly acidic. Otherwise, they may aid the release of existing hydrogen sulfide gas from wastewater streams.

It should be noted that some researchers have proposed equations for sulfide production that include a term for sulfate concentration. In particular, Thistlethwayte has proposed a model for sulfide generation which includes multiplication of the other terms by sulfate concentration raised to the 0.4 power (11).

3. High BOD Wastes

Industrial wastes with high BOD_5 concentrations (typically 300 mg/L to 1,000 mg/L or more) can greatly enhance sulfide production in wastewater systems. All equations used to predict sulfide build-up have BOD_5 as a basic parameter. Because sulfide production depends so much on anaerobic conditions, any waste which accelerates the consumption of oxygen from wastewater streams will tend to produce sulfide.

As an example, food processing wastes, such as those from canneries, often have BOD_5 concentrations as high as 1,000 mg/L. In a sewer with 1 part of such cannery waste to 2 parts of domestic waste (assumed BOD_5 of 200 mg/L), the BOD_5 of the mixture would be 433 mg/L. A properly designed sewer carrying this flow at 2. 5 ft/s (0.76 m/s) might experience little or no settling of solids in the pipe, yet the reaeration of the waste stream may not be adequate to maintain oxygen in the flow. Significant sulfide production will result.

Such a situation may be worse than the increase in BOD_5 alone would predict. For one thing, food processing wastewaters often have higher oxygen uptake rates than domestic wastewaters. Thus, the 5-day biochemical oxygen demand of such a waste may understate its shorter-term oxygen consumption.

4. Low pH Wastes

Many strong acids are employed in industry. The wastewaters from these industries may have pH values as low as 2 or 3. By lowering the pH of the wastewater in a sewer, such wastes may contribute greatly to sulfide production even if the industrial waste contains no sulfide, no sulfur compounds, and even little or no BOD_5.

Most sewering agencies have ordinances to prohibit the discharge of low pH wastes to their collection systems, and industries with such wastes are generally required to neutralize them with caustic (sodium hydroxide), lime (calcium oxide), or some other base. Most EPA pretreatment standards restrict industrial waste to a pH range of 6.0 to 9.0.Under some circumstances, a major industrial waste discharge with a pH of 6. 0 or even 6. 5 could cause considerable sulfide production. Conversely, a high pH industrial waste discharge (7. 5 to 9. 0 or higher) may cause a considerable reduction in sulfide production.

5. High Temperature Wastes

Industrial wastes frequently are warmer than the typical domestic wastes in the sewer system in a given area. Because sulfide production and sulfide release are both highly temperature dependent (increasing at high temperatures), high temperature industrial wastes may cause significant sulfide problems in wastewater systems.

An otherwise innocuous industrial waste can wreak havoc on a wastewater collection or treatment system if it is warm enough to raise the wastewater temperature by 5°C or more. Many sewer use ordinances place maximum temperature limits on industrial wastes.

6. Priority Pollutants

Increases in sulfide production should be anticipated as the discharges of industrial wastewaters containing heavy metals, pesticides, and other toxics are controlled as a result of local, state, or federal limitations. Industrial wastes containing significant quantities of metal ions can be expected to tie up sulfide ions as insoluble salts. However, these metal ions, as well as other toxic discharges, can act as biological inhibiting agents, for both the generation of sulfide through the reduction of sulfate, and the production of sulfuric acid as a result of the biological oxidation of hydrogen sulfide.

REFERENCES

1. Balden, A. R., Bossert, E. F., and Malhotra, S. K. "Joint Wastewater Treatment, City of Whitehall and Whitehall Leather Company," *Proceedings of the 34th Industrial Wastes Conference, Purdue University*, 1979, pp. 368-374.
2. Christiansen, J. A., and Spraker, P. W.,"Improving Effluent Quality of Petrochemical Wastewaters with Mutant Bacterial Cultures," *Proceedings of the 37th Industrial Wastes Conference, Purdue University,* 1982, pp. 567-576.
3. Higgins, T. E., and TerMaath, S. G., "Treatment of Plating Wastewaters by Ferrous Reduction, Sulfide Precipitation, Coagulation and Upflow Filtration," *Proceedings of the 36th Industrial Wastes Conference, Purdue University,* 1981, pp. 462-471.
4. Huber, C. V., and Doane, T. A., "A Case Study-Tannery Meets EPA Pretreatment Standards," *Proceedings of the 35th Industrial Wastes Conference, Purdue University*, 1980, pp. 95- 104.
5. Katz, P. B., "Control of Odors at a Tannery Wastewater Treat ment Plant," *Proceedings of the 34th Industrial Wastes Con ference, Purdue University*, 1979, pp. 360-367.
6. Kienow, K. E.,Pomeroy,R. D., and Kienow, K. K.,"Prediction of Sulfide Buildup in Sanitary Sewers," *Journal of Environmental Engineering*, ASCE, Vol. 109, No. EE6, Proc. Paper 17359-EE, 1983,pp. 1440-1447.
7. Metcalf & Eddy, Inc., *Wastewater Engineering: Collection, Treatment, Disposal,* McGraw-Hill, New York, NY, 1972, pp. 210-215.
8. Meyer, W. J. "Case Study of Prediction of Sulfide Generation and Corrosion in Sewers," *Journal of the Water Pollution Control Federation*, Vol. 52, No. 11, 1980, pp. 2666-2674.
9. Parkhurst, J. D., and Pomeroy, R. D., "Oxygen Absorption in Streams," *Journal of the Sanitary Engineering Division*, ASCE, Vol. 98, No. SA1, Proc. Paper 8701, Feb. 1972, pp. 101-124.
10. Paschke, R. A., Hwang, Y. S., and Johnson, D. W., "Catalytic Oxidation of Sulfide Wastewater from a Cellulose Sponge- Making Operation" *Journal of the Water Pollution Control Federation*, Vol. 49, No. 12, 1977, pp. 2445-2452.
11. Thistlethwayte, D. K. B., ed., *The Control of Sulfides in Sewerage Systems*, Ann Arbor Science Publishers, Ann Arbor, MI, 1972, pp. 14-20.
12. Tsai, K. C., Kamer, M. A., and Gray J. H. "Steam Stripping of Coal Liquefication Waste for Ammonia and Sulfide Removal," *Proceedings of the 37th Industrial Wastes Conference, Purdue University*, 1982, pp. 465-473.
13. U. S. Environmental Protection Agency. *Design Manual - Odor and Corrosion Control in Sanitary Sewerage Systems and Treatment Plants*, EPA/625/1-85/018, Chap. 4, 1985.
14. U. S. Environmental Protection Agency,*Process Design Manual for Sulfide Control in Sanitary Sewerage Systems* Chap. 3, 1974.
15. Warren, G. D., II, and Tchobanoglous, G.,*A Study of the Use of Concrete Pipe for Trunk Sewers in the City of Delano, California*, report prepared for the California Concrete Pipe Association, Sept. 1976.

CHAPTER 5

EFFECTS OF SULFIDE IN WASTEWATER SYSTEMS

A. INTRODUCTION

Hydrogen sulfide has wide-ranging effects in wastewater systems, most notably causing corrosion and odor problems. This chapter describes these effects, first for interceptor systems and then for treatment plants.

B. INTERCEPTOR SYSTEMS

Hydrogen sulfide problems typically do not occur in small-diameter gravity collector systems because sewage is usually fresh in these pipes. The most significant problems and effects from sulfide occur in trunk sewers, interceptors, partially filled force mains, and at the discharge from force mains. The problems to be considered are corrosion of sewer pipe materials, corrosion of appurtenances, odor, and toxic H_2S levels.

1. Corrosion of Sewer Pipe Materials

Corrosion of sewer pipe materials reduces the life of interceptors and thereby has a profound economic affect on many wastewater systems. The focus of this subsection is on the corrosion process, corrodible materials, distribution of corrosion, corrosion rates, and corrosion measurements.

Corrosion process.—Both cementitious pipe materials (such as concrete and asbestos cement) and ferrous pipe materials (such as cast iron, ductile iron, or steel) are subject to corrosion by sulfuric acid, which is formed from oxidation of hydrogen sulfide. Hydrogen sulfide (H_2S) is released from the surface of wastewater, enters the atmosphere of a sewer or other structure, and is oxidized on the pipe or structure surface as follows:

$$H_2S + 2\,O_2 \xrightarrow{\text{bacteria}} H_2\,SO_4 \qquad (5\text{-}1)$$

This reaction makes the moist walls of the pipe or structure acidic, and the reaction will proceed under extreme acidic conditions. One of the acting species under acidic conditions is *T. concretivorus* (6). In concrete pipe formed from granitic-based aggregate, the aggregate is not attacked by sulfuric acid because granitic materials are virtually inert. The acid attacks the cement bonding materials, producing a pasty mass of gypsum and inert products. The aggregate then drops into the wastewater,causing deposits of rocks in the sewer and exposing fresh cementitious material to acid attack. If calcareous aggregate is used in concrete, the acid attacks the aggregate as well as the cement, thus extending the time required to corrode a given thickness of wall material by a factor of up to 5. The corrosion rate of asbestos cement pipe is about twice the rate of concrete pipe made with

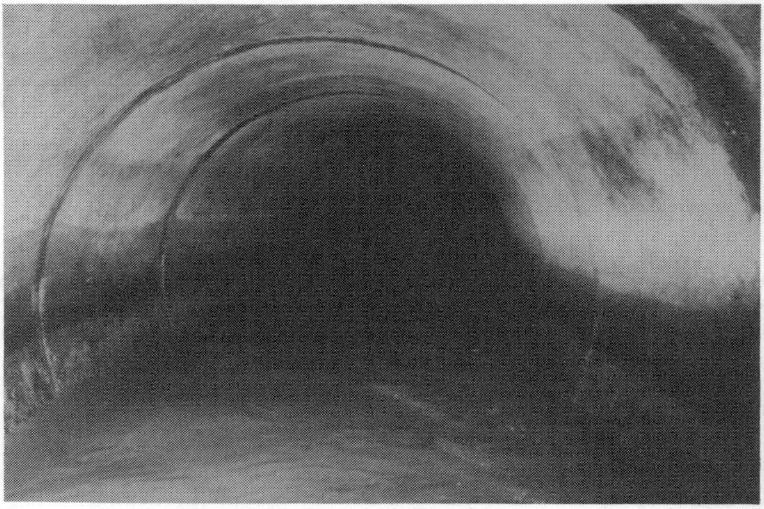

Figure 5-1. (a) 7-year–old 63-in. diameter reinforced concrete pipe, no sulfide in wastewater; (b) 20–year–old corroded 72-in. diameter reinforced concrete pipe, substantial sulfide in wastewater. (Courtesy: Los Angeles County Sanitation Districts)

TABLE 5-1. Pipe Materials Corrosion by H_2S or Sulfuric Acid

Material	Comment on corrosion potential
Cement and mortar	readily corroded
Mortar joints in all types of pipes	readily corroded
Mortar linings in all type of pipes	readily corroded
Type of pipe:	
Vitrified clay pipe and brick sewers	relatively inert (mortar between bricks is attacked)
Asbestos cement pipe	readily corroded
Concrete pipe	readily corroded, but the degree of corrosion depends on concrete alkalinity[a]
Plastic	relatively inert
Ferrous metals:	
Iron	attacked by acid and directly by H_2S
Steel	attacked by acid and directly by H_2S

[a] Concrete alkalinity depends primarily on the amount of calcareous aggregate in the concrete. Rate of corrosion will also depend on the distribution of cement and aggregate through the cross section, as this will vary with some types of pipe, i. e. , spun or cast.
Source: Ref. 3

calcareous aggregates, and about one-third the rate of granitic aggregate concrete pipe.

In the case of ferrous pipe, the acid attacks the metal directly. There is a second reaction involving direct combining of gaseous H_2S with iron to form ferrous and ferric sulfide, in addition to the normal oxidation (rusting) which occurs (3).

Corrosion of ferrous or cementitious materials does not occur below the waterline from this sulfide/sulfuric acid process. If corrosion is occurring below the water surface, it is from some other cause or mechanism, such as acidic wastewater.

If moisture is not present on the walls, or if the sewer atmosphere is devoid of oxygen, no acid is produced and no corrosion will result. However, it is rarely possible to avoid condensed moisture on walls of confined sewers and structures. It is equally difficult to maintain an oxygen-free sewer atmosphere.

Corrodible materials. —Table 5-1 lists materials which are attacked by the sulfide/sulfuric acid corrosion process described above. H_2S attacks some metals directly. Examples of typical corroded and uncorroded concrete pipe are shown in Fig. 5-1.

Distribution of corrosion.—Pomeroy (7) reports that the distribution of corrosion in a sewer is affected by air currents that control the rate of H_2S transfer to the pipe wall (see Fig. 5-2). The greatest corrosion is generally at the soffit of a manhole outlet, because of the high shear between the airstream and the pipe material. Items or structures extending into the sewer airstream will experience greater corrosion than

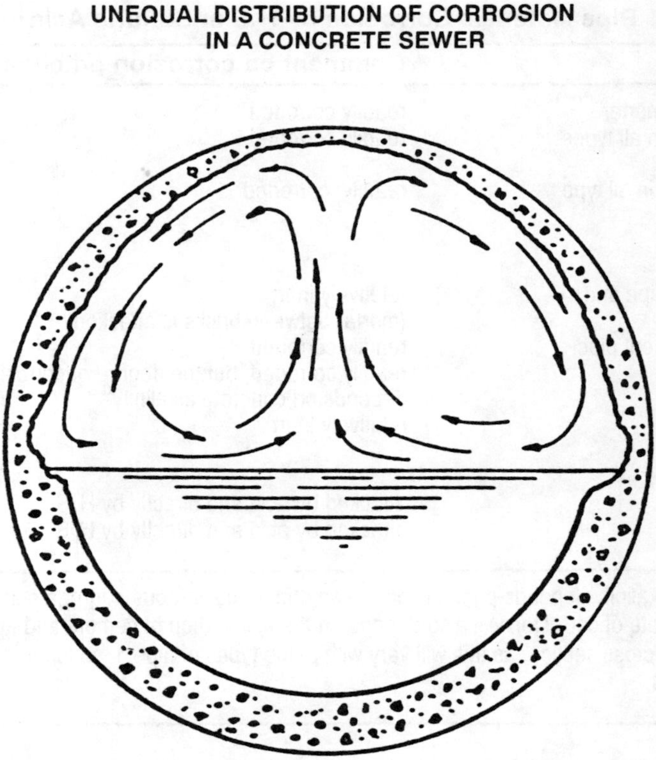

Figure 5-2. Sewer air movement and corrosion (7)

the pipe wall. Normally, airflow occurs down the sewer; this airflow is drawn by the movement of the wastewater. Higher than average corrosion occurs at the following points:

1. Crown of sewer. This is due to high H_2S transfer to the wall at this point,because of rising air currents at the sewer centerline.
2. Waterline. Wastewater flushes the corroded material away at the waterline, leaving fresh material to be corroded.
3. Turbulence points. All areas near wastewater turbulence points.

Corrosion rates.—Corrosion rates depend primarily on sulfide concentration in the sewer, rate of release of H_2S from the surface of the wastewater stream, and the type of material subject to corrosion. Chapter 6 explains the mathematical expressions which have been developed to describe sulfide build-up and corrosion processes.

Corrosion in sewer structures and areas of wastewater turbulence is dependent on factors which are not readily quantifiable. However, turbulence promotes higher than normal rates of hydrogen sulfide off-gassing. Multiplier factors are often used in estimating higher corrosion rates at turbulent locations. Turbulent conditions often require special protection for pipe and structures. See Chap. 6 of this manual for corrosion rate calculations.

Manholes corrode under the same principles that influence pipe corrosion. In some instances, inadequately protected manhole risers have been installed on adequately protected pipe. Manholes constructed with concrete produced with calcareous aggregates will have up to five times the corrosion resistance of manholes constructed with granitic aggregate concrete. Manhole risers in freeze/thaw climates may be subject to faster than normal sulfide deterioration, since cracks originating from freezing weather allow sulfuric acid attack to penetrate further into the material in warm weather (high sulfide) periods. Manhole steps are particularly subject to corrosion.

Corrosion measurements.—If corrosion has occurred in a pipe or structure to the extent that the original surface cannot be located, it is difficult to determine exactly the degree of deterioration. Several methods have been used to overcome this problem. Taking cross-section measurements inside a corroded sewer and comparing them to the original design provides a rough measurement of corrosion. Coring the wall of a pipe or structure has been done to determine remaining material or remaining cover over reinforcing steel (4). This method can also be used to estimate the amount of corrosion to date. Cores from the crown of the pipe can also be compared with cores taken beneath the waterline, i. e. , 10 to 15 degrees off the invert, where neither corrosion nor abrasion has reduced the original wall thickness. However, this may be a difficult and costly procedure.

2. Corrosion of Appurtenances

Often appurtenances, such as gates, control devices, and human access equipment (steps and platforms), are located within sewers and raw sewage structures. These appurtenances are subjected to a highly corrosive environment and should be designed accordingly. Many sewering authorities have discontinued the installation of rungs in manhole risers because of corrosion dangers.

Acid corrosion of ferrous metals.—Sulfuric acid attack in iron and steel occurs through the same processes as in cementitious materials. Higher density of iron and steel results in reduced rates of corrosion, but it takes relatively little corrosion to cause problems with metallic components in sewers. Pomeroy (3) reported that normal oxidation (rusting) of ferrous materials may be accelerated in the presence of sulfuric acid, and recommended that a corrosion rate 10 times the calculated corrosion rate due to acid alone should be used for ferrous materials when evaluating overall corrosion rate (due to both oxidation and acid attack).

Gates can be frozen in position if not properly maintained (see Fig. 5-3). Iron manhole covers can be frozen to the rings. Appurtenances often are located in moist atmospheres and turbulent locations, where H_2S off-gassing promotes high acid concentrations. The devices often protrude into the sewer airstream, so that maximum contact with H_2S is maintained, resulting in rapid acid attack. Stainless steel corrodes in closed sewers and raw wastewater structures.

Direct H_2S attack on metals.—Hydrogen sulfide in the atmosphere acts directly on most metals, including copper, iron, and silver. Cadmium is changed to yellow CdS. Cast iron and ductile iron turn to black FeS. Continued exposure destroys the metals. Electrical and instrumentation equipment that uses copper wiring and silver contacts should be kept out of hydrogen sulfide atmospheres.

Aluminum.—Aluminum is sometimes used for platforms, grates, handrails, and associated purposes in sewers. In closed sewers and structures, aluminum will not

Figure 5-3. (a) Corrosion in siphon box structure; (b) corrosion of siphon gates

be able to withstand the corrosion. In well ventilated areas, aluminum is used extensively.

3. Odor

The obnoxious odor of H_2S gas coming from pumping stations, manholes, and other structures causes distress to the public. Few large wastewater systems have escaped the problem of odor complaints at some time in their history. The number of locations where odorous air can be discharged from the interceptor system is limited, and they often are in residential, commercial, or other sensitive public-use areas. Citizen lawsuits against a sewering authority involving odor nuisance (5) have resulted in court orders to abate the odor nuisance or face stiff civil penalties. Subjects discussed in the following subsections are H_2S concentrations, raw sewage pump stations, manholes and structures, combined sewers, and interceptor maintenance operations. Odor control methods are discussed in Chap. 11.

H_2S concentrations.—Hydrogen sulfide is not the only odorous component of wastewater. However, it typically causes some of the worst odor problems, because it can be detected at low concentrations by humans (the practical limit of detectability is normally in the range of 1 to 10 parts per billion in air). With a dissolved sulfide concentration of 0. 5 mg/L in the wastewater and a wastewater pH of 7. 2, the atmosphere in equilibrium with the wastewater would have about 60 ppm of H_2S. This concentration would normally create a significant odor problem if discharged to the atmosphere. However, such a high concentration of H_2S is not usually reached in sewer atmospheres because equilibrium between the wastewater and the atmosphere is rarely achieved. Only in very turbulent conditions (e. g. , drop structures) would such equilibrium be likely to occur. Minimizing turbulence keeps hydrogen sulfide from off-gassing and creating high H_2S atmospheres.

Ventilation discharges containing 1 ppm or less of H_2S from pumping stations or other structures can cause odor problems. Releases from manholes at 1 ppm H_2S may also cause odor problems, depending on distance to human occupied areas.

Raw sewage pumping stations.—Pumping stations are some of the most common odor-causing facilities in a collection and interceptor system. Ventilation of wet wells in pumping stations is mandatory because the water surface elevation varies. This ventilation is handled in a variety of ways. In some cases, the ventilated air can be sent into a downstream gravity sewer.

For many pumping station wet wells, human entry is required for maintenance purposes. Providing a wet-well atmosphere safe for human entry requires in the range of 20 to 30 air changes per hour. This ventilated air often has H_2S concentrations above 1 ppm. If H_2S or odor control equipment is not installed, the foul air exhaust can cause a localized odor problem. The ventilation equipment, ductwork, and related facilities can corrode unless they are properly designed.

Manholes and structures.—Large gravity interceptors are often kept under a negative pressure, with fans pulling air out of the sewer at the treatment plant headworks. This practice effectively eliminates sewer air discharges from manholes and other structures along the interceptor system, but generally requires the installation of odor control equipment at the treatment plant. However, many interceptor systems are precluded from using this method because of the system configuration. In interceptors with no controlled sewer ventilation system, air is sometimes discharged at manholes and other structures due to surface wind conditions, changes in wastewater flow, and constricting features within the sewer, such as an inverted siphon or surcharging gravity sewer.

CORRODED DIGESTER GAS PIPING

CORRODED MANHOLE ON PLANT DRAIN LINE

CORRODED SUMP

PICTURES COURTESY OF LOS ANGELES COUNTY AND CITY OF OMAHA

Figure 5-4. Treatment plant corrosion problems. (Courtesy: Los Angeles County, California; City of Omaha, Nebraska)

Combined sewers.—Because storm inlets to combined sanitary and storm sewers must be available at all times to drain surface runoff, odors from these sewers are almost impossible to control. Hydrogen sulfide can be produced in combined sewers from normal sulfide production processes, from sulfide discharged by industries, and from sulfide produced in areas of grit and sludge build-up within the combined sewers. Warm weather, calm winds, and inversion conditions present the worst case for H_2S problems from storm drain inlets. H_2S-laden air wafts from the drains with relatively little atmospheric dilution and dispersion.

Interceptor maintenance operations.—When interceptors need maintenance work and human entry is necessary, sewer ventilation is required. Discharging this ventilation air in an urban area can cause odor problems; however, since it is of short-term duration, few agencies attempt to control the problem.

4. Toxic H_2S Levels

Atmospheric H_2S concentrations are lethal to humans at relatively low concentrations. Special equipment and procedures must be used by wastewater agencies to guard workers against H_2S levels. Toxicity and safety hazards of H_2S in the atmosphere of wastewater structures are discussed in Chap. 13.

C. TREATMENT PLANTS

Hydrogen sulfide causes corrosion, odor, safety, and process problems at wastewater treatment plants.

1. Corrosion of Cementitious Materials

Wastewater treatment plants have many kinds of conduits, tankage, and related structures which are subject to the same corrosion problems experienced in interceptor systems. Many of these facilities are made with concrete and are enclosed or partially enclosed. Moisture can collect on walls, and sulfuric acid attack can corrode the material. Figure 5-4 shows some examples of corrosion in treatment plants.

Plant headworks and grit tanks.—Plant headworks have the greatest potential of any area in the plant for sulfide corrosion, for the following reasons:

1. Wastewater sulfide concentration can be high at this point, due to sulfide build-up in the collection system.
2. Turbulence is often generated by flow measurement facilities, bar screens, wet wells, and grit tanks.
3. Headworks areas are often covered, to control odors and provide enclosed space for equipment.
4. Headworks are often underground where ventilation is poor.

Turbulence in headworks and grit tanks can cause mist or splash to occur on the wall area immediately above the water surface. The mist zone often receives the most corrosion. Areas above this zone are usually kept dry, if ventilation is provided, or if the facilities are open to the atmosphere.

Clarifiers and other facilities.—Areas of turbulence and splash are often corroded in primary clarifiers. This usually occurs at the weirs, launders, troughs, and outlet channels of clarifiers. Secondary clarifiers are typically not subject to H_2S corrosion, because sulfide concentrations in them are very low, usually unmeasurable.

Aeration facilities liberate H_2S due to the turbulence created and energy expended in oxygen transfer. This can create corrosion in the splashed areas and especially in confined, moist spaces, if the influent wastewater contains sulfide.

In-plant conduits and structures also experience substantial corrosion, depending on wastewater sulfide concentration, amount of turbulence, and ventilation within structures. Plant drain systems accept a variety of recycle flows within plants, some of which can be high in sulfide content. These pipelines and associated appurtenances can deteriorate quickly if adequate attention is not given to corrosion potential.

2. Corrosion of Equipment

Mechanical equipment in various areas of the plant (such as headworks, grit tanks, aeration, etc.) suffers substantial corrosion due to moist conditions, splashing, and H_2S release from wastewater. Electrical and instrumentation equipment can also undergo corrosion from H_2S.

Ventilation.—Equipment corrosion sometimes results directly from inadequately designed ventilation systems. If H_2S-laden air is brought through equipment rooms prior to discharge, these areas can suffer corrosion, odor, and safety problems.

Electrical and instrumentation equipment.—Electrical and instrumentation equipment is corrosion-sensitive because copper and silver can be directly corroded by H_2S in the air. Conversion to sulfuric acid is not necessary for this corrosion to occur. If electrical and instrumentation cabinets must be placed in areas of potential H_2S, then the equipment cabinets should be sealed and protected with a positive-pressure, clean, fresh air supply.

Other sensitive equipment.—Computer, laboratory, and other sensitive equipment must be kept free from H_2S environments. Computer and laboratory problems have been reported at several wastewater treatment plants. Electrical connection problems and circuit board malfunctions have been attributed to corrosion. These problems are primarily due to direct corrosion of materials by H_2S.

3. Odor

Odor problems at wastewater treatment plants are relatively common. A major odorant at many plants is H_2S. Various organic materials can also be responsible for odor problems, but if H_2S is controlled, the biggest portion of the problem is usually solved. Odor control methods and regulatory standards are discussed in Chap. 11.

Atmospheric H_2S discharges.—There are few established standards for H_2S concentrations in air discharged from treatment plants or industrial facilities. Specific conditions at each site determine the way in which the discharged air is transported and dispersed from the plant to off-site areas.

Level of control.—Because ventilation discharges and uncontrolled air emissions at treatment plants are often in the range of 1 to 20 ppm of H_2S, significant odor problems can exist. These problems must be analyzed carefully to determine how best to control the H_2S emission. These issues are considered in Chap. 11.

4. Liquid Wastewater Processes

Hydrogen sulfide can have adverse effects on wastewater treatment processes. The extent of these effects varies considerably among different treatment plants, different climates, and different wastes. Only limited information is available to explain the effects.

Sludge bulking.—Perhaps the most significant potential effect of sulfide is to increase the chances for sludge bulking in clarifiers. Sludge bulking occurs in secondary clarifiers, and is a condition where the biological sludge settles poorly, resulting in a relatively large sludge volume for a given mass of organisms. Filamentous organisms are generally responsible for sludge bulking. Researchers have found that septic wastewater causes sludge bulking (2) and H_2S specifically has been cited as the problem in some cases.

One thesis (8) states that H_2S or HS^- ties up trace elements (nutrients), such as certain metals, in the wastewater. Limiting the quantities of these key nutrients may cause predominance of filamentous bacteria, fungi, and actinomycetes. Controlled experiments (1) show that H_2S caused growth of the genus *Thiothrix*, which in turn caused sludge bulking in a pilot wastewater plant. *Thiothrix* forms are sulfur bacteria which oxidize sulfide to sulfur as an integral part of their metabolism.

There are many reasons for sludge bulking; presence of sulfide is only one of them. Also, the presence of sulfide will not always cause sludge bulking, since other factors may be necessary for growth of *Thiothrix* or other filamentous organisms.

Additional oxygen demand.—Sulfide exhibits oxygen demand in wastewater. The demand is typically not significant in total aeration requirements, because it normally represents a small portion of total wastewater BOD.

5. Sludge System Effects

Sulfide is formed in the anaerobic conditions in digesters and in anaerobic sludge deposits at the bottom of thickeners, clarifiers, conduits, and other structures and tanks. The wastewater in these tanks carries away the sulfide produced, thereby increasing sulfide concentrations in downstream processes.

Anaerobic digesters. —Wastewater sulfide concentrations of several milligrams per liter typically occur in anaerobic digesters. If sulfide concentration reaches 200 mg/L in digesters, it will be toxic to bacteria; this situation rarely develops. Even at 5 to 20 mg/L of sulfide in a digester—which is a typical range of concentration—recycle streams, such as digester supernatant, or sludge dewatering filtrates or centrates bring significant contributions of sulfide back to liquid processing streams. How and where these recycle streams are introduced to the liquid process are important considerations for corrosion, odor, and personnel safety.

Digester gas sulfide. —Hydrogen sulfide concentrations in digester gas can vary from less than 100 ppm to several thousand ppm. These concentrations typically cause no serious problems, because the gas is anaerobic and is not brought into contact with oxygen except in a combustion situation (engine, boiler, or waste gas burner). Gas leaks, however, can cause significant corrosion and odor. Gas leaks can occur in gas piping and along the gap between floating digester covers and digester walls, with resulting explosive hazards (from methane gas), corrosion, and potential exposure of personnel to toxic H_2S levels in the atmosphere.

REFERENCES

1. Farquhar, G. J. and Boyle, W. C. "Control of *Thiothrix* in Activated Sludge," *Journal of the Water Pollution Control Federation,* Vol. 44, No. 1, 1972.

2. Hudson, L. D., "Chlorination Controls Bulking Sludge," *Engineering News-Record*, Vol. 121, No. 242, 1938.
3. Kienow, K. K., and Pomeroy, R. D. , "Corrosion Resistant Design of Sanitary Sewer Pipe," ASCE Convention and Exposition, Chicago, IL, 1978.
4. Meyer, W. J., and Hall, G. H. , *Prediction of Sulfide Generation and Corrosion in Sewers,* prepared for the American Concrete Pipe Association, Brown and Caldwell, Sacramento, CA, April 1979.
5. Mountain View Community Improvement Association, et al., vs. Robert Fri, et al., Defendants, Civil Action 10171 in U. S. District Court, District of New Mexico.
6. Parker, C. D.,"Species of Bacteria Associated with the Corrosion of Concrete," *Nature*, Vol. 159, 1947, p. 439.
7. U. S. Environmental Protection Agency. *Process Design Manual for Sulfide Control in Sanitary Sewerage Systems,* Cincinnati, OH, Oct. 1974.
8. Wood, D. K., and Tchobanoglous, G., "Trace Elements in Biological Waste Treatment," *Journal of the Water Pollution Control Federation*, Vol. 47, No. 7, 1975.

CHAPTER 6

PREDICTING SULFIDE AND CORROSION

A. INTRODUCTION

The evaluation of a system's sulfide and corrosion potential can be very demanding from a resources standpoint. However, several valuable tools are available. Methods described in this chapter are: (1) The Pomeroy-Parkhurst equation, including graphical solutions and the SULF.BAS computer model; and (2) corrosion evaluation. Formulas and nomenclature are also discussed.

B. SULFIDE MODELING

When evaluating a system's sulfide potential, it may be necessary to simulate varied conditions. Inasmuch as the resources are not always available to perform a detailed in-line study and the condition desired cannot always be generated, a reliable quantitative model which accurately predicts sulfide levels is needed. For a detailed history of the development of sulfide evaluation criteria, see Ref. 26.

1. Pomeroy-Parkhurst Equation

In 1977, Pomeroy and Parkhurst presented a quantitative model to predict sulfide levels in partially-filled sanitary sewers (22). The model was based on research conducted in the wastewater collection systems of Los Angeles County, California, and other areas.

The model is applicable only when conditions are favorable for sulfide build-up, and misleading results may be obtained under other conditions. Sulfide build-up will not occur when the wastewater contains sufficient dissolved oxygen, or when insufficient nutrients are present, as indicated by a low BOD_5. Therefore, the model should not be used for systems when the dissolved oxygen level exceeds 0.5 mg/L. The model is represented by

$$\frac{d\,[S]}{dt} = M'\,[\mathrm{EBOD}]\,r^{-1} - N(sv)^{0.375}\,[S]d_m^{-1} \quad (6\text{-}1a)$$

Equation 6-1a can be rewritten as

$$\frac{d\,[S]}{dt} = 3.28M'\,[\mathrm{EBOD}]\,r^{-1} - 2.10N(sv)^{0.375}\,[S]d_m^{-1} \quad (6\text{-}1b)$$

in which

$\frac{d\,[S]}{dt}$ = rate of change of total sulfide, mg/L-hr;

3.28 = conversion from meters to feet;

M' = effective sulfide flux coefficient for sulfide generation by the slime layer in gravity sewers (empirical coefficient), m/hr;

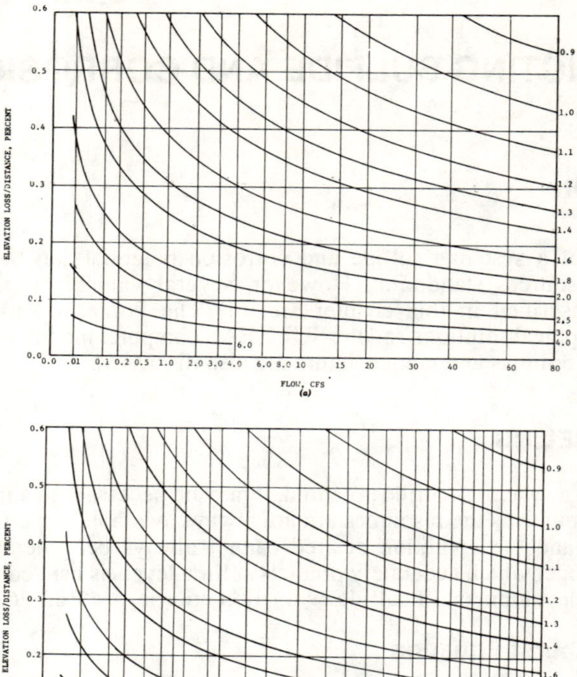

Figure 6-1. Plot of S_∞ d = 0.1 diam: (a) U.S.units; (b) S.I. units

EBOD = effective BOD, mg/L;

r = hydraulic radius, ft;

2.10 = $(3.28)^{5/8}$, conversion of meters to feet;

N = empirical factor accounting for sulfide losses by oxidation and escape to atmosphere;

$[S]$ = total sulfide concentration, mg/L;

s = slope of energy gradient, ft/ft;

v = mean wastewater velocity, ft/s; and

d_m = mean hydraulic depth (area of flow divided by surface width), ft.

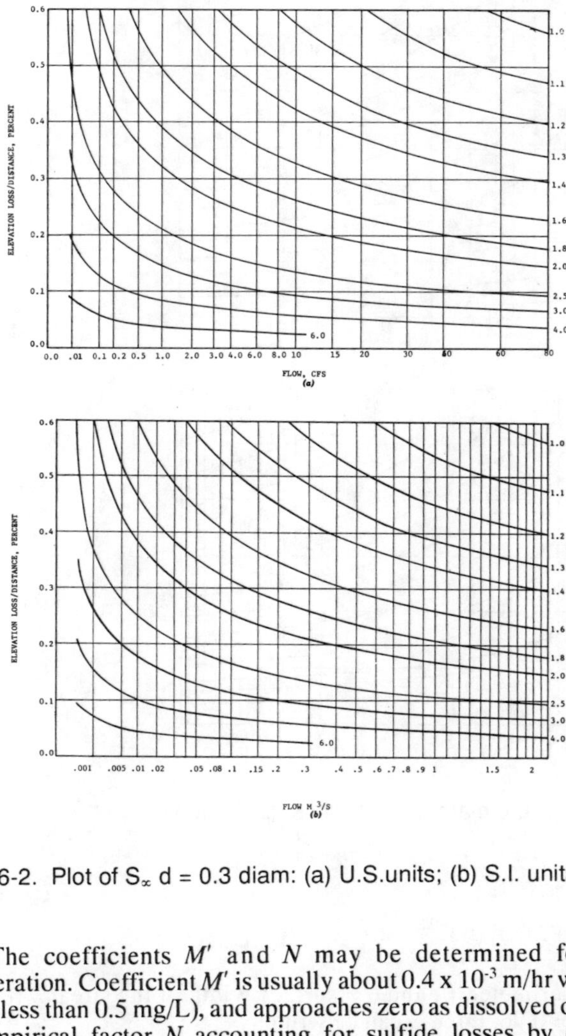

Figure 6-2. Plot of S_∞ d = 0.3 diam: (a) U.S.units; (b) S.I. units

The coefficients M' and N may be determined for the system under consideration. Coefficient M' is usually about 0.4×10^{-3} m/hr when dissolved oxygen is low (less than 0.5 mg/L), and approaches zero as dissolved oxygen levels increase. The empirical factor N accounting for sulfide losses by oxidation also varies. Pomeroy suggested in the discussion of the model that the value for M' should be 0.32×10^{3}, and N can be valued "conservatively" at 0.64 or "less conservatively" at 0.96. Analyses by J. B. Gilbert & Associates (5) and others (17, 18) demonstrated that the less conservative value of 0.96 for N adequately predicted maximum sulfide levels observed in the Sacramento, California Central Trunk sewer system.

The negative term in Eq. 6-1 is proportional to the sulfide concentration. At equilibrium, when the sulfide losses equal the sulfide generated, the rate of build-up, $d[S]/dt$ is zero. Thus, at equilibrium, Eq. 6-1 can be rewritten

$$3.28M' \,[\mathrm{EBOD}]r^{-1} = 2.10\, N[S](sv)^{0.375} d_m^{\;-1} \tag{6-2}$$

The limiting sulfide concentration $[S]_{\mathrm{lim}}$ is reached when $d[S]/dt = 0$. Therefore, $[S]_{\mathrm{lim}}$ is expressed as

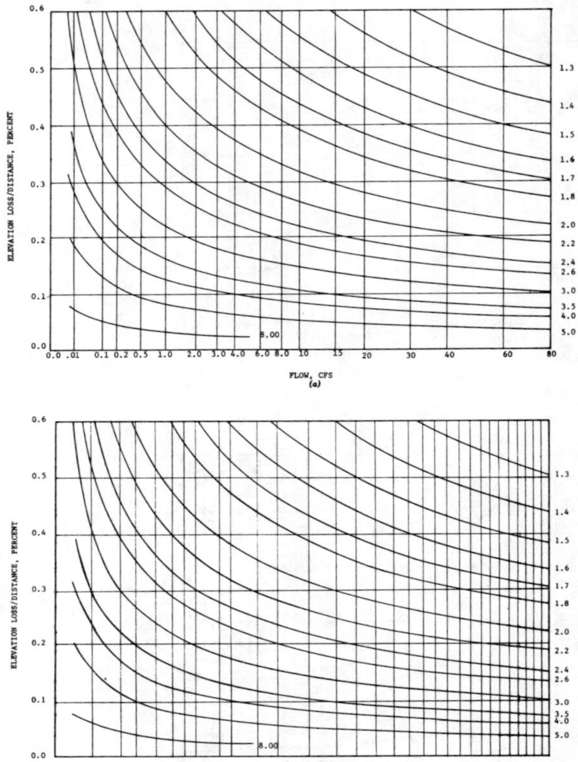

Figure 6-3. Plot of S_∞ d = 0.5 diam: (a) U.S.units; (b) S.I. units

$$[S]_{\text{lim}} = 1.56 \frac{M'}{N} [\text{EBOD}] (sv)^{-0.375} \times \frac{p}{b} \qquad (6\text{-}3)$$

in which: p = wetted perimeter, ft, and b = surface width of the stream, ft. Using $M' = 0.32 \times 10^{-3}$ and $N = 0.96$, Eq. 6-3 can be written

$$[S]_{\text{lim}} = \frac{(0.52 \times 10^{-3})\,[\text{EBOD}]}{(sv)^{0.375}} \times \frac{p}{b} \qquad (6\text{-}4)$$

When $M' = 0.32 \times 10^{-3}$ and $N = 0.64$, the numerical coefficient in Eq. 6-4 would be 0.78×10^{-3} rather than 0.52×10^{-3}.

Sulfide generation in a series of pipe sections with varying hydraulic conditions can be calculated by applying

$$t_2 - t_1 = \frac{1.10\, d_m}{N\,(sv)^{0.375}} \log \left[\frac{[S]_{\text{lim}} - S_1}{[S]_{\text{lim}} - S_2}\right] \qquad (6\text{-}5)$$

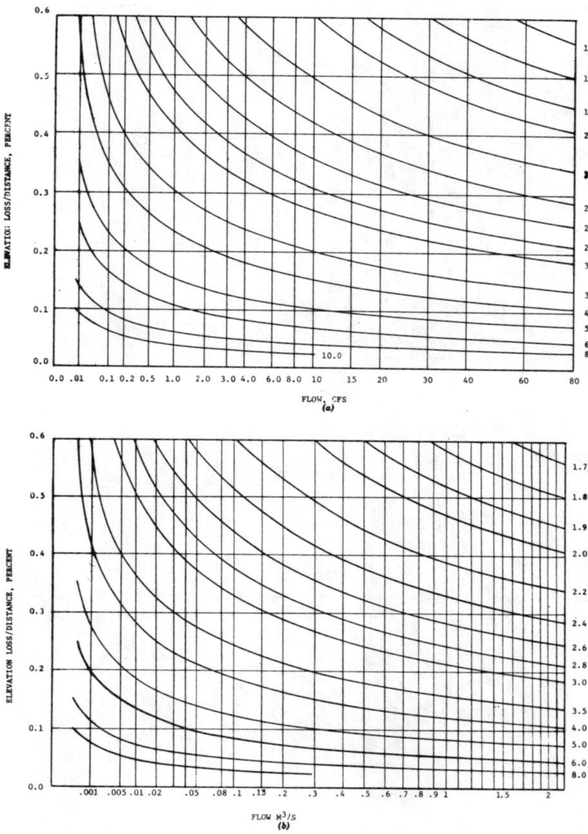

Figure 6-4. Plot of S_∞ d = 0.7 diam: (a) U.S.units; (b) S.I. units

in which $t_2 - t_1 = t$ = flow time in given reach with constant slope, diameter and flow, hr; S_1 = sulfide concentration at the upstream end of the reach, mg/L; and S_2 = sulfide concentration at downstream end of the reach, mg/L.

Using a value of 0.96 for N, Eq. 6-5 can be rearranged as follows:

$$S_2 = [\mathrm{S}]_{\mathrm{lim}} - \left[\frac{[S]_{\mathrm{lim}} - S_1}{\log^{-1}\left[\dfrac{(sv)^{0.375}\Delta t}{1.15 d_m}\right]} \right] \quad (6\text{-}6)$$

The caculated value of S_2 for the first reach will be used as S_1 for the second reach. Calculating $[S]_{\mathrm{lim}}$ and Δt, and knowing the hydraulic conditions for each reach of the system, it is possible to calculate the sulfide levels for a series of reaches. Sample calculations are presented in Appendix E.

When the characteristics of the wastewater in the tributary sewers vary significantly from those of the wastewater in the primary interceptor, the change in sulfide conditions occurring at major junctions should be considered. Each junction may be analyzed separately, and sulfide concentrations calculated using the mass balance method discussed in detail in Sec. D7. The resultant sulfide concentration

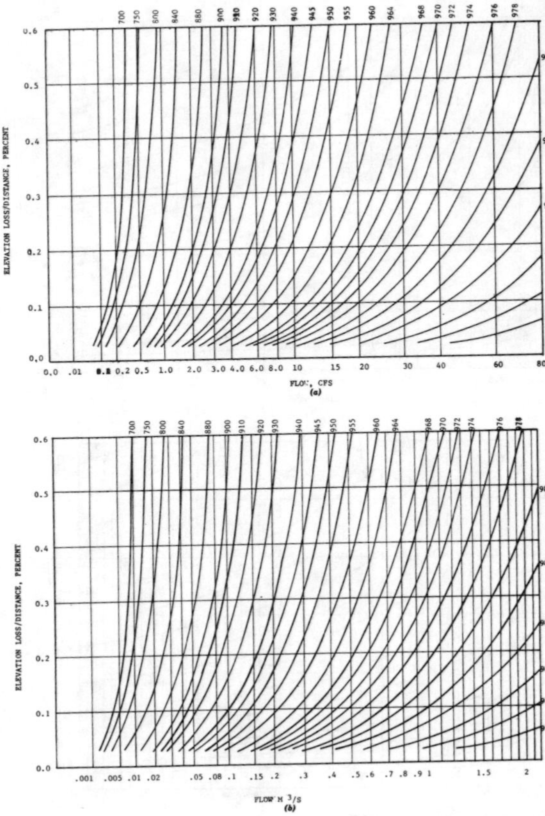

Figure 6-5. Plot of β (x 10 ^{3}) d = 0.1 diam: (a) U.S.units; (b) S.I. units

exiting the junction should in turn be used as the upstream sulfide level S_1 for the following reach.

The reliability of the Pomeroy-Parkhurst gravity model is discussed in Chap. 7. When a high percentage of the sewage is from nondomestic sources, or has a high soluble BOD (such as cannery wastes), or both, the Pomeroy-Parkhurst equation coefficients should be verified by tests on local sewer lines.

Graphical solution.—The application of the Pomeroy-Parkhurst model is tedious. Therefore, in 1982 Kienow, Pomeroy, and Kienow (13) reported the development of a graphical model intended to facilitate use of the Pomeroy-Parkhurst equation. The method uses a series of curves to evaluate the limiting total sulfide concentration S_{lim} and a factor (beta, β) combining the hydraulic variables. The curves are plotted on slope versus fiow grids for various depths, and are presented in both U.S. customary units and metric units. The curves were developed using the following values: EBOD = 500 mg/L, M'= 0.32, N = 0.96, and Manning roughness coefficient n = 0.013.

The basis of the graphical method is

$$S_{lim} - S_2 = (S_{lim} - S_1)\, \beta^{1/C_3} \qquad (6\text{-}7)$$

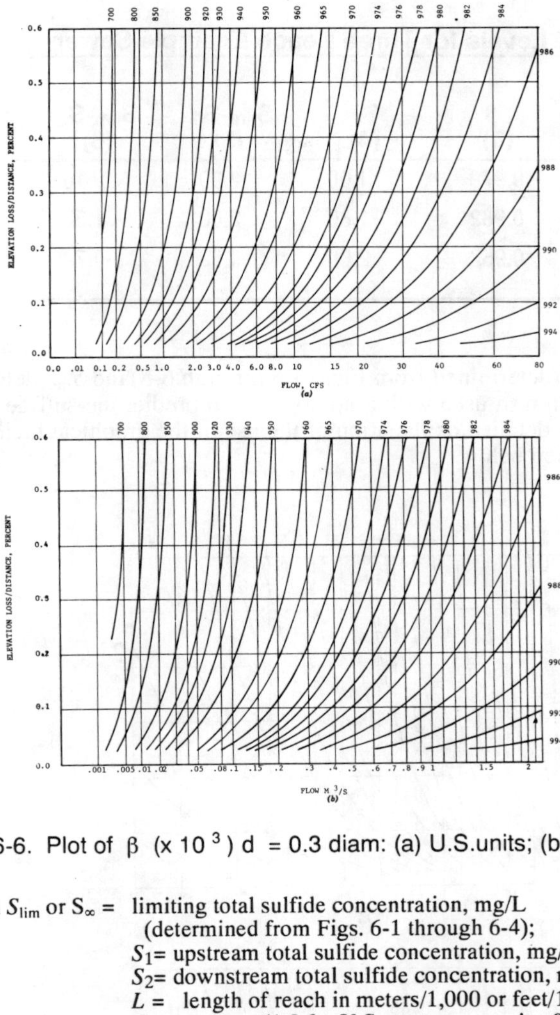

Figure 6-6. Plot of β (x 10 ^{3}) d = 0.3 diam: (a) U.S.units; (b) S.I. units

in which $S_{\lim}$ or S_∞ = limiting total sulfide concentration, mg/L (determined from Figs. 6-1 through 6-4);
S_1= upstream total sulfide concentration, mg/L;
S_2= downstream total sulfide concentration, mg/L;
L = length of reach in meters/1,000 or feet/1,000;
C_3= constant (1.0 for U.S. customary units, 0.3048 for metric units); and
β= factor combining hydraulic variables.

TABLE 6-1. Reach Data for Example Sewer

Reach number (1)	Q, cfs (2)	S, percentage (3)	d/D (4)	Length, ft (5)	S_1 mg/L (6)
1	8.0	0.09	0.50	13,650	1.0
2	10.0	0.09	0.50	4,650	
3	12.0	0.58	0.30	3,500	

TABLE 6-2. Sulfide Levels for Three Reach Example Sewer

Reach number (1)	S_{lim} (2)	β (3)	S_1 (4)	S_{lim}-S_1 (5)	S_{lim}-S_2 (6)	S_2 (7)
1	3.90	0.979	1.00	2.90	2.10	1.80
2	3.90	0.982	1.80	2.00	1.80	2.00
3	1.18	0.962	2.00	-0.82	-0.72	1.90

The values of β (determined from Figs. 6-5 through 6-8)and S_{lim}(determined using the curves) are in turn used with a nomograph to predict the sulfide level at the end of the reach. A detailed explanation of the use of the graphical method and an example are presented below.

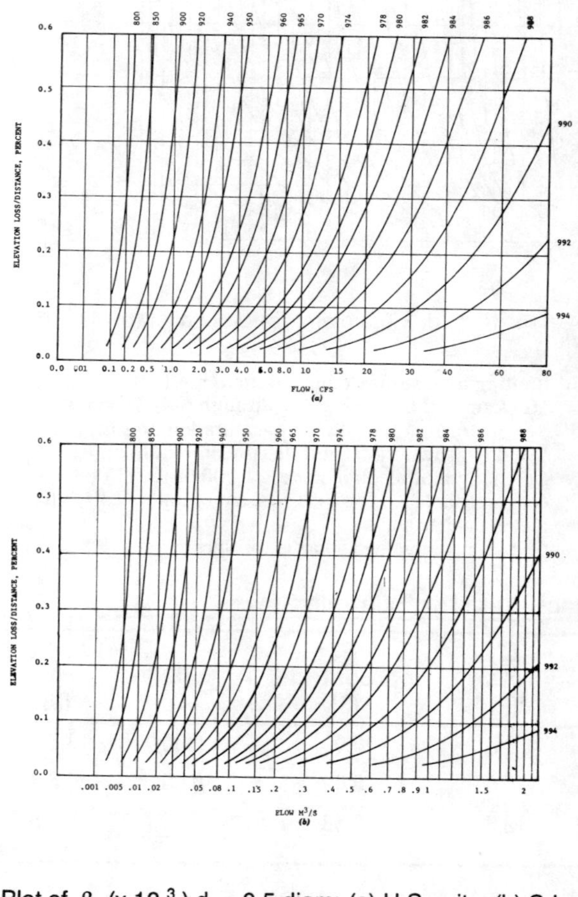

Figure 6-7. Plot of β (x 10 ^{3}) d = 0.5 diam: (a) U.S.units; (b) S.I. units

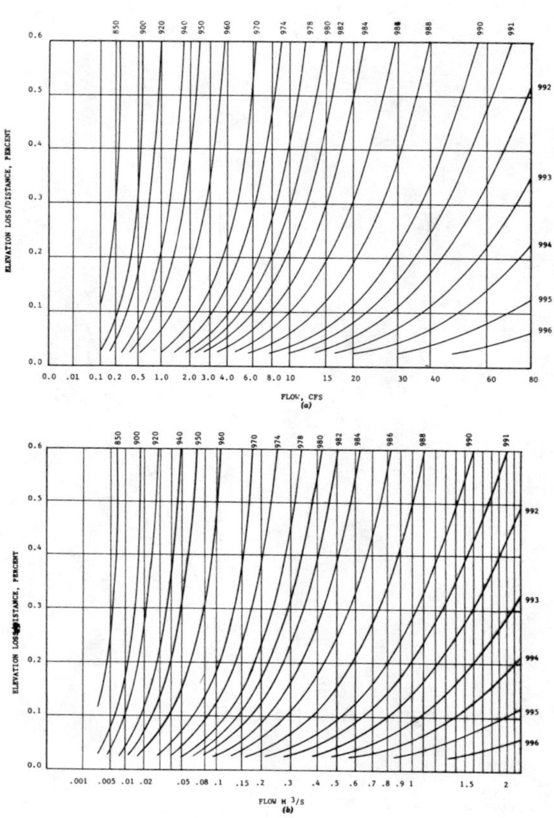

Figure 6-8. Plot of β (x 10^3) d = 0.7 diam: (a) U.S.units; (b) S.I. units

Use of prediction nomographs

1. Select the values of β and S_{lim} from the appropriate plot for the relative depth of flow.
2. Calculate the upstream sulfide deficit, S_{lim} - S_1.
3. For β values greater than 0.90, use Nomograph A (Fig. 6-9). For β values less than 0.90, use Nomograph B (Fig. 6-10). Locate β and L on the appropriate scales on the nomograph. Connect the β and L values with a straight line, extending the line to the unscaled pivot line.
4. Locate the upstream sulfide deficit S_{lim} - S_1 (determined in Step 2 above) on the left-hand scale. Draw a line from this point to the pivot point (determined in Step 3 on the unscaled line).
5. Read the downstream sulfide deficit S_{lim} - S_2 where the line drawn in Step 4 crosses the S_{lim} - S_2 scale.
6. Subtract the S_{lim} - S_2 value from S_{lim} to determine the downstream sulfide, S_2. (Note: If $S_1 > S_{lim}$, a negative value for S_{lim} - S_1 will be obtained, and sulfides will decline in the reach. The S_{lim} - S_2 quantity must be added to S_{lim} to determine S_2.)

Graphic solution design example.—Calculate the sulfide build-up for a trunk sewer consisting of three reaches (EBOD = 500 mg/L for all reaches; for EBOD values

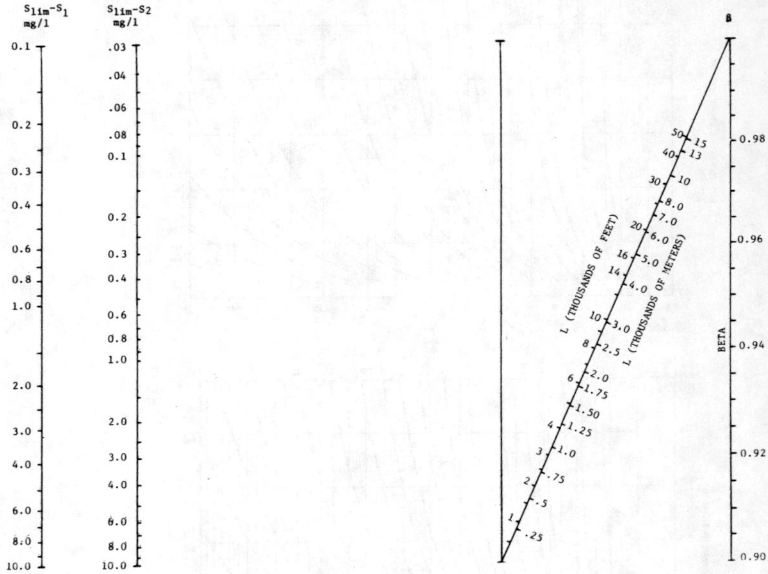

Figure 6-9. Prediction Nomograph A for β > 0.90

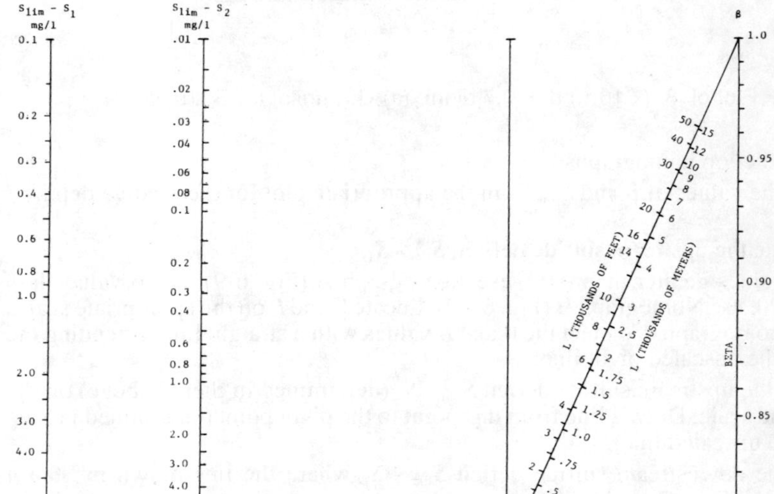

Figure 6-10. Prediction Nomograph B for β < 0.90

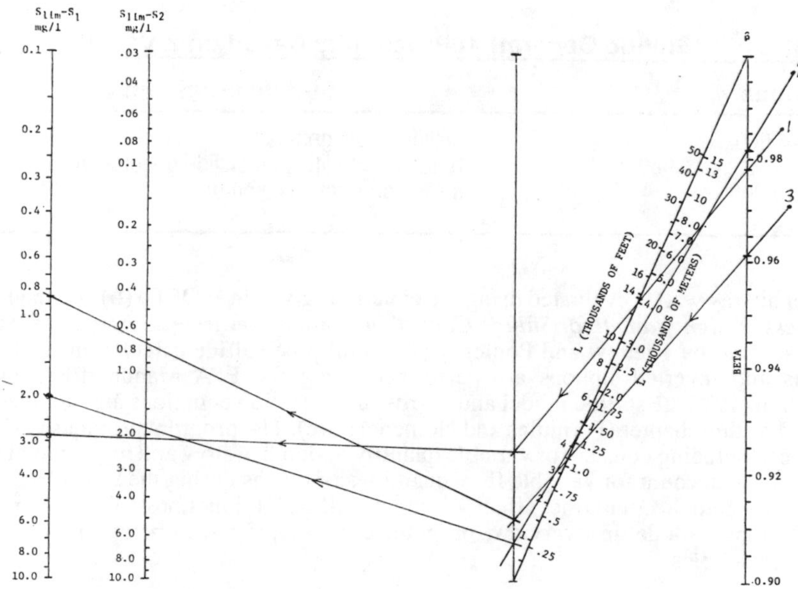

Figure 6-11. Prediction nomograph A example problem

other than 500, the plotted value should be multiplied by EBOD/500). Sulfide level entering Reach 1 is 1 mg/L. Reach data are given in Table 6-1.

The nomograph solution is shown on the example nomograph (Fig. 6- 11). For Reach 3, where a sulfide level decrease occurs, the algebraic signs for the S_{lim} - S_1 and S_{lim} - S_2 scales are changed to calculate the decrease. The calculations are tabulated in Table 6-2.

Total sulfide level at the downstream end of the three-reach sewer is predicted as 1.9 mg/L.

In actual trunk sewer systems, it is often necessary to take into account the effects of tributary flows at junctions. This may be done by computing the sulfide concentrations in the flows arriving at the junction, then calculating the sulfide leaving the junction using mass balance, as described in Sec. D7. If this is not done, the effect is the same as assuming that the sulfide concentration in a tributary flow is the same as in the main flow. Junctions also generally cause the dissolving of oxygen. This factor usually is ignored; it often causes sulfide concentrations to be somewhat lower than the calculations would indicate (5).

Computer modeling.—The nature of the Pomeroy-Parkhurst equation is such that it can easily be applied using a computer.

SULF.BAS is a Basic language computer model that predicts the sulfide build-up and corrosion protection required in a sanitary sewer system. The system may include both gravity flow and force main or full pipe sections. The prediction is based on the Pomeroy-Parkhurst model (Eqs. 6-4 and 6-6). The model is included in ASCE Manual of Practice No. 60, *Design and Construction of Gravity Sanitary Sewers* (6). The corrosion potential due to sulfide-induced acid formation on the

TABLE 6-3. Sulfide Generation Probability Based on Z Values

Z values	Sulfide condition
Z < 5,000	sulfide rarely generated
5,000 < Z < 10,000	marginal condition for sulfide generation
Z > 10,000	sulfide generation common

sewer airway wall is evaluated using the equations given in MOP 60 (6) and the EPA *Process Design Manual for Sulfide Control in Sanitary Sewerage Systems* (19), and as discussed by Kienow and Pomeroy (15). Full pipe sulfide calculations in force mains and inverted siphons are performed using the EPA Manual force main equation (19). The sulfide model and corrosion protection equations are included in Sec. D of this chapter (Formulas and Nomenclature). The program provides various options, including constant or variable quantity or depth of flow and incremental life analysis, to account for variable flow quantities or depths during the sewer life, and will take into account the effect of input sulfide at junctions. The following subsections include an overview, program options, program operation, and four design examples.

Overview—As previously indicated, the Pomeroy-Parkhurst gravity sulfide prediction model equations are used to predict downstream total sulfide level in the sewer on a reach-by-reach basis. When a high percentage of the sewage is from

TABLE 6-4. Hydraulic Properties

d/D	A/D^2	P/D	R/D	b/D	V/V_{Fa}	P/b	d_m/D	P′/D	b/P′
0.15	.074	0.795	0.093	0.714	0.517	1.11	.104	2.346	.304
0.20	.112	0.927	0.121	0.800	0.617	1.16	.140	2.214	.361
0.25	.154	1.047	0.147	0.866	0.702	1.21	.178	2.094	.414
0.30	.198	1.159	0.171	0.914	0.776	1.27	.217	1.982	.461
0.35	.245	1.266	0.194	0.954	0.845	1.32	.257	1.876	.509
0.40	.293	1.369	0.214	0.980	0.902	1.40	.299	1.772	.553
0.45	.343	1.471	0.233	0.995	0.954	1.48	.345	1.671	.595
0.50	.393	1.571	0.250	1.00	1.000	1.57	.393	1.571	.637
0.55	.443	1.671	0.265	0.995	1.040	1.68	.445	1.471	.676
0.60	.492	1.772	0.278	0.980	1.073	1.81	.502	1.369	.716
0.65	.540	1.876	0.288	0.954	1.099	1.97	.566	1.266	.754
0.70	.587	1.982	0.296	0.914	1.119	2.17	.642	1.159	.789
0.75	.6321	2.094	0.301	0.866	1.134	2.42	.730	1.047	.827
0.80	.674	2.214	0.304	0.800	1.139	2.77	.843	0.927	.863

[a] Based on Manning's n constant at all depths.Camp's variable n (Fig. 6-13) gives a closer approximation of V and Z at relative depths greater than 0.25D.

nondomestic sources or has a high soluble BOD, or both, the Pomeroy-Parkhurst equation coefficients should be verified by tests on local sewer lines. The full pipe (force main, inverted siphon, or surcharged gravity sewer) sulfide build-up calculation uses Eq. 6-13 given subsequently herein. Equations 6-27 and 6-28 are used to calculate the corrosion protection *Az* required for cementitious pipe material, such as concrete or asbestos cement pipe. The results may also be used to determine the required mortar lining thickness on ferrous pipe material, such as steel or ductile iron pipe. The use of the *Az* factor in designing corrosion protection for concrete, asbestos cement, ductile iron, or steel pipe, is discussed in Chap. 10. Example calculations for sulfide prediction and RCP corrosion protection are illustrated in Secs. B and C of Appendix E.

Program options.—The design period may be broken up into any number of increments to account for changing (generally increasing) flow conditions over the design life.

A crown corrosion factor CCF of 2.0 is used to calculate the required corrosion protection (Eq. 6-27). According to the 1974 EPA Manual (19), pipe crown corrosion may be 1-1/2 times the average unwetted perimeter corrosion, where uniform flow conditions exist. The result, from a practical standpoint, of using a CCF factor of 2.0 is that, at the end of the design life, 1/3 of the design corrosion protection will remain in the pipe crown. References 1 and 3 use an accelerated crown corrosion factor of 1.5 in the design examples. The use of 2.0 results in a slightly more conservative design, and assures that a reasonable amount of protection remains during the final years of design life.

To handle variations in flow quantity, four options are available. The program prompts for the following options:

1. Constant relative depth (Input Option 1): A constant relative depth of flow, i.e., 0.25 diameters (1/4 full), will be used for all reaches of the sewer during a given increment of the design life.
2. Variable relative depth (Input Option 2): A different relative depth value may be input for each reach of the sewer. If any reach consists of a force main or inverted siphon, this input option must be used, and 1.0 must be input for the relative depth of the force main or siphon reach.
3. Constant flow quantity (Input Option 3): A constant flow quantity will be used for all reaches of the sewer. Relative depths of flow will be calculated based on Manning's $n = 0.013$ and the slope and pipe diameter for the reach.
4. Variable flow quantity (Input Option 4): A different flow quantity may be input for each reach of the sewer. Relative depths of flow will be calculated for each reach as in Option 3.

Program operation.—For primary data input, sewage characteristics required are climactic BOD and sewage temperature as defined in Sec. 3.5.4 of Ref. 27; other primary input data are: (1) Design life, which may be broken into increments, with a different flow quantity or relative depth during each increment; (2) acid reaction factor k; (3) pH of the sewage; (4) upstream total sulfide level S_1, mg/L; (5) insoluble sulfides *IS*, mg/L; and the climactic ratio c. Climactic ratio is the ratio of annual average sulfide to the sulfide occurring during the climactic period when sulfide levels in the system are highest.The climactic ratio varies from 0.2 to 0.5, increasing with higher EBOD values. All of the input variables except climactic ratio are discussed in Refs. 1 and 3. Total sulfide must be greater than insoluble sulfide.

For reach data input, note that "reach" is a section of sewer with constant hydraulic conditions, i.e., slope, pipe size, and depth of flow. The program prompts first for the number of reaches to be analyzed; then for the pipe diameter in inches, slope, and length of the reach in feet for each reach in succession, beginning at the upstream end of the sewer. Beginning in Reach 2, the second reach downstream, the program prompts for intercepted sulfides. If an intercept occurs at the upstream end of Reach 2, the sulfide level in the intercepted flow in mg/L is input, followed by the quantity of flow intercepted, in ft^3/s. The program will mass balance the predicted sulfide level at the end of Reach 1 with the intercepted sulfide at the junction of Reaches 1 and 2 to determine S_1, the beginning sulfide level for Reach 2. The mass balance is calculated using the flow quantity in Reach 1 and the intercepted flow quantity. The number of reaches which can be input is limited only by the available data storage area in the particular computer used or by the "Dimension" statement on the variable $N(R)$.

To handle siphon or force main reach data, any reach except the first may be a full pipe (inverted siphon or force main). Input Option 2, described previously, must be used. The force main or siphon diameter and length are input, and zero is input for slope. The force main flow quantity Q is taken as the flow quantity of the previous reach. A relative depth of 1.0 must be input for the force main. Because S_{lim} is not defined for a full pipe, it is set to 999.99. A full pipe will not be subject to sulfide-induced acid corrosion; therefore, the corrosion protection Az is set to zero.

The program's output includes reach number and increment of design life if an incremental life has been used. For each increment of the design life, the relative depth of flow as the ratio of depth to diameter, quantity of flow in cubic feet per second, and velocity in feet per second are tabulated. The limiting sulfide value for the reach S_{lim}, the upstream sulfide level S_1 in the reach (taking into account intercepted sulfides, if any), and the predicted downstream sulfides S_2 are output. The required corrosion protection factor Az for the reach is listed for each increment of design life. After each increment of life is analyzed and the incremental output listed, a summary line is printed which includes the pipe diameter for the reach, the slope of the reach, and the corrosion protection Az required for the total design life. (The sum of incremental life periods in years equals the design life, and the sum of the corrosion protection values required for each increment of life is the corrosion protection required for the design life of the pipe.)

If an incremental life analysis has been performed, the life increments in years are printed. The final line of the output indicates the effective biochemical oxygen demand for the climactic BOD and sewage temperature input ($EBOD_c$).

Design examples.—Three design examples are given in this subsection.

For Example 1, using Input Option 1 (constant relative depth), the sewage characteristics and climactic ratio are given below:

Example No. 1 PRIMARY DATA INPUT
(Incremental Life Analysis with 4 Increments)

Your System Choice:1
CONSTANT RELATIVE DEPTH
START PRIMARY INPUT

Bio-chemical Oxygen Demand?	= 250 mg/L
Climactic Temperature?	= 28 deg C
Design Life, increment 1?	= 5 years
Design Life, increment 2?	= 5 years
Design Life, increment 3?	= 10 years
Design Life, increment 4?	= 30 years
Acid-Reaction Factor k?	= 0.75

Sewage pH	= 7.4
Climactic Ratio C?	= 0.4
Upstream Sulfide Level?	= 1 mg/L
Insoluble Sulfides?	= 0.2 mg/L

PRIMARY INPUT COMPLETED
REACH DATA INPUT
Number of Reaches = 3

Design Rel. Depth, increment 1?	= 0.2
Design Rel. Depth, increment 2?	= 0.3
Design Rel. Depth, increment 3?	= 0.45
Design Rel. Depth, increment 4?	= 0.6

START REACH DATA

Pipe Diameter 1?	= 36 inches
Slope of Reach 1?	= 0.0015 ft/ft
Length of Reach 1?	= 5300 ft
Pipe Diameter 2?	= 39 inches
Slope of Reach 2?	= 0.0018 ft/ft
Length of Reach 2?	= 10500 ft

The system to be analyzed consists of pipe diameters, slopes, and reach lengths. The design life is 50 years, with a depth of flow (at average dry-weather flow) in the interceptor line estimated to vary as follows:Years 1-5, d = 0.2 diam; years 6-10, d = 0.3 diam; years 11-20, d = 0.45 diam; and years 21-50, d = 0.6 diam.

In order to design a sulfide control system, the anticipated sulfide levels along the sewer during each of the flow periods is needed. The required corrosion protection is to be determined for each reach of the line, assuming that no sulfide control system is constructed.

The program output is listed below:

Pipe Diameter 3?	= 42 inches
Slope of Reach 3?	= 0.00155 ft/ft
Length of Reach 3?	= 10500 ft

REACH DATA INPUT COMPLETED
Computer Output:
ASCE EXAMPLE NO. 1
Project Engineer: K. K. KIENOW 4-2-85 Page 1

RCH, INC	REL DEPTH (d/DIA)	FLOW QNITY (cfs)	FLOWS VLCTY (fps)	SULFIDE LIMIT (mg/L)	U.STRM SULF S1 (mg/L)	DN.STRM SULF S2 (mg/L)	Az FACTOR
1 , 1	0.20	2.27	2.25	2.19	1.00	1.37	0.01
1 , 2	0.30	5.07	2.84	2.19	1.00	1.22	0.01
1 , 3	0.45	10.79	3.50	2.36	1.00	1.15	0.03
1 , 4	0.60	17.40	3.93	2.77	1.00	1.12	0.10
Pipe Diam 1=36 in. Slope = 0.0015 ft/ft 50 Year Life Az = 0.146							
2 , 1	0.20	3.08	2.60	1.93	1.37	1.64	0.01
2 , 2	0.30	6.88	3.29	1.93	1.22	1.44	0.01

2 , 3	0.45	14.63	4.04	2.09	1.15	1.32	0.04
2 , 4	0.60	23.60	4.54	2.45	1.12	1.29	0.13

Pipe Diam 2=39 in. Slope = 0.0018 ft/ft 50 Year Life Az = 0.196

3 , 1	0.20	3.48	2.54	2.07	1.64	1.83	0.01
3 , 2	0.30	7.78	3.20	2.07	1.44	1.62	0.02
3 , 3	0.45	16.54	3.94	2.23	1.32	1.47	0.04
3 , 4	0.60	26.68	4.43	2.62	1.29	1.43	0.14

Pipe Diam 3 =42 in. Slope = 0.00155 ft/ft 50 Year Life Az = 0.208
LIFE INCREMENTS (YRS)
INC 1 = 5 INC 2 = 5 INC 3 = 10 INC 4 = 30
Effective Biochemical Oxygen Demand = 430 mg/L

From the output sheet, it can be seen that the maximum downstream sulfide will be 1.8 (1.83) mg/L during the initial five-year low flow period, and 1.6 (1.62) mg/L during the next five years. For years 11 to 20, sulfide of about 1.5 mg/L should be anticipated. When design flow is reached, at 20 years, sulfide will be on the order of 1.4 mg/L during the climactic period (the high-flow six hours of the day during the warmest three months of the year). The required corrosion protection *Az* for each reach is 0.146, 0.196, and 0.208, for the required 50-year design life.

Example 2 is a force main problem. The analysis in Example 1 is repeated, except that Reach 2 is an 18 in. diam force main 3,000 ft long. The output is listed below:

Example No. 2 PRIMARY DATA INPUT
(Incremental Life Analysis with 4 Increments)

Your System Choice:2
VARIABLE RELATIVE DEPTHS
START PRIMARY INPUT

Bio-chemical Oxygen Demand?	= 280 mg/L
Climactic Temperature?	= 28 deg C
Design Life, increment 1?	= 5 years
Design Life, increment 2?	= 5 years
Design Life, increment 3?	= 10 years
Design Life, increment 4?	= 30 years
Acid-Reaction Factor k?	= 0.75
Sewage pH?	= 7.4
Climactic Ratio C?	= 0.4
Upstream Sulfide Level?	= 1 mg/L
Insoluble Sulfides?	= 0.2 mg/L

PRIMARY INPUT COMPLETED
REACH DATA INPUT
Number of Reaches = 3
START REACH DATA INPUT

Pipe Diameter 1?	= 36 inches
Slope of Reach 1?	= 0.0015 ft/ft
Length of Reach 1?	= 5300 ft

Design Rel. Depth for Reach 1, increment 1? = 0.2
Design Rel. Depth for Reach 1, increment 2? = 0.3
Design Rel. Depth for Reach 1, increment 3? = 0.45
Design Rel. Depth for Reach 1, increment 4? = 0.6

Pipe Diameter 2? = 18 inches
Slope of Reach 2? = 0 ft/ft
Length of Reach 2? = 3000 ft

Design Rel. Depth for Reach 2, increment 1? = 1
Design Rel. Depth for Reach 2, increment 2? = 1
Design Rel. Depth for Reach 2, increment 3? = 1
Design Rel. Depth for Reach 2, increment 4? = 1

Pipe Diameter 3? = 42 inches
Slope of Reach 3? = 0.00155 ft/ft
Length of Reach 3? = 10500 ft

Design Rel. Depth for Reach 3, increment 1? = 0.2
Design Rel. Depth for Reach 3, increment 2? = 0.3
Design Rel. Depth for Reach 3, increment 3? = 0.45
Design Rel. Depth for Reach 3, increment 4? = 0.6

REACH DATA INPUT COMPLETED
COMPUTER OUTPUT:
ASCE Example No. 2
Project Engineer: K. K. KIENOW 4-2-85 Page 1

RCH , INC	REL DEPTH (d/DIA)	FLOW QNITY (cfs)	FLOW VLCTY (fps)	SULFIDE LIMIT (mg/L)	UP.STRM SULF S1 (mg/L)	DN.STRM SULF S2 (mg/L)	Az FACTOR
1 , 1	0.20	2.27	2.25	2.45	1.00	1.45	0.01
1 , 2	0.30	5.07	2.84	2.45	1.00	1.27	0.01
1 , 3	0.45	10.79	3.50	2.65	1.00	1.18	0.03
1 , 4	0.60	17.40	3.93	3.10	1.00	1.15	0.10
Pipe Diam 1=36 in. Slope = 0.0015 ft/ft 50 Year Life Az = 0.151							
2 , 1	1.00	2.27	1.28	999.99	1.45	4.67	0.00
2 , 2	1.00	5.07	2.87	999.99	1.27	2.71	0.00
2 , 3	1.00	10.79	6.11	999.99	1.18	1.86	0.00
2 , 4	1.00	17.40	9.85	999.99	1.15	1.57	0.00
Pipe Diam 2=18 in.Slope = 0 ft/ft 50 Year Life Az = 0.000							
3 , 1	0.20	3.48	2.54	2.31	4.67	3.62	0.04
3 , 2	0.30	7.78	3.20	2.31	2.71	2.60	0.03
3 , 3	0.45	16.54	3.94	2.50	1.86	1.96	0.05
3 , 4	0.60	26.68	4.43	2.93	1.57	1.72	0.17

Pipe Diam 3=42 in. Slope=0.00155 ft/ft 50 Yr. Life Az = 0.290
LFE INCREMENTS (yrs)
INC 1 = 5 INC 2 = 5 INC 3 = 10INC 4 = 30
Effective Biochemical Oxygen Demand = 481 mg/L

The sulfide build-up in the force main during the early low-flow period is significant, increasing from 1.45 to 4.67 mg/L in the 3,000-ft force main. Sulfide build-up in the force main after design flow is reached is 0.42 mg/L. The required corrosion protection *Az* for Reach 3 is 0.29.

Example 3 provides results for variable flow quantity with intercepted sulfide. A variable flow quantity (Option 4) analysis, with intercepted sulfides, is desired for the same sewer analyzed in Example 1. The flow quantity is 15 ft^3/s for Reaches 1 and 2. An intercept of 5.0 ft^3/s, with 3.5 mg/L total sulfide in the intercepted flow, occurs at the junction of Reaches 2 and 3. The intercept data is entered as part of the reach data input as shown below:

Example No. 3 PRIMARY DATA INPUT

Your System Choice: 4
VARIABLE FLOW QUANTITIES
START PRIMARY INPUT

Bio-chemical Oxygen Demand?	= 280 mg/L
Climactic Temperature?	= 28 deg C
Design Life?	= 50 years
Acid-Reaction Factor k?	= 0.75
Sewage pH?	= 7.4
Climactic Ratio C?	= 0.4
Upstream Sulfide Level?	= 1 mg/L
Insoluble Sulfides?	= .2 mg/L

PRIMARY INPUT COMPLETED
REACH DATA INPUT
Number of Reaches = 3
START REACH DATA INPUT

Pipe Diameter 1?	= 36 inches
Slope of Reach 1?	= 0.0015 ft/ft
Length of Reach 1?	= 5300 ft
Design Flow Quantity for Reach 1?	= 15 cfs
Pipe Diameter 2?	= 39 inches
Slope of Reach 2?	= 0.0018 ft/ft
Length of Reach 2?	= 10500 ft
Design Flow Quantity for Reach 2?	= 15 cfs
Pipe Diameter 3?	= 42 inches
Slope of Reach 3?	= 0.00155 ft/ft
Length of Reach 3?	= 10500 ft
Design Flow Quantity for Reach 3?	= 20 cfs
Intercepted Sulfide at start of Reach 3?	= 3.5 mg/L
Intercepted Flow Quantity at start of Reach 3?	= 5 cfs

REACH DATA INPUT COMPLETED
COMPUTER OUTPUT
ASCE Example No. 3
Project Engineer:K. K. KIENOW 4-2-85 Page 1

RCH	REL DEPTH (d/DIA)	FLOW QNITY (cfs)	FLOW VLCTY (fps)	SULFIDE LIMIT (mg/L)	UP.STRM SULF S1 (mg/L)	DN.STRM SULF S2 (mg/L)	Az FACTOR
1	0.55	15.00	3.80	2.90	1.00	1.15	0.16
2	0.46	15.00	4.07	2.35	1.15	1.37	0.19
3	0.50	20.00	4.14	2.62	1.90	2.01	0.30

Effective Biochemical Oxygen Demand = 481 mg/L

The effect of the high sulfide intercept is to increase sulfide from 1.37 to 1.90 mg/L at the junction of Reaches 2 and 3. The program performs the mass balance for junctions, as described in Secs. B1 and D7.

Important note.—The design *Az* calculated is that which should be specified for the main line pipe. The EPA corrosion rate formula (Eq. 6-27) calculates the average perimeter corrosion for uniform hydraulic conditions; therefore, some additional protection is required downstream of those structures where non-uniform flow (turbulence) may occur, such as major intercepts, angle points, or grade changes.

The Sacramento central trunk study (5) showed that, where moderate turbulence occurs, the pipe crown corrosion could be from 2 to 5 times the average (uniform flow) perimeter corrosion given by the EPA corrosion rate formula. It is a common design practice to specify double the calculated main-line corrosion protection *Az* for a distance of 10 diameters, or 50 ft minimum, downstream of such structures. Doubling the computed main-line design *Az* provides an accelerated corrosion factor (CCF x TCF) of 4.0 on the calculated uniform-flow average unwetted perimeter corrosion when the design *Az* required is calculated using a factor of 2 (CCF) on average perimeter corrosion, as is done in this program. The upstream or downstream *Az*, whichever is larger, should be doubled and applied downstream of the structures where turbulence occurs. For additional discussion of turbulence effects on corrosion rates, see Sec. D9 of this chapter. Where high turbulence occurs, either an inert pipe material or an inert lining should be provided to protect the pipe if the design cannot be modified to eliminate the turbulence.

C. CORROSION EVALUATION

Two important aspects of corrosion evaluation are physical inspection of corroded sewer lines and calculation of a sewer's remaining life expectancy.

1. Physical Inspection

Corrosion is the graphic end result of the presence of significant levels of hydrogen sulfide in the collection system. The degree of corrosion can yield valuable insight into the condition of the sewer and its expected life. Data gathered during the in-line study addresses the corrosion experienced near manholes. The degree of corrosion experienced between manholes can be observed by internal inspection of the sewer using closed circuit television. When corrosion of reinforced

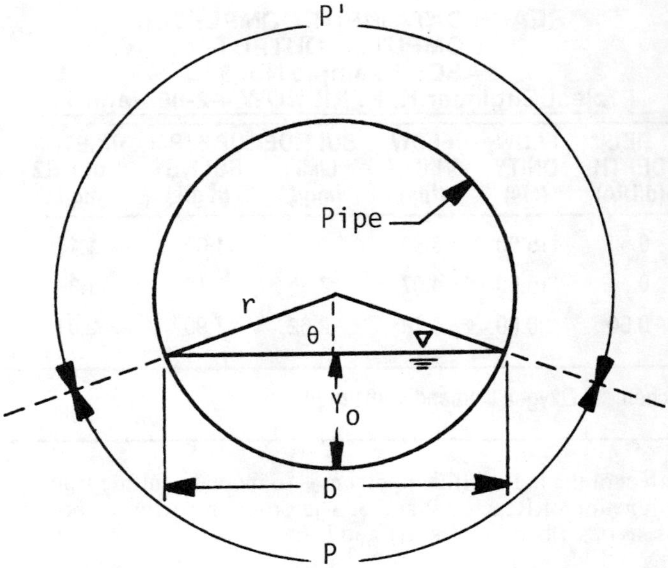

Figure 6-12. Hydraulic factors

concrete pipe has progressed to the point that the reinforcing steel is exposed, the corrosion is regarded as serious.

While visual inspections give good indications as to the presence and degree of corrosion, the most reliable method to determine the degree of corrosion is to take core samples. The samples can then be analyzed, and dimensions compared to the original pipe design or to core samples taken below the flow line, to define the exact degree of corrosion (see Chap. 8).

2. Calculations

To evaluate the useful life of a sewer, the rate of corrosion and current condition of the sewer must be determined. The current condition of the sewer is defined as outlined previously. Determining the future rate of corrosion, however, is much more involved.

Corrosion rate determination was addressed by the Environmental Protection Agency (19) in 1974. The EPA Manual indicated that the corrosion of a pipe depends on the pipe material, as well as on the rate of release of hydrogen sulfide from the wastewater stream. Several equations were presented, and the conditions defining their use are presented in Secs. D8, D9, and D10.

The flux of hydrogen sulfide to the sewer wall is calculated as follows:

$$\phi_{sw} = 0.45\,(sv)^{3/8}\,J\,[DS]\,\frac{b}{p'} \tag{6-8}$$

in which

ϕ_{sw} = flux of hydrogen sulfide to the sewer wall, g/m^2-hr;

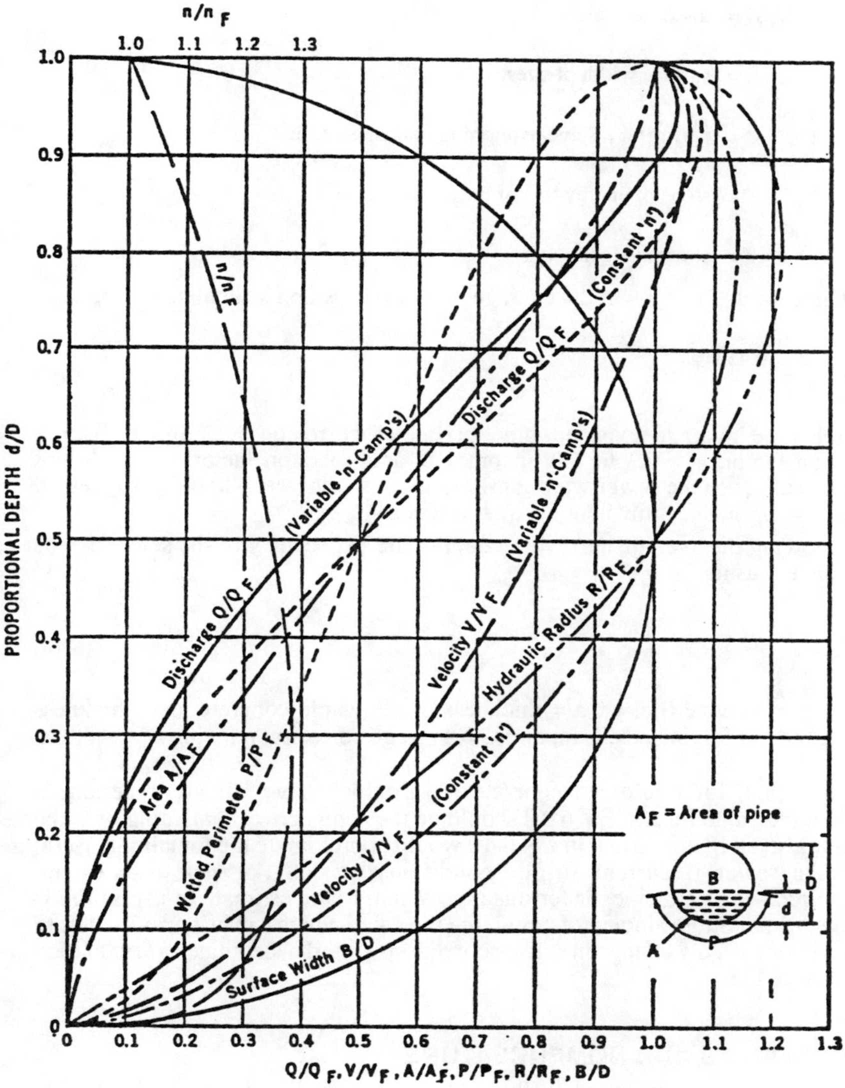

Figure 6-13. Hydraulic elements of circular sewers running partly full

$0.45 = (3.28)^{-3/8}(0.7)$, conversion factor from meters to feet;

J = factor relating fraction of dissolved sulfide as hydrogen sulfide to wastewater pH;

$[DS]$ = dissolved sulfide, mg/L;

b = surface width of stream, ft;

P' = perimeter of pipe exposed to atmosphere, ft;

s = slope of energy line of the stream; and

v = stream velocity, ft/s.

Once the ϕ_{sw} is known, the average rate of corrosion is calculated using

$$C_{avg} = \frac{0.45\, k\, \phi_{sw}}{A} \tag{6-9}$$

in which C = average exposed pipe perimeter corrosion rate, in./yr; 0.45 = conversion factor; k = 0.3 to 1.0, incomplete acid reaction factor; ϕ_{sw} = flux of hydrogen sulfide to the sewer wall, g/m^2-hr; and A = concrete alkalinity, expressed as calcium carbonate, equivalent decimal fraction.

Knowing the average corrosion rate, the life expectancy of the sewer can be approximated using

$$L = \frac{z}{C} \tag{6-10}$$

in which L = service life, yr; z = thickness of allowable concrete loss (thickness covering the reinforcing steel), in; and C = average corrosion rate, in./yr.

Note that, for reinforced concrete pipes which have already experienced corrosion, the value for z in Eq. 6-10 should be the actual cover remaining over the reinforcing steel. In this case L in Eq. 6-10 would approximate the remaining useful life of the sewer if current sulfide conditions persist. For asbestos cement, nonreinforced concrete pipe, or ferrous pipe material, the estimate of useful life is based on the maximum amount of remaining pipe wall which can be corroded before failure will occur. Any change in sulfide conditions would necessitate a reevaluation of the service life.

D. FORMULAS AND NOMENCLATURE

The 11 terms considered in this section are EBOD, Pomeroy Z formula, quantitative sulfide forecasting (both filled pipe and partially filled pipe conditions), hydraulic factors, climactic EBOD, sulfide calculation at junctions (mass balance), corrosion calculations, peak corrosion rates, *Az* design equations, and units of measurement used in odor control.

1. EBOD

Effective biochemical oxygen demand is defined by

$$\mathrm{EBOD} = \mathrm{BOD}_5 \times 1.07^{(T_C - 20)} \tag{6-11}$$

in which EBOD is in milligrams per liter; BOD_5 = five- day BOD, in milligrams per liter; T_c= sewage temperature, in degrees Celsius; and 1.07 = empirical coefficient which may vary with wastewaters having uncommon characteristics (see discussion of percentage soluble BOD in Chap. 3).

2. Pomeroy Z Formula

One indicator of the likelihood of sulfide build-up in relatively small gravity sewers (not over 600-mm or 24-in. diameter) is afforded by the Z formula (25):

$$Z = \frac{\mathrm{EBOD}}{S^{0.50}\, Q^{0.33}} \, \frac{P}{b} \tag{6-12}$$

in which Z = defined function; S = hydraulic slope; Q = discharge volume, in cubic feet per second; P = wetted perimeter, in feet; and b = surface width, in feet. Sulfide generation probability is approximated in Table 6-3.

Regarding Table 6-3, note that both the 1970 Pomeroy (25) Z Formula (Eq. 6-12) and later the EPA (19) flow-slope relationship (EPA Fig. 3-13) were presented as crude interim qualitative guidelines, before research had advanced to a point at which a rational predictive equation could be developed. The use of either of the older qualitative predictive tools should be discouraged, and new systems should be modeled using the 1977 Pomeroy-Parkhurst (22) equation (Eq. 6-1). For a detailed history of the development of sulfide evaluation and predictive criteria, see Ref. 26.

3. Quantitative Sulfide Forecasting for Filled Pipe

In completely filled sewer pipes (force mains), an estimate of sulfide build-up can be obtained using the following equation (20, 15):

$$\frac{d[S]}{dt} = 3.28(M)\,[\mathrm{EBOD}]\,(1 + 0.48r)\,r^{-1} \tag{6-13}$$

in which S = total sulfide, mg/L; t = flow time, hr; M = specific sulfide flux coefficient, m/hr; [EBOD] = effective BOD_5, mg/L; and, r = hydraulic radius, ft = $D/4$ for full pipe.

The use of M equal to 0.3 mm/hr (0.001 ft/hr) is intentionally conservative. Actual build-up will be less than predicted in about 80% of the cases principally because of dissolved oxygen in the wastewater before it enters the main. The use of M = 0.3 mm/hr (0.001 ft/hr) is advised unless there is good reason to believe that there will be a substantial delay in the onset of sulfide generation, or if sulfide build-up studies indicate another value is appropriate.

4. Quantitative Sulfide Forecasting for Partially Filled Pipe

In 1977, Pomeroy and Parkhurst presented a quantitative method for sulfide prediction in gravity (partly filled) sanitary sewers (Eq. 6-1), based on continuing research on the wastewater collection systems of Los Angeles County and other data (22). The method assumes that all conditions are favorable for sulfide build-up, Conservative (high) sulfide predictions may be obtained under other conditions. Sulfide build-up will not occur when the wastewater contains significant dissolved oxygen, or when sufficient nutrients are absent, as indicated by a low BOD.

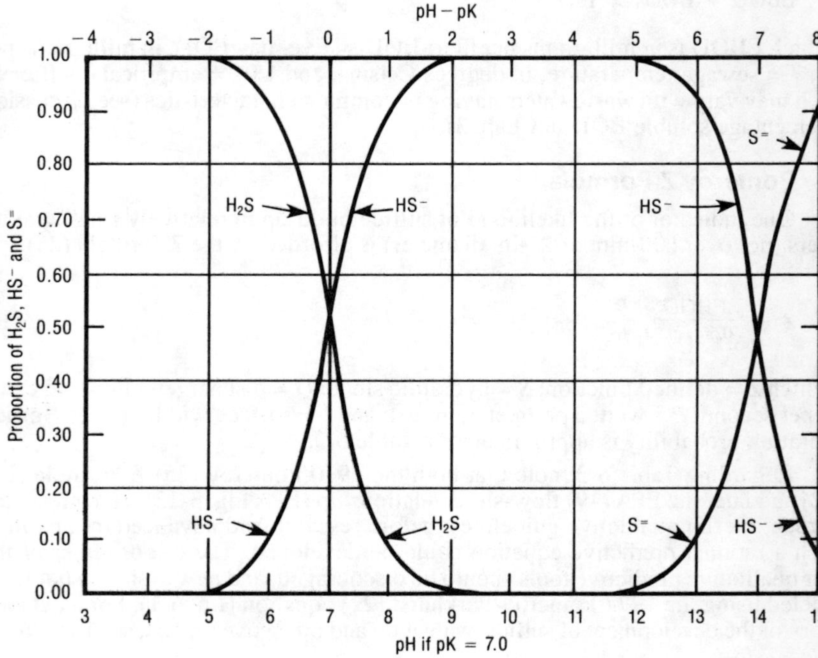

Figure 6-14. Distribution of Hydrogen sulfide components in aqueous solution at 25° c

Therefore, the method denoted by Eq. 6.1a in this chapter will predict too high sulfide levels for systems where dissolved oxygen exceeds 0.5 mg/L:

$$\frac{d[S]}{dt} = M' \,[\mathrm{EBOD}]\, r^{-1} - N(sv)^{0.375}\,[S] d_m^{-1} \tag{6-1}$$

The coefficients M' and N may be established to suit the conditions of the system under consideration. As noted previously, however, M' is usually about 0.4 x 10^{-3} m/hr (1.3 x 10^{-3} ft/hr) when dissolved oxygen is low (less than 0.5 mg/L) and approaches zero as dissolved oxygen concentrations increase. The sulfide loss coefficient N is a parameter for sulfide losses due to oxidation and escape of H_2S to the atmosphere. Pomeroy has suggested values for M' of 0.32 x 10^{3} m/hr (1.0 x 10^{-3} ft/hr) and, for N, either a "conservative" value of 0.64 or a "less conservative" value of 0.96. Analysis of several in-line studies led to the view that the less conservative value for N was adequate to predict maximum sulfide levels where major sanitary sewers of several miles in length were involved (13).

The negative term of Eq. 6-1 is proportional to the sulfide concentration. The sulfide concentration approaches a limiting concentration $[S]_{\mathrm{lim}}$. When $d[S]/dt = 0$, and taking N equal to 0.96, the equation may be rearranged for the sulfide limiting condition, as shown by

$$[S]_{\lim} = \frac{0.52 \times 10^{-3}\,[\text{EBOD}]}{(sv)^{0.375}} \times \frac{d_m}{r} \tag{6-4}$$

in which

$$d_m = \frac{\text{area of flow}}{\text{width of flow surface}} = \frac{A}{b}$$, and units are in feet or meters as appropriate.

To calculate the sulfide generation in a long reach with uniform slope and flow, the integrated forms of Eq. 6-4 may be applied as shown by

$$S_2 = [S]_{\lim} - \left[\frac{[S]_{\lim} - S_1}{\log^{-1} \dfrac{(sv)^{0.375} \Delta t}{1.15 d_m}} \right] \tag{6-6}$$

in which t = time of flow from upstream location to downstream location; S_1 = sulfide concentration at upstream location, mg/L; and S_2 = sulfide concentration at downstream location, mg/L.

5. Hydraulic Factors

The hydraulic factors which are involved in sulfide prediction and corrosion calculations are shown in Fig. 6-12.

In Fig. 6-12

$$\text{Area} = 3.14r^2\,(\theta/180) - b/2\,(r - Y_0)$$

in which r = diameter/2; A = area = $3.14r^2$ (6-14)

$$A = \cos^{-1} \frac{r - Y_O}{r} \tag{6-15}$$

Y_o = depth of flow; width of flow surface

$$b = 2[r^2 - (r - Y_O)^2]^{0.5} \tag{6-16}$$

$$\text{wetted perimeter } P = 2(3.14)\ r - P\,(\theta/180) \tag{6-17}$$

$$\text{unwetted perimeter } P' = 2(3.14)\,r - P \tag{6-18}$$

$$R = \frac{\text{area}}{P} \tag{6-19}$$

$$\text{velocity of flow } v = \frac{1.486}{n} R^{2/3} S^{1/2} \tag{6-20}$$

and mean depth

$$d_m = \frac{\text{area}}{b} \tag{6-21}$$

Table 6-4 lists commonly required hydraulic relationships for various d/D values, while Fig. 6-13 illustrates hydraulic elements of partly filled pipes.

For gravity sewer calculations, the assumption of a constant n results in:

1. An overestimate of V and Z of up to 25%.
2. A conservative (high) ϕ_{sw} and corrosion rate estimate for a given sulfide level, due to the overstated velocity.
3. An unconservative (lower) rate of sulfide build-up using the Pomeroy-Parkhurst model due to the higher indicated reaeration rate resulting from the higher velocity.
4. The effects of 2 and 3 above will tend to offset each other so that the refinement of using Camp's variable *n* probably is not necessary.

6. Climactic EBOD

The temperature of the wastewater in a sewer follows an annual cycle, BOD and quantity of flow have diurnal cycles, and flows usually have long-time trends. These variations are discussed in detail in Chap. 3. It is useful to define a climactic condition as the combination of the average BOD and flow for the highest six-hour flow period of the day and the average wastewater temperature for the warmest three months of the year. Where diurnal BOD curves have not been made, it may be assumed (19) that the BOD for the 6-hr high-flow period is 1.25 times the BOD of a flow-proportioned 24-hr composite. The climactic EBOD is defined by

$$[\text{EBOD}]_c = [\text{BOD}]_c \times 1.07^{T_c-20} \tag{4-1}$$

for which the symbols are as defined in Chap. 4.

7. Sulfide Calculation at Junctions (Mass Balance)

The sulfide exiting the structure S_C, can be calculated from the flow quantities entering the structure Q_A and Q_B, and the sulfide entering the structure S_A and S_B, as follows:

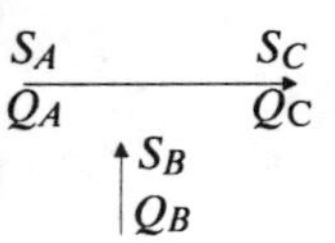

(6-22)

$$Q_C = Q_A + Q_B$$

Thus, quantity of flow at C, Q_C, in cubic feet per second, is $Q_A + Q_B$.

To calculate sulfide level S_C at the downstream side of the junction structure

$$S_C = \frac{S_A \times Q_A + S_B \times Q_B}{Q_C} \tag{6-23}$$

in which S_C = total sulfide leaving the structure at point C, mg/L.

8. Corrosion Calculations - Predicting Rates of Corrosion

The rate of corrosion of cementitious pipes can be predicted theoretically by equations. The corrosion rate depends upon the rate of sulfuric acid production, the alkalinity of the pipe material, and the hydraulic conditions. Sulfuric acid production is related to the mass emission of sulfide from the wastewater. The corrosion rate of cement bonded pipe can be estimated by assuming the rate at which H_2S will reach the pipe wall and the amount reaching the wall that will be oxidized and available for reaction. Thirty-two grams of sulfide are required to produce the sulfuric acid to dissolve 100 g of alkalinity expressed as $CaCO_3$. The corrosion rate equation developed by Pomeroy (19) has been presented herein as

$$C_{avg} = \frac{0.45\, k\, \phi_{sw}}{A} \tag{6-9}$$

in which C_{avg} = average rate of penetration, mm/yr or in./yr; k = coefficient of efficiency for acid reaction considering the estimated fraction of acid remaining on the wall (may be as low as 0.3 and will approach 1.0 for a complete acid reaction); ϕ_{sw} = flux of H_2S to the pipe wall, g/m^2-hr; and A = alkalinity of the cement bonded material, expressed as $CaCO_3$ equivalent. The alkalinity is approximately 0.17 to 0.20 for granitic aggregate concrete, 0.9 to 1.1 for calcareous aggregate concrete, 0.4 for mortar linings, and 0.5 for asbestos cement. For ferrous pipe, the equivalent

TABLE 6-5. Proportions of Dissolved Sulfide Present As H_2S (J Factors)

pH if pK = 7.0 (1)	Proportion of H_2S (2)	pH if pK = 7.0 (3)	Proportion of H_2S (4)
5.0	0.99	7.5	0.24
5.2	0.98	7.6	0.20
5.4	0.975	7.7	0.17
5.6	0.96	7.8	0.14
5.8	0.94	7.9	0.11
6.0	0.91	8.0	0.091
6.1	0.89	8.1	0.074
6.2	0.86	8.2	0.059
6.3	0.83	8.3	0.048
6.4	0.80	8.4	0.039
6.5	0.76	8.5	0.031
6.6	0.72	8.6	0.025
6.7	0.67	8.7	0.020
6.8	0.61	8.8	0.016
6.9	0.56	8.9	0.013
7.0	0.50	9.0	0.010
7.1	0.44	9.5	0.003
7.2	0.39	10.0	0.001
7.3	0.33		
7.4	0.28		

alkalinity A in the corrosion calculation may be assumed as 0.50 (15). The predicted corrosion rate will include direct corrosion by H_2S as well as acid corrosion.

Sulfide flux to the pipe wall ϕ_{sw} may be calculated using

$$\phi_{sw} = 0.45\, J\, [DS] \frac{(b)}{p'} (sv)^{0.375} \qquad (6\text{-}8)$$

in which s = energy gradient of wastewater stream, m/m or ft/ft; v = stream velocity, m/s or ft/s; J = fraction of dissolved sulfide present as H_2S as a function of pH (see Table 6-5 and Fig. 6-13); $[DS]$ = average annual concentration of dissolved sulfide in the wastewater, mg/L; and, b/p' = ratio of width of wastewater stream at surface to exposed perimeter of the pipe wall above the water surface. Rates of corrosion are usually expressed in inches or millimeters of penetration per year measured into the pipe wall from the original interior profile.

Note that the dissolved sulfide concentration used in Eq. 6-9 is the annual average sulfide concentration, not the peak or climactic concentration. Average annual dissolved sulfides may be only 25% to 50% of the climactic values.

The choice of the coefficient of efficiency for the acid reaction k is a matter of engineering judgment, because k will approach unity when the rate of acid production is very slow. In fact, k may be as low as 0.3 to 0.4 if acid production is rapid and if much condensate is formed, as with warm wastewater flowing in a cold pipe. In large pipes with slow to moderate rates of acid production, much of the acid will react, and k will most likely be in the range of 0.9 to 1.0.

In Table 6-5, the relative proportions of H_2S and HS^- dissolved in wastewater are predicted by

TABLE 6-6. Observed Results in Sacramento Study (5)

Pipe diameter, in. (1)	Predicted corrosion, w/k = 0.7, in. (2)	Observed maximum corrosion at manhole, in. (3)	CCF x TCF (4)
30	0.11	0.25	2.3
36	0.15	0.75	5.0
39	0.22	0.50	2.3
42	0.29	0.75	2.6
48	0.40	1.0	2.5
54	0.49	1.0	2.0
60	0.43	1.5	3.5

Note: The Sacramento study averaged the corrosion in all the manholes in a given pipe diameter, rather than comparing calculated corrosion to the worst manhole in a given diameter.

$$\log \frac{(HS^-)}{(H_2 S)} - pH - pK^1 \qquad (6\text{-}24)$$

in which (HS^-) = molar concentration of the hydrosulfide ion; (H_2S) = molar concentration of hydrogen sulfide; pH = negative logarithm of the activity of the hydrogen ion; and pK^1 = negative logarithm of the practical ionization constant in water (average value of 7.0 is adequate for design estimates).

9. Peak Corrosion Rates

Pipe crown corrosion rate where uniform flow conditions exist, may be calculated using

$$C_{\max_u} = C_{ave} \times \text{CCF} \times \text{TCF} \qquad (6\text{-}25)$$

in which CCF (crown corrosion factor) may be on the order of 1.5, and in extreme cases 2.0, and TCF = 1.0.

Calculate pipe crown corrosion rate where nonuniform or turbulent flow conditions exist using

$$C_{\max_T} = C_{avg} \times \text{CCF} \times \text{TCF} \qquad (6\text{-}26)$$

in which TCF (turbulence corrosion factor) may vary from 1 to 2.5 for well designed junction structures, angle points, or grade changes where nonuniform flow conditions exist. Table 6-6 shows the values of CCF times TCF observed in pipe downstream of junctions in the Sacramento study (5).

Where turbulence is moderate to severe, for example at drops, hydraulic jumps, or turbulent junctions with high energy loss, pipe crown corrosion rates may be as high as 10 or 20 times the calculated average uniform flow corrosion (TCF = 5 to 10).

10. Az Design Equations for Cementitious or Ferrous Materials

For uniform flow conditions in cementitious materials

$$Az = 0.45k\, \phi_{sw}\, L \text{ CCF TCF} \qquad (6\text{-}27)$$

in which CCF = 1.5 to 2.0; TCF = 1.0; and, L = design life, years to steel exposure.

For nonuniform flow conditions in cementitious materials, the Az is calculated using Eq. 6-27 but with TCF = 1.5 to 2.0. This Az value applies 10 pipe diameters or at least 50 ft downstream of structures with nonuniform flow, and may apply upstream if air flow is impaired in the sewer.

For ferrous pipe materials, A may be assumed to be 1.79, and the wall thickness loss, Z may be calculated from Eq. 6-27 (15). The calculated corrosion rate will reflect both normal oxidation, direct reaction with H_2S, and acid corrosion.

11. Units of Measurement Used in Odor Control

The sanitary engineer must be very careful when interpreting concentrations given in the literature for threshold odor (TO), dilution threshold, recognition threshold, etc. The literature gives concentrations in terms of milligrams per liter (mg/L); parts per million on a weight to weight basis (ppm, w/w); and in parts per million on a volume to volume basis (ppm, v/v). Authors often do not clarify which basis is being used. Worse still, several publications in sanitary engineering convert mg/L directly to ppm on the assumption that the terms are equivalent. This may result in extremely erroneous values.

Parts per million is equivalent to a weight fraction for solids and liquids, because the total amount of material is of a much higher order of magnitude than the amount of solute. In water solutions and mixtures, 1 mg/L = 1 ppm (w/w). That is never true in the case of gases, where the concentration of a given component in ppm (v/v) is essentially a mole fraction. In gases, where the air fraction is of a much higher order of magnitude than the amount of solute of concern, concentrations can be expressed as a weight or volume fraction of air. By convention, the molecular weight of pure air is 28.9, and the specific weight at standard temperature and pressure is 0.080 lb/cf.

Making the conversion from one set of standard conditions to another is not a problem, provided the person making the conversion is aware of the system being used. In gases, concentrations in mg/L are not the same as in ppm (w/w) or ppm (v/v). In the case of H_2S for example, at Universal Scientific standard conditions (32°F and 760 mm Hg), 1 ppm (w/w) H_2S in air is the same as 0.8 ppm (v/v) and 0.0013 mg H_2S/L air (see following subsections for detailed calculations).

Confusion is very likely when using threshold odor (TO) concentrations available in the literature. For example, the TO concentration of H_2S is commonly given as 0.0011 ppm (at 25°C and 760 mm Hg), when in fact what is really meant is 0.0011 mg/L. The correct concentrations are 0.0011 mg H_2S/L of air = 0.93 ppm (w/w) = 0.8 ppm (v/v)(see following subsections for detailed calculations).

The calculations that follow apply to cases when the air density is expressed on the basis of the Universal Scientific standard of 32°F and 760 mm Hg, where 1 g mole = 22.4 L. There are other standard conditions of temperature and pressure, commonly used in the literature and not always clearly stated, which must be recognized. Some of these are:

1. Natural gas industry (60°F and 760 mm Hg) Sp. Wt. = 0.076.
2. Fan industry (68°F and 14.7 psia) Sp. Wt. = 0.075.
3. Industrial ventilation (70°F and 760 mm with 50% humidity) Sp. Wt = 0.075.

Calculations to Convert Parts per Million by Weight.—Use the following equation for 1 ppm (w/w) H_2S in air, at 32°F and 760 mm Hg:

$$\left(\frac{1\text{ g H}_2\text{S}}{10^6\ g\text{ air}}\right)\left(\frac{22.4\text{ L H}_2\text{S}}{34.08\text{ g H}_2\text{S}}\right)\left(\frac{28.9\text{ g air}}{22.4\text{ L air}}\right) = \left(\frac{0.848\text{ L H}_2\text{S}}{10^6\text{ L air}}\right) = 0.848\text{ ppm (v/v)}$$

To convert ppm (w/w) to mg/L, use

$$\left(\frac{1\text{mg H}_2\text{S}}{10^6\text{ mg air}}\right)\left(\frac{28.9\text{ g air}}{22.4\text{ L air}}\right)\left(\frac{1000\text{ mg}}{\text{g}}\right) = 0.0013\text{ mg H}_2\text{S/L air}$$

Calculations to Convert Threshold Odor Concentration of H_2S.—The TO concentration is commonly given as 0.0011 mg/L at 25°C and 760 mm Hg. This concentration is also equal to

$$\left(\frac{0.0011\text{ mg H}_2\text{S}}{1\ L\text{ air}}\right)\left(\frac{22.4\text{ L air}}{28.9\text{ g air}}\right)\left(\frac{298°\text{K}}{273°\text{ K}}\right)\left(\frac{1{,}000\text{ g}}{\text{kg}}\right)$$

$$= 9.93\text{ ppm (w/w)}$$

and also equal to

$$\left(\frac{0.0011 \text{ mg } H_2S}{1 \text{ L air}}\right)\left(\frac{22.4 \text{ L } H_2S}{34.08 \text{ g } H_2S}\right)\left(\frac{298°K}{273° \text{ K}}\right)\left(\frac{\text{g } H_2S}{1{,}000 \text{ mg } H_2S}\right)$$

$$= \frac{7.9 \times 10^{-7}\ H_2S}{\text{L air}}$$

$$= \frac{0.8 \text{ L } H_2S}{10^6 \text{ L air}}$$

$$= 0.8 \text{ ppm } (^v\!/_v)$$

Calculations to Convert Threshold Odor Concentrations Given in Milligrams per Liter.—To convert to parts per million by weight, use

$$\text{TO, ppm } (^w\!/_w) = \text{TO, mg/L}\left(\frac{22.4 \text{ L}}{28.9}\right)\left(\frac{T^* + 273°K}{273°K}\right)\left(\frac{1 \text{ atm}}{P^*}\right)(10^3)$$

To convert TO concentration given in mg/L to ppm (v/v), use

$$\text{TO, ppm } (^v\!/_v) = \text{TO, mg/L}\left(\frac{22.4 \text{ L}}{\text{M.W.}}\right)\left(\frac{T^* + 273°K}{273°K}\right)\left(\frac{1 \text{ atm}}{P*}\right)(10^{-3}\text{g/mg})$$

in which T* = temperature at which TO is given, °C; P* = pressure at which TO is given, atm; and M.W. = molecular weight of substances, g.

REFERENCES

1. *Concrete Pipe Handbook*, American Concrete Pipe Association, Vienna, VA, 1980.
2. Davy, W. J., "Influence of Velocity on Sulfide Generation in Sewers," *Sewage and Industrial Wastes*, Vol. 22, No. 9, Sept. 1950.
3. "Design and Construction of Gravity Sanitary Sewers," *Manual of Practice No. 60*, ASCE, New York, NY, 1982.
4. Foster, W. S., and Sullivan, R. H., *Sewer Infiltration and Inflow Control Product and Equipment Guide*, Environmental Protection Agency, Cincinnati, OH, July 1977.
5. Gilbert, J. B. & Associates, *A Case Study of Sulfide Generation and Corrosion in Sewers*, American Concrete Pipe Association, April 1979.
6. *Gravity Sanitary Sewer Design and Construction*, ASCE Manuals and Reports of Engineering Practice No. 60, WPCF Manual of Practice No. FD-5, ASCE and Water Pollution Control Federation, 1982.
7. Hall, G. H., McLaren, F., and Hyde, W. S., "Analysis and Design for Sulfide and Corrosion Control in the Sacramento Regional Wastewater Collection System," *Bulletin, California Water Pollution Control Federation*, Vol. 17, No. 3, Jan. 1981.
8. *Handbook for Sewer System Evaluation and Rehabilitation*, Environmental Protection Agency, Cincinnati, OH, Aug. 1973.
9. *Handbook for Sewer System Evaluation and Rehabilitation*, Environmental Protection Agency, Dec. 1975.
10. Kienow, K. K., "Concrete Interceptor Sewer Corrosion Protection-A State of the Art Report," *Proceedings, Third International Conference on the Internal and External Protection of Pipes*, British Hydromechanics Research Association Fluid Engineering, Cranfield, Bedford, England, Sept. 1979.
11. Kienow, K. K., "Protecting Reinforced Concrete Sewers," *Water and Sewage Works*, Oct. 1975.

12. Kienow, K. K., and Kienow, K. E., "Computer Predicts Pipe Corrosion," *Water and Sewage Works*, Sept. 1978.
13. Kienow, K. E., Pomeroy, R. D., and Kienow, K. K., "Prediction of Sulfide Buildup in Sanitary Sewers," *Journal of the Environmental Engineering Division*, ASCE, Vol. 108, No. EE5, Proc. Paper 17359-EE, Oct. 1982.
14. Kienow, K. E., Pomeroy, R. D., and Kienow, K. K., closure to "Prediction of Sulfide Buildup in Sanitary Sewers,"*Journal of the Environmental Engineering Division*, ASCE, Vol. 109, No. EE6, Dec. 1983.
15. Kienow, K. K., and Pomeroy, R. D., "Corrosion Resistant Design of Sanitary Sewer Pipe," *Water and Sewage Works*, Reference Edition, April 1979.
16. Kienow, K. E., and Kienow, K. K., "Predicting Sulfide in Force Mains," *Water and Sewage Works*, Dec. 1978.
17. Meyer, W. J., "Case Study of Prediction of Sulfide Generation and Corrosion in Sewers," *Journal of the Water Pollution Control Federation*, Vol. 52, No. 11, Nov. 1980
18. Meyer, W. J.,and Hall, G. H., "Prediction of Sulfide Generation and Corrosion in Concrete Gravity Sewers, A Case Study," J. B. Gilbert & Associates, A Division of Brown & Caldwell, Sacramento, CA, April 1979.
19. Pomeroy, R. D., "Process Design Manual for Sulfide Control in Sanitary Sewerage Systems," EPA 625/1-7-005, U. S. Environmental Protection Agency, Oct. 1974.
20. Pomeroy, R. D., "Generation and Control of Sulfide in Filled Pipe," *Sewage and Industrial Wastes*, Vol. 31, No. 9, 1959.
21. Pomeroy, R. D., "The Problem of Hydrogen Sulfide in Sewers," Clay Pipe Development Association Ltd., London, England, 1978.
22. Pomeroy, R. D., and Parkhurst, J. D., "The Forecasting of Sulfide Buildup Rates in Sewers," *Prog. Wat. Tech.*, Vol. 9, Pergamon Press, London, England, 1977, pp. 621-628.
23. Pomeroy, R. D., and Bowlus, F. D., "Progress Report on Sulfide Control Research," *Sewage Works Journal*, Vol. 18, No. 4, 1946.
24. Pomeroy, R. D., "Flow Velocities in Small Sewers," *Journal of the Water Pollution Control Federation*, Vol. 39, No. 9, Sept. 1967.
25. Pomeroy, R.D., "Sanitary Sewer Design for Hydrogen Sulfide Control," *Public Works*, Oct. 1970.
26. Pomeroy, R. D., and Bryant, C. W., "A History of Sulfide Evaluation and Prediction Criteria," presented at the ASCE National Conference on Environmental Engineering held at Cincinnati, OH, July 1986.
27. *Process Design Manual for Sulfide Control in Sanitary Sewerage Systems*, Environmental Protection Agency Technology Transfer, Oct. 1974.
28. Rogers, E. H., "West Fresno Herndon Avenue Interceptor Sulfide Buildup Tests," *California Precast Concrete Pipe Association*, San Rafael, CA, Oct. 1978.
29. *Standard Methods for the Examination of Water and Wastewater*, 16th ed., American Public Health Association, American Water Works Association, and Water Pollution Control Federation, 1985.
30. Thistlethwayte, D.K.B., *Control of Sulfides in Sewerage Systems*, Ann Arbor Science Publishers, Ann Arbor, MI, 1972.
31. Warren, G. D., II, and Tchobanoglous G., "A Study of the Use of Concrete Pipe for Trunk Sewers in the City of Delano, California," Department of Civil Engineering, University of California at Davis, CA, Sept. 1976.

CHAPTER 7

RELIABILITY OF SULFIDE GENERATION AND CORROSION PREDICTION MODELS

A. INTRODUCTION

Case studies comparing the results predicted by sulfide generation and corrosion equations with observed conditions in sewers establish the reliability of the forecasting models. Summaries of several such studies are reported in this chapter. Work continues on collecting and reporting data to improve understanding.

Because the factors controlling sulfide generation in sewers are so complex, it would be unrealistic to expect that sulfide concentrations can be accurately predicted on an hour-by-hour basis. Even predictions of average sulfide conditions in a sewer are not considered precise, but they will be adequate for many design and operation purposes. The Pomeroy equations that have been devised have coefficients that can be modified to meet the objectives of the engineer, giving results that will approximate average performance of all sewers represented by any given set of parameters or that will give results with varying degrees of conservatism. Thus, a sulfide prediction may show a concentration that will rarely be exceeded in any sewer, or one that will be exceeded only part of the time in any sewer or in few sewers, or one that will be an average value where septic conditions prevail. The level of understanding of sulfide generation mechanisms and corrosion of both cement bonded and ferrous materials allows a relatively accurate assessment of anticipated conditions in sewer systems and the cost-effective design of control measures.

B. CORRELATIVE STUDIES

Correlation studies summarized in this chapter include evaluations of the Delano trunk sewers, the Sacramento Central Trunk, the San Diego-West Point Loma interceptor, and the Mesa Trunk sewer system. Comparisons are made of observed sulfide and corrosion in sewers and one or more of the following predictive models:

- The EPA Flow-slope relationship as a qualitative guide for evaluating sulfide build-up potential in sewers with flow depths not exceeding two-thirds of the pipe inside diameter (11).
- The Z formula as an indicator of the likelihood of sulfide build-up in relatively small gravity sewers (10).
- Pomeroy's quantitative sulfide forecasting model for filled pipe condition (force mains) (3, 13).
- Pomeroy's sulfide forecasting model for partially filled pipe conditions where dissolved oxygen is less than 0.5 mg/L and the effective slope of the sewers is less than 0.6% (3, 12).

TABLE 7-1. Wastewater Characteristics [a]

Trunk (1)	BOD_5, mg/L (2)	pH (3)	Temperature °F (4)	DO, mg/L[c] (5)	SO_4, mg/L (6)	Flow mgd[d] (7)
Cecil	123	7.44	b	0.5	79	1.85
Stradley	117	7.49	b	0.5	89	0.94

[a] Data in the table represent average values measured from the field study, plant monitoring sheets, and lab work done by Braun, Skaggs, Kevorkian, and Simons. Approximately 60 BOD_5 and SO_4 tests, and 100 pH, temperature and DO measurements are included in this table.

[b] Temperature data were compiled for a year. The yearly wastewater temperature range was 69°F to 83°F (20.5°C to 28.3°C). Maximum yearly temperatures occur during the six-month period of May to October and the average temperature during this period was 81°F (27.2°C).

[c] Average dissolved oxygen levels were consistently less than 0.5 mg/L from May through July.

[d] Flow data were compiled from the monitoring reports and the lab report. Flow data on Stradley are not available. Value shown is the average flow to Plant No. 2. 1 mgd = 3,785,000 L/day.

Source: Ref. 15.

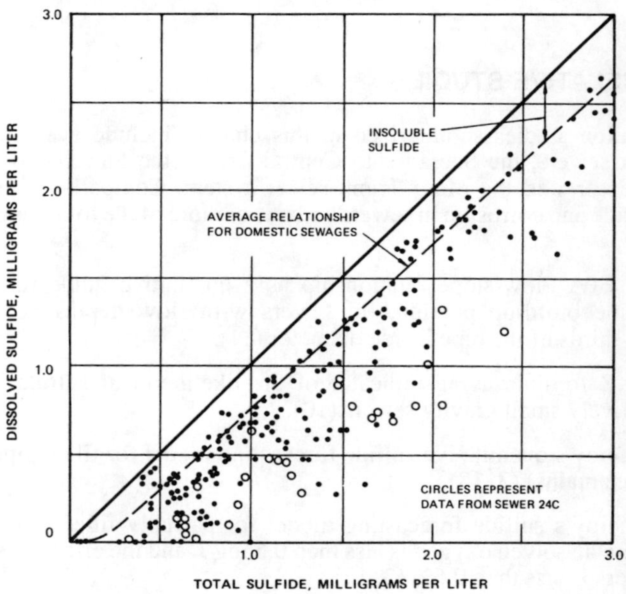

Figure 7-1. Total–dissolved sulfide relationship. (*Source:* Ref. 11)

TABLE 7-2. Sulfide Levels With M′= 0.35 X 10^{-3} and N = 0.6

Run (1)	$(S)_{lim}$, mg/L (2)	Transit time t, hr (3)	$t/t_{1/2}$ (4)	TS, mg/L[a] (5)	TS, mg/L[b] (6)
A	3.62	0.502	0.114	0.28	0.3
B	3.95	0.495	0.108	0.30	0.25
C	3.94	0.405	0.110	0.30	0.3
D	3.83	0.522	0.121	0.32	0.3
E	3.88	0.512	0.119	0.32	0.25
F	3.86	0.512	0.119	0.31	0.45

[a]Calculated total sulfide levels at the end of the test run (MH 8).
[b]Actual total sulfide build-up at MH 8. Values were zeroed at MH 1.
Source: Ref. 15.

- Thistlethwayte's sulfide forecasting model for partially filled pipe conditions (14, 15).
- The EPA rate of corrosion equation for cementitious materials (3, 11).
- The "life factor" design equation for required corrosion protection *Az* (3).

1. Delano Trunk Sewers

Warren and Tchobanoglous (15) compared both the Pomeroy model (12) and the model proposed earlier by Thistlethwayte (14) for predicting sulfide build-up in gravity trunk sewers with sulfide levels observed in Delano, California.

System description.—Two distinct wastewater collection systems conveying sewage to two treatment plants were visually inspected for sulfide corrosion. Based on these inspections, a 4,800-ft (1,460 m) long reach of the 21-in. (53-cm) Cecil Avenue concrete trunk line and a 2,800-ft (850-m) long reach of the 18-in. (45- cm) Stradley Avenue vitrified clay trunk line were selected for the correlation study. The slopes of the Cecil Avenue and Stradley Avenue trunk lines were 0.00225 and 0.0008, respectively. The friction factors for the same trunk lines were 0.0145 and 0.0147, respectively. Each reach had no junctions, drops, or structures that would alter the energy gradient of the flow.

In 1963, Pomeroy (9) had found very low sulfide levels in the Delano wastewater collection system. Subsequent visual inspections for this study in 1976 found minor corrosion [1/16-in. (1.6-mm) to 1/8-in. (3-mm) depth] in the Cecil Avenue trunk line only, in the manholes closest to the wastewater treatment plant. The Stradley Avenue trunk line was in excellent condition. Overall, the main trunk lines in the City of Delano showed little or no evidence of hydrogen sulfide corrosion.

The wastewater characteristics that were monitored in the trunk lines included five-day biochemical oxygen demand (BOD_5), pH, temperature, dissolved oxygen

TABLE 7-3. Predicted Sulfide Levels Using Thistlethwayte's Equation

Run (1)	TS, mg/L[a] (2)	TS, mg/L[b] (3)
A	0.03	0.3
B	0.14	0.25
C	0.10	0.3
D	0.05	0.3
E	0.04	0.25
F	0.05	0.45

[a] Calculated total sulfide values at MH 8 using Thistlethwayte's equation.
[b] Measured total sulfide levels at MH 8, data zeroed at MH 1.

(DO), sulfate (SO_4), total sulfide, and flow. Table 7-1 summarizes the collected data. The relationship between the total and dissolved sulfide for the Delano system is shown in Fig. 7-1. The insoluble sulfide levels varied from 0.05 to 0.3 mg/L.

Results of the in-line sulfide testing found levels consistently low. Sulfide generation was first noted in the Stradley Avenue trunk line in mid-May and coincided with a sewage temperature rise of 4°F (2.2°C) from 76°F (24.4°C) in late April to 80°F (26.6°C) in mid-May. The net sulfide build-up in the test reach of the Stradley Avenue trunk line varied from zero to 0.7 mg/L. The total sulfide levels in the Cecil Avenue trunk line were never greater than 0.1 mg/L. Because of the low sulfide levels in the Cecil trunk line, slug doses of sodium sulfide were added to provide information on the sulfide build-up characteristics of the line. Total sulfide measuring as much as 1.9 mg/L was reduced to zero 4,800 ft downstream, probably as a result of reaeration due to the steep slope (0.00225) and corresponding wastewater velocity of approximately 2.5 ft/s (0.76 m/s). It was concluded that the Cecil Avenue trunk line had a net sulfide loss rather than a build-up. As pointed out by Warren and Tchobanoglous (15), the sulfide levels found in the Delano wastewater collection system could have been qualitatively predicted by use of Pomeroy's flow-slope relationship (11) and Z formula (10). Figure 7-2 incorporates the data from the trunk lines in Cecil and Stradley Avenues; and further, the Z function (Eq. 6-12) is less than 5,000 and greater than 10,000 for the respective trunk lines.

Sulfide Prediction by Pomeroy's Equation.—An excellent correlation between the sulfide calculated using Eq. 6-6 and the measured sulfide levels was obtained for the Stradley Avenue trunk line. Table 7-2 summarizes sulfide levels predicted by the first trial, assuming the constants M' and N in Pomeroy's equation to be $M' = 0.35 \times 10^{-3}$ m/hr, and $N = 0.6$. If the computed sulfide levels had not correlated well with the observed values in the field, the constants M' and N would have been adjusted accordingly and another trial computation made.

The calculated and actual average total sulfide values for the six runs were identical (0.31 mg/L). Because of this excellent correlation, no further trials were

TABLE 7-4. Hydraulic Characteristics of the Central Trunk

Diameter, in. (1)	Slope (2)	Length, ft (3)	Depth,[a] in. (4)	Q,[b] cfs (5)	V,[a] fps (6)	t, hours (7)
27	0.001	2,699	12	3.8	2.5	0.30
30	0.0007	4,785	14	4.8	2.5	0.53
36	0.0005	9,159	16	5.8	2.6	0.98
39	0.0005	5,834	19	8.6	2.3	0.70
42	0.0006	15,873	22	12.4	2.4	1.84
48	0.0007	12,148	23	17.4	2.4	1.41
54	0.0007	9,071	23	18.4	3.4	0.74
60	0.0005	12,263	26	23.2	3.1	1.10

[a]Measured during 1976 study.
[b]Calculated.
Source: Ref. 8.

necessary, and the empirical constants for Pomeroy's equation for the Stradley Avenue trunk line were found to be $M' = 0.35 \times 10^{-3}$ m/hr, and $N = 0.6$. Pomeroy's equation (12) was then used to predict sulfide levels in the new Delano interceptor.

A correlation with values measured in the Cecil Avenue trunk line was not made because of the net sulfide loss characteristics of the line.

Sulfide Prediction by Thistlethwayte Equation.—Essentially no correlation was found between the values predicted using Thistlethwayte's equation (14) and the actual observed data. The results calculated using Thistlethwayte's equation are compared with measured values in Table 7-3.

2. The Sacramento Central Trunk

A correlation study completed in 1979 by Meyer and Hall (8) compared sulfide generation and corrosion predictive models to observed conditions in the Central Trunk sewer system of Sacramento County, California, where data had been accumulated since the system was placed in operation in 1965.

System description.—Approximately 13 miles (21 km) of the trunk line consists of reinforced concrete pipe varying from 27 in. (68.6 cm) to 60 in. (152 cm) in diameter. The slope varies from 0.18% in the upper reaches to 0.05% in the lower end of the system. Table 7-4 shows the hydraulic characteristics of the trunk, and Table 7-5 summarizes the wastewater characteristics used in the predictive equations.

Both total and dissolved sulfide as well as other water quality parameters were measured at 28 tributary laterals, 14 daily sampling points in the trunk, and 4 diurnal sampling points. The dissolved sulfide was about 75% of the total sulfide.

Sulfide prediction by Pomeroy's equation.—Table 7-6 shows the characteristics for the Central trunk that were used in calculating the limiting sulfide concentrations, $S_{\lim}$, and the corresponding S_2 calculations.

TABLE 7-5. Summary of Wastewater Quality Parameters of the Central Trunk (*Source:* Ref. 8.)

Reach diameter, in.	BOD,mg/L				SO_4,mg/L				Average temperature,°C	pH			
	10-7-76		10-8-76		10-7-76		10-8-76			10-7-76		10-8-76	
	Max.	Aver.	Max.	Aver.	Max.	Aver.	Max.	Aver.		Max.	Aver.	Max.	Aver.
(1)	(2)	(3)	(4)	(5)	(6)	(7)	(8)	(9)	(10)	(11)	(12)	(13)	(14)
27	1,400	762	600	548	33	34	22	21	25.0	7.5	7.3	7.4	7.3
30	600	505	510	495	53	49	32	27	24.4	7.2	7.2	7.3	7.2
36	600	432	540	473	74	63	35	32	26.1	7.4	7.3	7.5	7.4
39	570	410	420	342	24	21	34	29	25.0	7.4	7.3	7.5	7.4
42	630	437	260	245	62	51	55	45	27.2	7.3	7.2	7.3	7.3
48	410	395	340	225	45	27	230	131	26.7	7.2	7.2	7.3	7.3
54	320	295	310	260	85	59	83	64	28.9	7.1	7.0	7.2	7.2
60	360	313	270	243	52	29	53	49	26.7	7.1	7.0	7.3	7.1

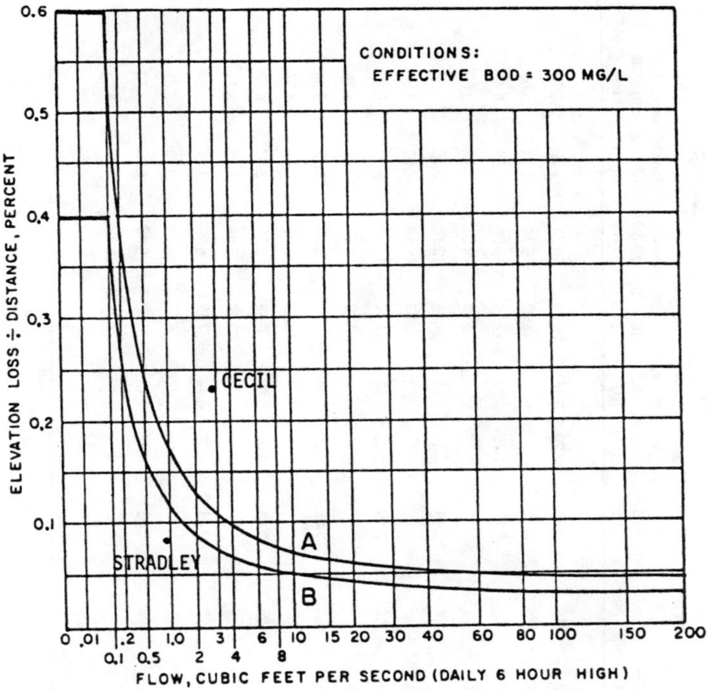

Figure 7-2. EPA flow–slope relationship for Cecil and Stradley Trunk (*Source*: Ref. 15)

Pomeroy suggests the use of both conservative and less conservative empirical coefficients in his predictive equation (12). Use an M' value of 0.32×10^{-3}, with an N value of 0.64 for a conservative prediction, or an N value of 0.96 for a less conservative prediction. The less conservative value of $N = 0.96$ was used in this study. Table 7-7 presents the calculated sulfide levels, using Pomeroy's sulfide prediction equation without the effects of junctions, compared with the observed sulfide levels in the Central Trunk.

The ratios of predicted to observed sulfide concentrations are also shown. In the upper reaches of the system, the average predicted sulfide concentrations were 4 to 5 times the observed levels. However, in the lower reaches, the predicted sulfide concentrations were only 1 to 2 times the observed levels. Figures 7-3 and 7-4 compare the observed values for the average total dissolved sulfides for October 7 and 8, 1976, with the values predicted using Pomeroy's equation with both the conservative and less conservative constants. In summary, measured sulfide concentrations correlate well with the sulfide concentrations predicted by Pomeroy's less conservative equation.

Sulfide prediction by Thistlethwayte's equation.—Table 7-8 shows the observed sulfide data compared to the predicted sulfide concentrations using Thistlethwayte's equation (14). The table shows that the predicted concentrations are very high compared to the observed data. This is caused by the linear progression of the sulfide build-up using the equation. Thistlethwayte's model predicted 3 to 20 times the actual sulfide build-up observed.

TABLE 7-6. Summary of Calculations for S_{lim} for Central Trunk (*Source:* Ref. 8)

Reach diameter, in. (1)	Condition (2)	Slope percentage[a] (3)	V,[a] fps (4)	BOD5,[b] mg/L 10-7-76 (5)	BOD5,[b] mg/L 10-8-76 (6)	EBOD,[c] mg/L 10-7-76 (7)	EBOD,[c] mg/L 10-8-76 (8)	Depth, d[a] in. (9)	d/D (10)	Width, b, ft (11)	Wetted perimeter, ft (12)	S_{lim}, mg/L Conservative[d] 10-7-76 (13)	S_{lim}, mg/L Conservative[d] 10-8-76 (14)	S_{lim}, mg/L Less Conservative[d] 10-7-76 (15)	S_{lim}, mg/L Less Conservative[d] 10-8-76 (16)
27	maximum	0.001	2.5	1,400	600	1,964	842	12	0.44	2.20	3.26	25.02	10.72	14.29	6.13
	average			762	548	1,069	769					13.62	9.79	7.78	5.60
30	maximum	0.0007	2.5	600	510	808	687	14	0.47	2.49	3.77	12.00	10.21	6.86	5.83
	average			505	495	680	667					10.10	9.91	5.77	5.66
36	maximum	0.0005	2.6	600	540	907	816	16	0.44	2.94	4.35	14.66	13.19	8.38	7.54
	average			432	473	653	715					10.56	11.56	6.03	6.61
39	maximum	0.0005	2.3	570	420	799	589	19	0.49	3.25	5.00	14.17	10.45	8.10	5.97
	average			410	342	575	480					10.20	8.51	5.83	4.86
42	maximum	0.0006	2.4	630	320	1,025	521	22	0.52	3.50	5.64	17.36	8.82	9.92	5.04
	average			437	275	711	448					12.04	7.59	6.88	4.34
48	maximum	0.0007	2.4	410	340	645	535	23	0.48	4.00	5.96	9.92	8.23	5.67	4.70
	average			395	225	622	354					9.57	5.45	5.47	3.11
54	maximum	0.0007	3.4	320	310	584	566	23	0.43	4.38	6.44	7.12	6.90	4.07	3.94
	average			295	260	539	475					6.57	5.79	3.75	3.31
60	maximum	0.0005	3.1	360	270	566	425	26	0.43	4.87	7.15	8.62	6.48	4.93	3.70
	average			313	243	493	382					7.51	5.82	4.29	3.33

[a] From Table 4-1.
[b] From Table 4-2.
[c] EBOD = $BOD_5 \times (1.07)^{(T-20)}$, T = °C from Table 4-2.
[d] Conservative N = 0.64; less conservative N = 0.96.

TABLE 7-7. Comparison of Predicted Total Sulfide With Observed Data for Pomeroy's Equation

Reach diameter in. (1)	Condition (2)	Observed Sulfide, mg/L 10/7/76 (3)	Observed Sulfide, mg/L 10/8/76 (4)	Pomeroy's sulfide equation, mg/L Less conservative[a] 10/7/76 (5)	Less conservative[a] 10/8/76 (6)	Conservative[b] 10/7/76 (7)	Conservative[b] 10/8/76 (8)	Ratio[c] predicted/observed 10/7/76 (9)	Ratio[c] predicted/observed 10/8/76 (10)
	maximum[d]	0.20	0.10	1.19	0.51	1.34	0.57	5.9	5.1
27	average[e]	0.02	0.10	0.65	0.47	0.73	0.52	3.25	4.7
	maximum	0.40	0.40	1.75	1.04	2.02	1.19	4.4	2.6
30	average	0.30	0.33	1.16	0.99	1.33	1.13	3.9	3.0
	maximum	0.60	0.90	2.70	1.98	3.20	2.31	4.5	2.2
36	average	0.52	0.73	1.86	1.80	2.19	2.10	3.6	2.5
	maximum	0.70	1.10	3.16	2.32	3.81	2.76	4.5	2.1
39	average	0.65	0.68	2.20	2.06	2.63	2.45	3.4	3.0
	maximum	1.50	2.40	4.53	1.87	5.62	3.57	3.0	0.8
42	average	1.45	1.53	3.15	2.52	3.89	3.14	2.2	1.6
	maximum	2.50	1.40	4.71	3.16	6.06	4.05	1.9	2.3
48	average	2.12	1.35	3.52	2.61	4.48	3.38	1.7	1.9
	maximum	1.80	2.70	4.64	3.24	6.13	4.25	2.6	1.2
54	average	1.65	2.43	3.54	2.68	4.62	3.54	2.1	1.1
	maximum	2.50	1.60	4.67	3.29	6.32	4.41	1.9	2.1
60	average	2.35	1.53	3.63	2.76	4.84	3.72	1.5	1.8

[a] N = 0.96
[b] N = 0.64
[c] Using less conservative predicted sulfide concentrations.
[d] Maximum values for BOD and temperature during sampling period; assumes initial sulfide concentration equals zero.
[e] Average values for BOD and Temperature during sampling period; assumes initial sulfide concentration equals zero.
Source: Ref. 8.

In short, there was no correlation found between the measured and predicted total sulfide using the Thistlethwayte equation. (It is important to note that predicted sulfide in Tables 7-7 and 7-8 does not account for increases or decreases in sulfide concentration due to tributary sources.)

Effect of junctions.—Meyer and Hall (8) determined the effect of junctions, considering only the quantity and quality of tributary wastewater, on the predicted sulfide concentrations using Pomeroy's equation. Table 7-9 is a summary of the sulfide data collected on the Central Trunk tributaries. At each junction, a new starting sulfide concentration (S_1) can be determined with a simple flow mass balance procedure (see Chap. 6, Section D7), using the calculated sulfide build-up

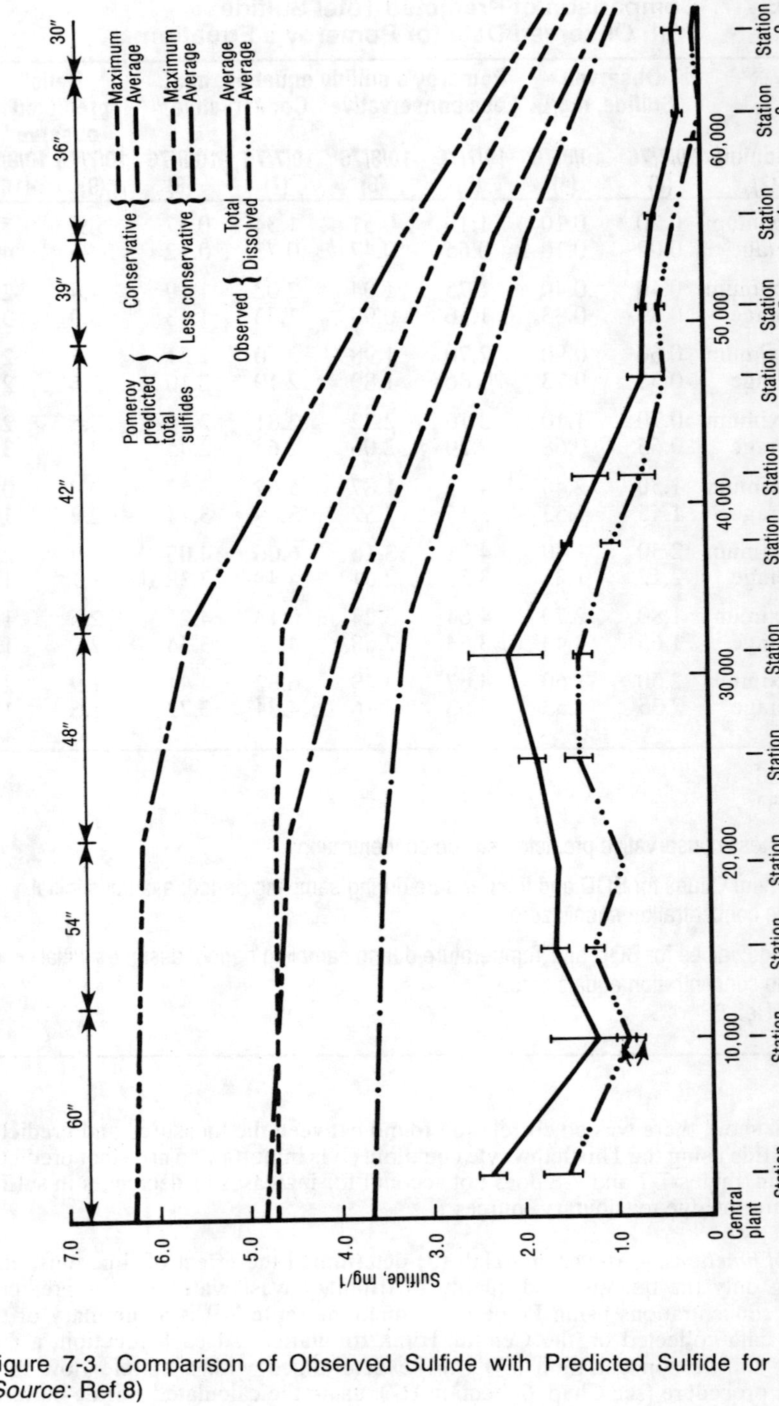

Figure 7-3. Comparison of Observed Sulfide with Predicted Sulfide for 10/7/76 (*Source*: Ref.8)

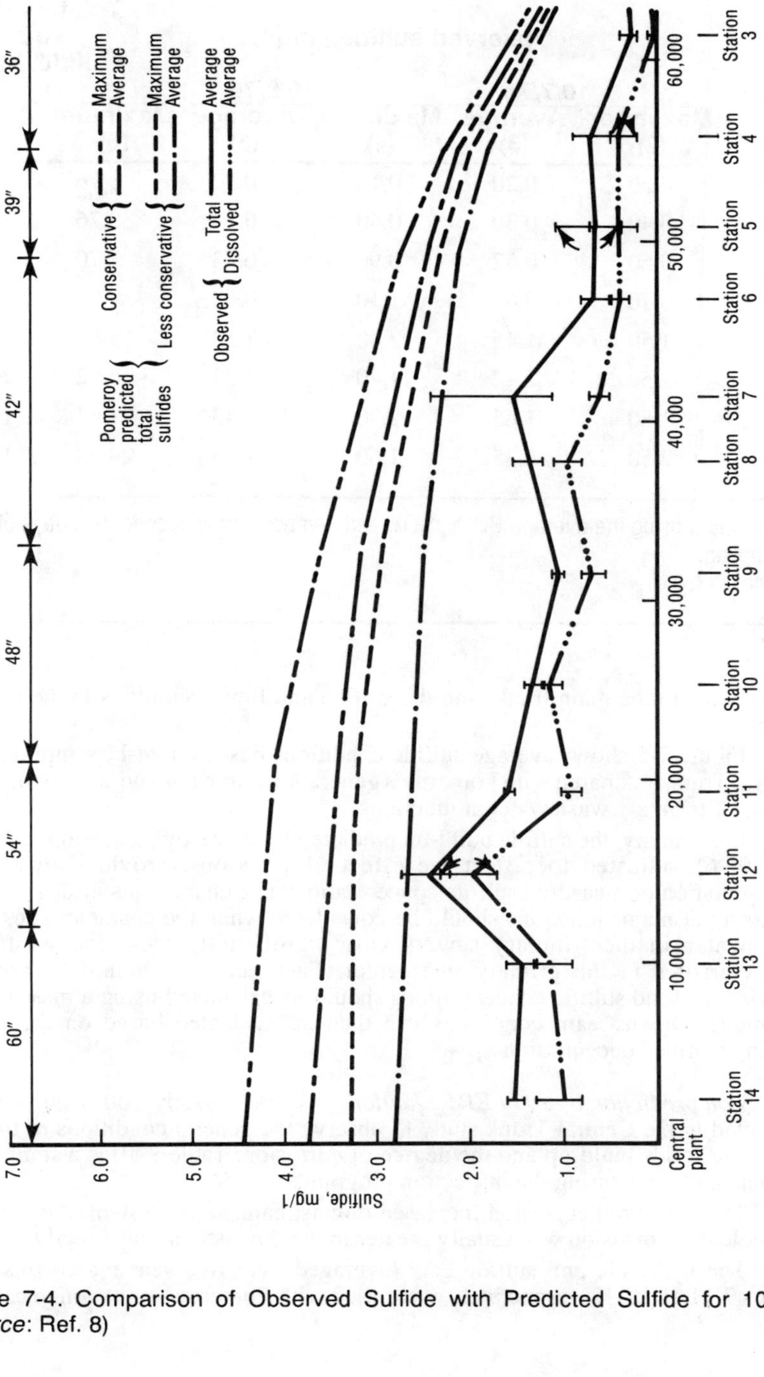

Figure 7-4. Comparison of Observed Sulfide with Predicted Sulfide for 10/8/76 (*Source*: Ref. 8)

TABLE 7-8. Comparison of Predicted Total Sulfide With Observed Data for Thistlethwayte's Equation

Reach, in. (1)	Observed sulfide, mg/L 10/7/76 Maximum (2)	Average (3)	10/8/76 Maximum (4)	Average (5)	Thistlethwayte equation sulfide, mg/L[a] Maximum (6)	Average (7)
27	0.20	0.20	0.10	0.10	1.76	1.03
30	0.40	0.30	0.40	0.33	3.76	2.52
36	0.60	0.52	0.90	0.73	7.70	5.72
39	0.70	0.65	1.10	0.68	9.01	6.53
42	1.50	1.45	2.40	1.53	15.66	9.28
48	2.50	2.12	1.40	1.35	18.92	11.09
54	1.80	1.65	2.70	2.43	21.61	12.95
60	2.50	2.35	1.60	1.53	23.48	14.26

[a] Calculated using the average BOD_5, SO_4, and temperature values for two days of sampling.
Source: Ref. 8.

S_2, the flow in the main trunk, and the sulfide and flow conditions in the tributary sewer.

Figure 7-5 shows average sulfide conditions based on 24-hr sampling in the Central Trunk compared with Pomeroy's generation curve both with and without the effects of tributary wastewater at junctions.

In summary, the sulfide build-up predicted by Pomeroy's equation (assuming $N = 0.96$), adjusted to reflect the effect of junctions, provides an accurate representation of measured sulfide concentration. The change in sulfide conditions occurring at major junctions should be considered when the characteristics of the wastewater in the tributary sewers vary significantly from the wastewater characteristics in the primary interceptor. Each junction should be analyzed individually, and sulfide concentrations should be calculated using a mass balance technique. Downstream conditions can then be evaluated based on the revised starting sulfide concentration S_1.

Corrosion prediction by 1974 EPA equation.—Approximately 100 manholes were inspected in the Central Trunk study to observe the general conditions of the line, including solids build-up and the degree of corrosion. Table 7-10 is a summary of the data gathered during the inspection program.

The measured corrosion increased downstream in the system. Also, at each manhole, the corrosion was usually greater in the downstream pipe section.

The hydraulic and sulfide data (averaged over five-year increments) from Tables 7-11 and 7-12, respectively, were used with Pomeroy's equation to calculate

TABLE 7-9. Sewers Tributary to Central Trunk

Diameter, in. (1)	Station, ft (2)	Average flow,[a] cfs (3)	Average total sulfide, mg/L (4)	Average dissolved sulfide, mg/L (5)
27	95+00	1.4	0.78	0.52
33	123+00	2.4	0.52	0.32
39	123+00	3.7	0.22	0.10
24	161+00	1.1	0.26	0.06
12	161+00	0.2	0.35	0.15
10	213+00	0.2	0.25	0.10
18	213+00			
27	249+00	1.4	0.10	0
12	289+00	0.2	0	0
18	306+00	0.6	0.10	0
24	306+00	1.1	0.10	0
15	335+00	0.4	0	0
15	362+00	0.4	0.15	0
15	373+00	0.4	0.10	0
10	393+00			
12	399+00	0.2	0	0
15	423+00	0.4	0	0
21	467+00			
21	493+00	1.1	0	0
15	553+00	0.4	0.22	0.10
18	584+00	1.2	0.30	0.12
10	614+00	0.2	0.13	0
18	643+00	0.6	0	0
12	659+00	0.2	0.10	0
15	691+00	0.4	0.13	0
15	718+00	0.4	0.13	0.10
15	776+00	0.4	0.60	0.20
8	820+00	0.3	0	0

[a]Calculated using measured depth of flow
Source: Ref. 8.

TABLE 7-10. Data from Manhole Inspection Program

Reach diameter, in. (1)	No. of manholes sampled (2)	Average pH Crown (3)	Average pH Springline (4)	Solids build-up (5)	Penetration, in. Crown Aver. (6)	Penetration, in. Crown Max. (7)	Penetration, in. Springline Aver. (8)	Penetration, in. Springline Max. (9)
27	5	5.4	5.8	no	0	0	0.02	0.13
30	9	4.8	5.2	no	0.04	0.25	0.04	0.25
36	14	2.7	3.7	no	0.33	0.75	0.35	0.50
39	8	1.8	3.5	no	0.33	0.50	0.33	0.50
42	19	2.8	4.1	yes	0.59	0.75	0.61	0.75
48	19	3.4	4.2	yes	0.80	1.0	0.80	1.0
54	14	3.4	4.6	no	0.70	1.0	0.70	0.75
60	24	3.5	3.9	no	0.83	1.5	0.78	1.0

(*Source*: Ref. 8)

the predicted corrosion for five-year periods and total predicted corrosion from 1962 to 1976.

The predicted corrosion is compared with the observed corrosion in Table 7-13 (see discussion of corrosion rates in Chap. 6, Secs. D8 and D9). Equation 6-28 was used for the calculations, with $k = 0.7$. In general, the maximum crown corrosion at manholes where turbulence occurs is 2 to 5 times greater than the predicted average corrosion. The high rate of corrosion in the 60-in. line was probably caused by the turbulent junction where tributaries number 26 and 27 enter the manhole and the trunk changes from a 54-in. to 60-in. diameter line (see Fig. 7-5).

The hydrogen sulfide in the atmosphere was measured in the 60-in. line, and these values were used in Pomeroy's equation to calculate $\phi/_{sw}$ shown in Table 7-14. The hydrogen sulfide in the atmosphere was measured in the other reaches to check the predicted values using Pomeroy's equation. Good correlation between the measured and calculated hydrogen sulfide in the atmosphere was found.

In summary, the observed crown corrosion at structures is 2 to 5 times greater than the average uniform flow predicted value.

Increased corrosion due to turbulence.—The predicted corrosion is compared to the maximum observed crown corrosion downstream of manholes, to determine the required accelerated corrosion factor, as shown in Table 7-15.

The maximum corrosion in structures or in pipes immediately downstream of those structures where turbulence occurs can be 2 to 5 times the average C calculated if $k = 0.7$ is used. Further, the increased rate of corrosion applies only to those areas in the vicinity of induced turbulence. It does not address the question of a crown corrosion factor required for those reaches of pipe not in the vicinity of structures, i.e., the majority of pipe, where uniform hydraulic conditions exist.

A minimum factor of 1.5 must be applied to the average exposed perimeter corrosion C, to account for crown corrosion in a reach "well removed from points of turbulence,"as stated in the EPA design manual (11). This is documented in the EPA manual in the Joint Outfall B case history. Failure to use a crown corrosion

TABLE 7-11. Hydraulic Characteristics, 1962 to 1976

Reach diameter, in. (1)	Year (2)	Q[a] cfs (3)	s (4)	V,[b] fps (5)	$SV^{0.375}$ (6)	b,[c] ft (7)	P,[d] ft (8)	b/P′ (9)
	1962-66	0.58		1.4	0.085	1.7	5.1	0.33
27	1966-71	1.84	0.001	1.9	0.095	2.1	4.4	0.48
	1971-76	4.24		2.3	0.102	2.2	3.4	0.65
	1962-66	0.76		1.3	0.072	1.9	5.7	0.33
30	1966-71	2.41	0.0007	1.9	0.083	2.3	4.9	0.47
	1971-76	6.62		2.3	0.090	2.5	3.9	0.64
	1962-66	0.90		1.2	0.062	2.3	6.8	0.34
36	1966-71	2.87	0.0005	1.6	0.069	2.8	5.8	0.48
	1971-76	6.62		2.0	0.075	3.0	4.5	0.67
	1962-66	1.34		1.3	0.064	2.5	7.4	0.34
39	1966-71	4.25	0.0005	1.8	0.072	3.0	6.3	0.48
	1971-76	9.80		2.2	0.078	3.2	5.1	0.63
	1962-66	1.92		1.5	0.072	2.7	7.9	0.34
42	1966-71	6.09	0.0006	2.1	0.082	3.2	6.8	0.47
	1971-76	14.04		2.5	0.087	3.5	5.3	0.66
	1962-66	2.68		1.7	0.080	3.1	9.1	0.34
48	1966-71	8.51	0.0007	2.4	0.091	3.7	7.8	0.47
	1971-76	19.60		2.9	0.098	4.0	6.2	0.65
	1962-66	2.86		1.9	0.083	3.5	10.2	0.34
54	1966-71	9.08	0.0007	2.6	0.094	4.1	8.8	0.47
	1971-76	20.90		3.1	0.100	4.5	7.4	0.61
	1962-66	3.62		1.7	0.071	3.9	11.3	0.35
60	1966-71	11.49	0.0005	2.3	0.079	4.6	9.7	0.47
	1971-76	26.48		2.8	0.085	5.0	8.2	0.61

[a]Average annual flows measured at the Central Treatment Plant.
[b]Calculated based on hydraulic conditions.
[c]Surface width of flow.
[d]Exposed perimeter.
Source: Ref. 8.

TABLE 7-12. Dissolved Sulfide Concentrations, 1962 to 1976[a]

Reach diameter, in.	Year 1962-66	1966-71	1971-76
27	0.07	0.18	0.32
30	0.02	0.28	0.62
36	0.32	0.52	0.90
39	0.45	0.74	1.29
42	0.52	0.84	1.47
48	0.15	0.60	2.38
54	0.41	0.67	2.98
60	0.60	0.49	1.79

[a]Adjusted for average daily conditions (1.4 x observed average).
Source: Ref. 8

TABLE 7-13. Comparison of Predicted Corrosion with Observed Corrosion, 1962 to 1976

Pipe diameter, in. (1)	Predicted corrosion (2)	Average observed[a] (3)	Worst case[b] (4)
30	0.11	0	0.25
36	0.15	0.3	0.75
39	0.22	0.3	0.50
42	0.29	0.6	0.75
48	0.40	0.8	1.0
54	0.49	0.7	1.0
60	0.43	0.8	1.5

[a]Average of all manholes in this diameter.
[b]Worst manhole in this diameter.
Source: Ref. 8.

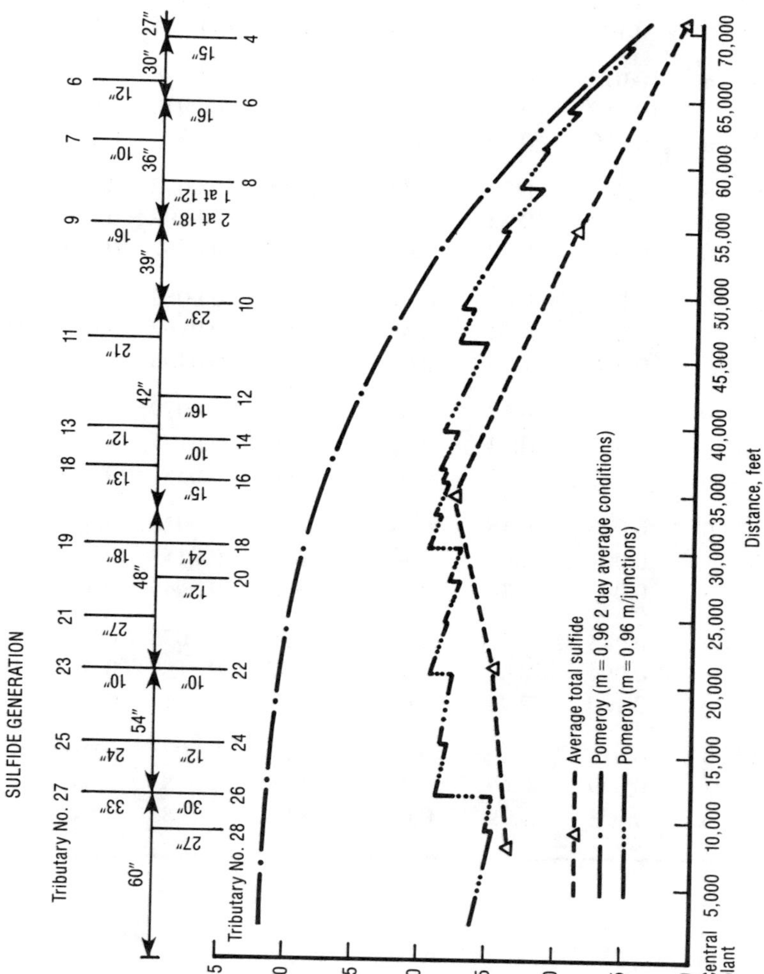

Figure 7-5. Effect of tributaries on Pomeroy predicted sulfide generation. (*Source*: Ref. 8)

TABLE 7-14. Predicted Corrosion for Central Trunk (*Source:* Ref. 8)

Reach diameter, in.	Year	$sv^{0.375}$	[DS] (mg/L)	b/P′	ϕ/sw	Average corrosion, in. C′ (in./yr)	5-yr	15-yr
	1962-66	0.085	0.07	0.33	0.0004	0.0011	0.005	
27	1966-71	0.095	0.18	0.48	0.0018	0.0051	0.026	0.10
	1971-76	0.102	0.32	0.65	0.0048	0.0135	0.068	
	1962-66	0.072	0.20	0.33	0.0010	0.0028	0.014	
30	1966-71	0.083	0.83	0.28	0.0025	0.0070	0.035	0.16
	1971-76	0.090	0.90	0.62	0.0080	0.0225	0.113	
	1962-66	0.062	0.32	0.34	0.0015	0.0042	0.021	
36	1966-71	0.069	0.52	0.48	0.0039	0.0109	0.054	0.22
	1971-76	0.075	0.90	0.67	0.0100	0.0281	0.141	
	1962-66	0.064	0.45	0.34	0.0022	0.0062	0.031	
39	1966-71	0.072	0.74	0.48	0.0057	0.0162	0.080	0.31
	1971-76	0.078	1.29	0.63	0.0140	0.0406	0.203	
	1962-66	0.072	0.52	0.34	0.0028	0.0079	0.040	
42	1966-71	0.082	0.84	0.47	0.0070	0.0197	0.099	0.42
	1967-76	0.087	1.47	0.66	0.0196	0.0551	0.276	
	1962-66	0.080	0.15	0.34	0.0010	0.0028	0.014	
48	1966-71	0.091	0.60	0.47	0.0056	0.0158	0.079	0.57
	1971-76	0.098	2.38	0.65	0.0336	0.0945	0.473	
	1962-66	0.083	0.41	0.34	0.0025	0.0070	0.035	
54	1966-71	0.094	0.67	0.47	0.0067	0.0188	0.094	0.70
	1971-76	0.100	2.98	0.61	0.0406	0.1142	0.571	
	1962-66	0.071	0.60	0.35	0.0034	0.0085	0.048	
60	1966-71	0.079	0.49	0.47	0.0065	0.0183	0.091	0.79
	1971-76	0.085	1.79	0.61	0.0458	0.1288	0.640	

factor in calculating corrosion rates for main line pipe results in a design which will last for only two-thirds of the desired service life. The 1.5 crown corrosion factor discussed in the EPA manual relates crown corrosion to average exposed perimeter corrosion and has nothing to do with the additional corrosion which will occur at points of nonuniform flow, such as might occur in pipe downstream of structures with intercepted flows.

TABLE 7-15. Predicted and Observed Corrosion, Sacramento Central Trunk

Pipe diameter, in. (1)	Predicted corrosion[a] (2)	Observed maximum corrosion (3)	Accelerated corrosion factor[b] (4)
30	0.11	0.25	2.3
36	0.15	0.75	5.0
39	0.22	0.50	2.3
42	0.29	0.75	2.6
48	0.40	1.0	2.5
54	0.49	1.0	2.0
60	0.43	1.5	3.5

[a] Using k = 0.7.

[b] Accelerated corrosion factor = observed maximum/predicted (downstream of structures).Note: See Eqs. 6-30 and 6-31. Accelerated corrosion factor = CCF x TCF.

Source: Ref. 8.

3. The San Diego-West Point Loma Interceptor

An 8,000-ft (2,440-m) long, 114-in. (290-cm) diameter reinforced concrete pipe installed in 1963 in the City of San Diego's wastewater collection system has been the subject of intensive study surrounding its shortened life due to sulfide attack. Kienow (6) has reviewed the data collected over 13 years, as well as the reports prepared by Graham (4), Koebig and Koebig (7), and others who have studied the sulfide corrosion and control measures in this interceptor's system. He has correlated observed conditions in the interceptor with predictive models to draw some significant conclusions regarding the reliability of the models.

System description.—The San Diego-West Point Loma interceptor was constructed in 1963, and includes some 22 miles (35.4 km) of interceptor sewer and force mains, ranging in pipe sizes from 42 to 114 in. (106 to 290 cm) in diameter. The last reach of the interceptor is an 8,000 ft gravity flow section of 114-in.- diameter reinforced concrete pipe. The original designers recognized the potential for high sulfide generation in the force mains and corrosion potential throughout the gravity flow sections. Two types of pipe-protection systems were installed: (1) Sacrificial concrete with granitic aggregate; and (2) PVC-lined RCP for up to 5 diameters upstream and 10 diameters downstream of all junction structures. In addition, provisions were made to inject air into each force main to control hydrogen sulfide build-up. The sulfide control system was to limit the maximum sulfide concentration in the air at 2 ppm. The pipe design called for 2 in. (5 cm) of concrete cover over the steel at the invert, 2-1/2 in. (6.35 cm) at the springline, and 3 in. (7.6 cm) at the crown. Pomeroy's quantitative formula for predicting sulfide generation in pipes flowing full (force mains) was used in the design (4).

At the time of construction of the interceptor, an annual inspection of the system was recommended. The air injection system was inoperative most of the first year; sulfide corrosion was noted in the pipe. By the end of the second year, the 114-in.-diameter RCP was undergoing severe corrosion. In late 1966, the air injection system was causing problems in the force mains, and the City of San Diego abandoned air injection (4). Annual inspections of the interceptor system continued.

In 1969, consultants were retained to study the 114- in.-diameter pipe and to report on its life expectancy, based on observed corrosion and predicted future sulfide levels. Koebig and Koebig (7) estimated the 114-in. interceptor had an additional 15 years before pipe failure might be anticipated.

Inspection of the 114-in.-diameter pipe in 1971 revealed that the reinforcing steel was exposed in one-third of the pipe length. In 1978, another consultant evaluated the extent of sulfide damage within the interceptor, and concluded that some sections were in serious danger structurally.

Sulfide prediction by Pomeroy's equation.—Kienow (6) reviewed possible causes for the rapid corrosion and failure of the 114- in.-diameter section of the San Diego-West Point Loma interceptor. One theorized possible cause for early failure was that the sulfide prediction methodology was faulty. The initial design of the interceptor was based on an expected summertime BOD of 300 mg/L with a corresponding sewage temperature of 84°F (28.9°C). For the initial average wastewater flow rate of 63 mgd, the total sulfide generated in the force main preceding the 114-in. gravity section is calculated from Pomeroy's equation to be 7.7 mg/L. This is well above the levels of dissolved sulfide concentration that cause severe corrosion (3). The same equation predicted that sulfide generation in the force main would range from 1.90 mg/L to 7.07 mg/L for wastewater flows existing in 1975 (65 ft^3/s minimum, and 244 ft^3/s maximum). Thus, conditions predicted by Pomeroy's equation were right for severe sulfide generation to occur.

Corrosion prediction by 1974 EPA equation.—For flows prevailing in the 114-in. interceptor in 1964, the EPA corrosion prediction equation estimates the reinforcing steel in the pipe would be exposed in about nine years. Based on the 1974 flows, the reinforcing steel would be exposed in about six years,according to the same equation. Actually, the reinforcing steel was exposed in some of the pipe after seven years of service. Thus, corrosion forecasts using the EPA equation correlate well with the observed corrosion level (6).

Life factor correlation.—The Az value for the 114-in. interceptor [with z equal to 3 in. (7.6 cm) of granitic aggregate concrete cover over the reinforcing steel at the crown, and alkalinity A equal to 0.19] is 3 x 0.19 = 0.57. The calculated life factor for the 114-in. interceptor (without providing sulfide control in the force mains) is 3.0, or more than 5 times the level of protection that was specified and furnished. Kienow (6) concluded that the life factor design method would have predicted that, without sulfide controls, a very short life for the 114-in.-diameter granitic aggregate pipe should have been anticipated.

4. Mesa Trunk Sewers

A more recently completed study (1) of observed sulfide build-up in the Mesa, Arizona, trunk sewer system demonstrated good correlation between predicted sulfide levels using Pomeroy's model for gravity sewers and observed sulfide build-up in the system.

The Thistlethwayte model, on the other hand, predicted 12 to 14 mg/L total sulfide concentration, while the observed values were 2 to 3 mg/L. In this case, the

Thistlethwayte model over-predicted sulfide concentration by a factor of 6. The total flow time in the system was on the order of five hours.

C. CONCLUSIONS

Correlation studies published to date (1988) substantiate the reliability (and unreliability) of predictive models and qualitative guides available to designers for forecasting sulfide build-up and corrosion of cementitious materials in wastewater collection and treatment systems.

Flow-slope relationship.—The flow-slope relationship (11, 3) can serve as a qualitative guide for evaluating sulfide build-up potential in sewers with flow depths not exceeding two-thirds of the pipe inside diameter. There cannot be any pressure mains upstream of the reach being analyzed unless measures are taken to offset their oxygen-depleting and sulfide-producing potentials. It must be emphasized that the flow-slope relationship cannot be used to predict sulfide conditions accurately. The purpose is to indicate an apparent trend in the effect of flow-slope relationship and sulfide build-up.

The Z formula.—The Z formula (10, 3) has generally been successful in indicating the likelihood of sulfide problems in relatively small gravity sewers [not over 24 in. (61 cm) in diameter], but a degree of uncertainty still exists. For example, sulfide has been found to be present in a sewer with a Z value as low as 6,000, and absent in one system with a Z value as large as 28,000 (2).

Pomeroy's model for sulfide generation in gravity sewers.—Sulfide build-up predicted by Pomeroy's equations, assuming a less conservative coefficient for sulfide losses of $N = 0.96$, correlates well with measured sulfide concentrations. The change in sulfide conditions occurring at major junctions should be considered when the characteristics of the wastewater in tributary sewers vary significantly from the wastewater characteristics in the primary sewer.

The Pomeroy model, when used to forecast sulfide levels for gravity domestic sanitary sewers, results in a reasonably accurate and conservative sulfide prediction.

1974 EPA equation for sulfide corrosion in cementitious sewers and structures.—Actual corrosion at manholes and downstream of those structures where turbulence occurs can be up to 5 times the average corrosion predicted using the EPA equation, because turbulence causes more outgassing of hydrogen sulfide, making it available for rapid attack of the sewer. The distance turbulence effects extend downstream of structures such as intercepts, change in direction, grade changes, drop manholes, etc., varies, but a distance of 10 pipe diameters is not unusual.

The corrosion at the crown of pipes may be 50% greater than the average corrosion over the exposed unwetted perimeter. Thus, a minimum crown corrosion factor of 1.5 should be applied to the average exposed perimeter corrosion C calculated by Eq. 6- 25 for a reach with uniform flow conditions.

In summary, the corrosion of sewer pipe made of cement- bonded materials is not uniform around the exposed pipe perimeter, due to factors affecting the rate of transfer of hydrogen sulfide to the pipe wall. Corrosion observations and measurements are usually greater at the crown and springline of the sewer. The use of equations (Eqs. 6-25 and 6-26) for prediction of average corrosion provides a useful tool in the design of sewers. Proper factors (CCF and TCF) must be applied to the predicted average sulfide corrosion. These factors are discussed in section D9 of Chap. 6.

Pomeroy's predictive equation for sulfide generation in force mains.—See Eq. 6-13 in Chap. 6. Good correlation can be anticipated between observed sulfide levels in force mains and levels predicted by Pomeroy's force-main sulfide prediction equation. Correlation data is presented in Ref. 11.

Thistlethwayte's model for sulfide generation in gravity sewers.—Poor correlation has been found between measured and predicted sulfides using Thistlethwayte's equations. These equations result in sulfide predictions which are much higher than those observed for longer reaches. For short reaches, the Thistlethwayte model will predict lower sulfides than observed. In summary, use of the Thistlethwayte model to predict sulfide levels in gravity sewers will often result in overpredicting or, worse, underestimating sulfides by a factor of 10 to 20 or more.

In conclusion, engineers designing wastewater collection and treatment systems should view the models for sulfide generation and corrosion as useful tools to indicate sulfide trouble spots. The models are not accurate within 5% to 10%. Therefore, emphasis by designers should be on redesign of the predicted sulfide trouble spots, to further reduce the generation of sulfide, and incorporation of protective measures into the system to minimize hydrogen sulfide attack, odor nuisance, and toxic hazards to personnel.

REFERENCES

1. Brown and Caldwell Consulting Engineers, *Sewer System Sulfide Study, City of Mesa*, Tucson, AZ, June 1983.
2. *Concrete Pipe Handbook*, American Concrete Pipe Association, Vienna, VA, 1980.
3. *Design and Construction of Gravity Sanitary Sewers*, Manual of Practice 60, ASCE, New York, NY, 1982.
4. Graham, R. E., "First Year of Operation of the San Diego Metropolitan Sewerage System," *CWPCA Bulletin*, July 1965.
5. Kienow, K. K., "Calculations on Sacramento Trunk," personal communication, Aug. 7, 1984.
6. Kienow, K. K., "The San Diego-West Point Loma 114-Inch Interceptor Failure," *Water and Sewage Works*, May 1980.
7. Koebig & Koebig, *Special Study of San Diego Metropolitan Interceptor and Point Loma Wastewater Treatment Plant*, Los Angeles, CA, March 30, 1970.
8. Meyer, W. J., and Hall, G. H., *Prediction of Sulfide Generation and Corrosion in Concrete Gravity Sewers: A Case Study*, J. B. Gilbert & Associates, A Division of Brown and Caldwell, Sacramento, CA, April 1979.
9. Pomeroy, Johnson, and Bailey, Consulting Engineers, *Investigation of the Conditions of Sewers of the City of Delano, California*, Pasadena, CA, June 28, 1963.
10. Pomeroy, R. D., "Sanitary Sewer Design for Hydrogen Sulfide Control," *Public Works*, Oct. 1970.
11. Pomeroy, R. D., *Process Design Manual for Sulfide Control in Sanitary Sewerage System*, EPA 625/1-7-005, U. S. Environmental Protection Agency, Cincinnati, OH, Oct. 1974.
12. Pomeroy, R. D., and Parkhurst, J. D., "The Forecasting of Sulfide Buildup Rates in Sewers," *Prog. Wat. Tech.*, Vol. 9, Pergamon Press, London, England, 1977, pp. 621-628.
13. Pomeroy, R. D., "Generation and Control of Sulfide in Filled Pipe," *Sewage and Industrial Wastes*, Vol. 31, No. 9, 1959.

14. Thistlethwayte, D.K.B., *Control of Sulfides in Sewerage Systems*, Ann Arbor Science Publishers, Ann Arbor, MI, 1972.

15. Warren, G. D., II, and Tchobanoglous, G., *A Study of the Use of Concrete Pipe for Trunk Sewers in the City of Delano, California*, Department of Civil Engineering, University of California at Davis, CA, Sept. 1976.

CHAPTER 8

EVALUATING EXISTING COLLECTION SYSTEMS

A. INTRODUCTION

Many wastewater collection and interceptor systems generate sulfide. In some cases, the extent of sulfide generation is sufficient to cause corrosion, odor, and safety problems, or to indicate to the managing organization that a problem may be developing. An evaluation of the existing system may be warranted, and, if required, the scope and detail of that evaluation must be determined. The purpose of this chapter is to: (1) Provide guidance to wastewater system managers, engineers, and system operators; (2) outline the fieldwork and analytical work used to define the problem and the cause of the problem; and (3) review methods to evaluate the sulfide production potential of a collection and interceptor system.

Once the problem, its extent, and the source of the sulfide are confirmed, alternative sulfide control measures can be assessed and a solution to the problem developed. Without a sound understanding of the problem and the source of the sulfide, it is doubtful that cost-effective control measures could be devised. Wastewater treatment plants may need to be included in sulfide evaluations. Much of the material in this chapter can be used to evaluate treatment plants as well as interceptor systems.

B. PRELIMINARY PROBLEM ASSESSMENT

Before a major field assessment of wastewater facilities, or the organization of wastewater sampling programs or other work programs, a preliminary assessment of the problem should be undertaken. The two primary objectives of this assessment are to define whether the problem is sufficient to warrant evaluation and —assuming further work is warranted—to develop the scope of the evaluation work.

1. Uncovering Sulfide Problems

The problem may be obvious. That is, concrete corrosion at wet wells and manholes may be readily seen, or the rotten-egg odor of hydrogen sulfide may be easily recognizable. However, in many cases, the problem is not obvious. Perhaps limited access to an interceptor allows only minimal and cursory examination of potential pipe corrosion. Also, hydrogen sulfide concentrations may be successfully controlled within the atmospheres of the interceptors, and the problem may only surface when the wastewater reaches the treatment plant. In many instances, the first evidence of a sulfide problem upstream in the collection system is an odor problem at the plant headworks or at a pump station.

Sulfide corrosion problems can also be uncovered by maintenance inspections. Maintenance personnel may inspect a structure or manhole that has not been examined for several years, and find a problem; or a sudden increase in the corrosion rate of a structure may be identified by operations and maintenance staff.

Another way problems can be uncovered is through gas monitoring. Safety procedures (see Chap. 13) require gas monitoring for hydrogen sulfide in locations classified as "confined spaces," where there is potential for hydrogen sulfide to be present, and where operating personnel have access. Asking operating and maintenance supervisors to identify the locations where, because of safety considerations, they do not allow staff access, can be a first step in investigating a sulfide problem.

2. Assessing the Problem

An initial problem assessment should be sufficiently limited so that it can be completed within a few days or, at most, a few weeks. The information provides the system manager with enough understanding of the seriousness of the problem to decide whether to undertake a more detailed evaluation. Table 8-1 suggests information sources and limited fieldwork that can be completed quickly to help make decisions.

If wastewater sulfide concentrations (dissolved or total sulfide) are above about 0.5 to 1.0 mg/L, there is strong potential for problems, although odor problems have been experienced at concentrations below this range. If hydrogen sulfide gas concentrations are measured above about 1 ppm (v/v), there is a strong indication of an odor problem, although, again, problems can occur with hydrogen sulfide gas concentrations well below this. (The odor detection limit of hydrogen sulfide is on the order of 0.001 ppm.)

3. Developing Strategy

The initial assessment should provide some limits or bounds on the problem. The reaches or portions of the collection system (or portions of the treatment plant) subject to problems should be identified so that further work can be concentrated where it will be most productive. Is the problem one of corrosion, odor, or safety? Or is it a combination of these three, with emphasis on one or two? Is it a year-round or seasonal problem, or does it have other frequency?

The answers to these questions will help formulate a strategy or an approach. Advice and assistance may be necessary from knowledgeable people in the industry, from staff in governmental agencies, consulting firms, or from universities with experienced personnel.

TABLE 8-1. Initial Sulfide Problem Assessment

- Discuss situation with operations and maintenance personnel, both collection system and plant staff.
- Conduct brief visual inspection of several manholes and key structures.
- Take a field sulfide kit on the inspection trip, and collect and analyze a few wastewater samples. Determine the dissolved or total sulfide concentration (or both) in the wastewater.
- Take a hydrogen sulfide gas detector along and measure H2S concentration inside structures and in ventilation exhausts, if possible.
- Undertake a quick check of maintenance records of existing facilities.
- Determine if there are known locations of sluggish wastewater flow or sludge deposits in the interceptors.

TABLE 8-2. Common Field Information Needed to Evaluate Sulfide Problems

- Note the conditions of concrete and other materials, especially the extent of corrosion. Quantify the amount of corrosion which has occurred, if possible.
- Determine the pH of moisture on wall or pipe crown. A pH of less than 2 indicates active corrosion potential (7).
- Take pictures.
- Collect scrapings of corroded materials, if possible.
- Collect samples of solids deposits below the waterline, if they exist.
- Drill concrete core samples at key locations.
- Determine or estimate wastewater flow and velocity at various locations in the system.
- Collect wastewater samples (usually collected without entering confined spaces). Conduct field analysis or preserve samples for laboratory analysis, or both. Since preservation of sulfide samples involves precipitation of the sulfide, dissolved sulfide measurements must be made in the field.

Note: See Appendix D for detailed sampling and analytical procedures.

C. FIELD EVALUATION

Field evaluation programs involve entering and inspecting wastewater facilities to gather information. Because the work almost always requires entering confined spaces, safety procedures must be observed. Typical information collected during these evaluations and inspections is listed in Table 8-2.

1. Planning and Preparation

Prior to the start of actual fieldwork, a good understanding of the wastewater system is mandatory. Available information on wastewater flows, wastewater quality data (including any previous sulfide measurement), other conditions likely to be encountered, and physical constraints in entering structures needs to be evaluated. As-built drawings should be examined to fully acquaint the field crew with expected situations and to plan for equipment requirements.

Substantial preparation and planning are needed to collect accurate information for locations which are, at best, uncomfortable for the field personnel, and, in many instances, dangerous. Contingency planning is mandatory because unlikely events have a habit of occurring when least expected. (Consult Chap. 13 of this manual and other sources for safety planning prior to any fieldwork.)

2. Inspection Work

Typically, the data gathered during the physical inspection are documented on a form designed for that purpose. The content and format of the forms vary from

Figure 8-1. Collecting information within sewers: (a) Drilling concrete core; (b) taking wall pH measurement. (*Courtesy*: Los Angeles County Sanitation Districts, California)

study to study (see Appendix F for sample inspection form). Figure 8-1 shows a field crew collecting corrosion data in an interceptor.

The field measurement of the depth of corrosion is important information. This information is needed to evaluate the structural integrity as well as the remaining life of the structure. If the original surface of the material can be observed at a nearby location, then a baseline exists, and an estimate of the depth of corrosion at the worst location can usually be made (3). Such measurements are often hampered by the lack of a baseline, i.e., extensive corrosion has obliterated the original surface. In this case, sometimes the inside dimensions of the corroded pipe or structure can be compared with construction documents, or with an uncorroded section of similar pipe made under the same construction contract (7). If direct measurement methods fail, coring of the material may be the only method available to determine the extent of corrosion and to define remaining cover over reinforcing steel.

Locations chosen for inspection need to be evaluated carefully to try to assess the worst conditions, as well as more typical areas. Since physical human entry is often difficult and time-consuming, more extensive pipeline coverage can often be accomplished using closed-circuit television. Television inspection may provide only limited details on extent of corrosion, but television results can be compared with sections physically inspected to provide qualitative judgments.

3. Concrete Coring

Concrete cores can be drilled in pipe and the walls and ceilings of various structures (3). Gaining access to the best location for drilling can sometimes be difficult. Drilling directly down through the pipe crown from above the ground surface has been successfully undertaken; the drilled pipe material in this case was dropped onto a raft moored inside the pipe (7).

Drilled cores supply important information on the condition of the concrete. The remaining cover over reinforcing steel can be measured directly. Also, the alkalinity of the concrete used to manufacture the pipe can be determined through analytical testing of a sample of concrete from the core (1). Figure 8-2 shows a core taken from a corroded interceptor in Sacramento County, California.

4. Hydraulic Assessment

Wastewater flow and velocity information for gravity interceptors at different locations in the system are important in assessing sulfide production. Predicted flows and velocities, based on design data, can be substantially different from actual field conditions due to pipe settlement, solids build-up in pipes, inflow and infiltration, and other factors. Relationships of flow, velocity, pipe roughness, pipe slope, pipe size, depth of flow, hydraulic radius, and other factors can be determined from basic hydraulics texts (4, 2). Manning's formula for open channel flow is discussed in detail in these references.

A wide variety of equipment is available which provides instantaneous and long-term measurement of velocity, depth of flow, and flow rate. The United States Environmental Protection Agency has published a report on this equipment (6). Prior to purchasing or using such equipment, prospective users should determine its capability, applicability, and maintainability.

Tracer or dye measurements or other methods can be used to check the velocity of flow. Tracer measurements involve the injection of a tracer into the wastewater and subsequent monitoring of its movement through the system. Commonly used tracers are dyes (8), salt solutions (12), and radioactive solutions.

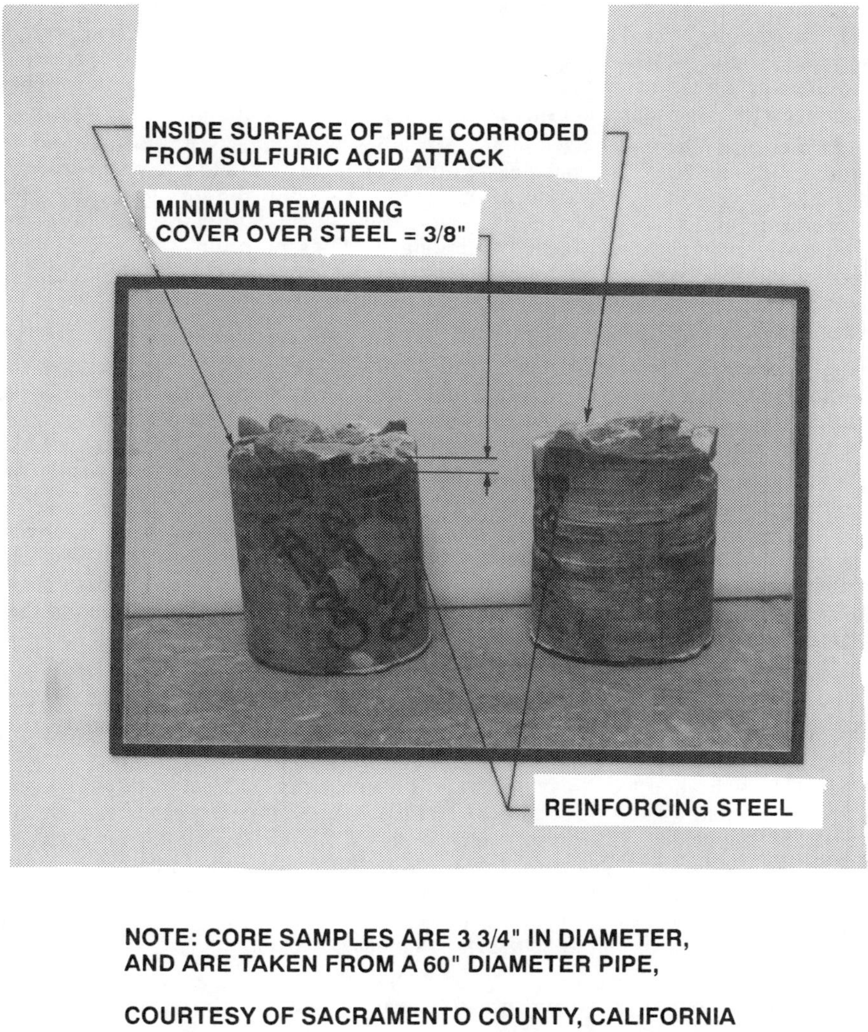

Figure 8-2. Corroded concrete core samples: samples are 3 ¾ in. diameter, from 60-in. diameter pipe. (*Courtesy*: Sacramento County, California)

The user should carefully review the manufacturer's data on tracers to assure that they are properly handled and that they provide the results desired.

D. ANALYTICAL WORK

As described in earlier chapters of this manual, several parameters affect the presence and concentration of sulfide. To evaluate the relationships in a given situation, the parameters need to be quantified. The typical parameters are BOD_5, pH, temperature, dissolved oxygen (DO), total and dissolved sulfide, and perhaps sulfate, soluble BOD_5, oxidation-reduction potential, and others. Appendix D provides details on sampling and analytical procedures and should be examined together with this chapter to establish a field analytical program.

Because there are often many variables associated with sulfide and its production, sampling, and analysis, a primary element of a field analytical program should be to eliminate many of the extraneous variables through following good procedures and through properly organizing and establishing the program. Objectives should be clearly stated; sampling locations, procedures, and frequencies chosen with care; and quality assurance checks made on analytical techniques.

1. Sulfide Cycling and Peaking

When a sulfide corrosion problem is being investigated, annual averages and probably seasonal peaks of the listed parameters are most helpful. However, if the sulfide problem is primarily odor or safety related, instantaneous peaks (or perhaps the maximum events in a 1-minute or 5-minute interval) will be of most interest. The sampling and analytical program, therefore, needs to be developed within the framework of the specific problem.

A key relationship may be the diurnal sulfide cycle. In many cases, this cycle shows minimal amounts of sulfide during the early morning, low-flow period. The sulfide concentration usually rises shortly after the flows begin to increase in the morning. After reaching a peak, sulfide levels usually decline somewhat and then plateau for several hours (12).

The cycle is often less distinct in large sewers. In these sewers, the daily rise in sulfide levels may lag two or three hours behind the rise of flow, as the wave of increased flow travels faster than the actual velocity of the wastewater itself. Also, the plateau may last 8 to 12 hours (12).

The cycles defined above will vary if the system is subject to surcharging or receives discharge from force mains. The discharges from force mains with low velocities [especially those less than 1.2 to 1.4 ft/s (0.36 to 0.43 m/s)] can be extremely troublesome, because septic solids that settle in a force main at low velocity can be washed downstream during periods of higher flow. Also, solids deposits, which encourage sulfide generation, are disturbed at higher flows, thus releasing slug loads of sulfide to the flowing waste stream.

Seasonal sulfide cycles are typically the result of wastewater temperature changes. However, these cycles can also be heavily influenced by seasonal industrial discharges, which cause substantial increase in BOD_5, or, more importantly, in soluble BOD_5.

Erratic peaks are often the result of industrial waste discharge. Sometimes these discharges are not adequately controlled through pretreatment regulations or other source control programs of the municipal authority. The differences between weekday and weekend sulfide concentrations can often be tied to changes in industrial discharges.

Sampling programs aimed at tracking these cycles can be developed, if necessary. Some sampling programs involve hourly (or even more frequent) grab samples, round the clock, for short periods, to adequately define sulfide concentration variations. Other programs involve sampling less frequently, perhaps only during daylight working hours.

2. Choice of Sampling Locations

Sampling locations should be easily accessible and should, preferably, be located to allow the sampling crew to concentrate its attention on the sampling work, rather than on extraneous problems, such as confined space, safety requirements or traffic problems. The locations should be well documented for future reference and proper evaluation of data.

The wastewater flow stream at sampling locations needs to be well mixed. Therefore, sampling immediately downstream from a junction structure or other combination of flows needs to be examined carefully. Sampling the interceptors upstream from a junction structure may help to determine which interceptor is contributing most of the sulfide to the downstream reach.

E. SULFIDE GENERATION INDICATORS FOR SEWERS

Previous chapters defined the mechanism whereby hydrogen sulfide is generated in sanitary sewers. An understanding of that mechanism has been used to develop some basic relationships to evaluate the potential for sulfide generation. These relationships can be used to undertake desk top evaluations where analytical data are lacking. The sulfide generating potential in gravity sewers is intimately tied to velocity of flow, slope of pipe, and other hydraulic parameters.

1. Limiting Velocity

Before 1946, the relationships of the various parameters which affect sulfide generation and resulting corrosion were not well understood. The common theory was that sulfide was generated solely in the solids deposits within the sewer and that minimizing the deposition of solids would minimize the generation of sulfide. Thus, it was suggested that a minimum velocity of 2.0 ft/s (0.61 m/s) would be adequate to avoid solids deposits and eliminate sulfide generation.

In 1946, Pomeroy and Bowlus published an article (10) which gave a new perspective on the generation of sulfide. The article presented what was to become an important temperature - BOD relationship, defined as follows:

$$\text{EBOD} = \text{BOD}_5 \times 1.07^{(T-20)} \tag{3-7}$$

in which the symbols are as defined in Chap. 3. The EBOD was, in turn, related to the velocity by

$$\text{marginal } V = \sqrt{\text{BOD}_5 \times 1.07^{(T-20)} \times 0.018} \tag{8-1}$$

in which V = wastewater velocity required so that no sulfide is generated, ft/s, and 0.018 = empirical constant. Thus, the "limiting velocity" became a useful concept. (See Fig. 4-3 in this manual for a pictorial representation of the effect of velocity on solids deposition.)

2. Z Formula

In 1950, Davey (5) published a more complete formula relating the Reynolds number, EBOD, cross-sectional area of the flow, and the surface width of the flow. Davey's formula was subsequently modified by Pomeroy and reformulated into what became known as the *Z* formula (11). The *Z* formula relates EBOD and the hydraulic properties of the sewer. It applies to small gravity sewers (less than 24 in. or 600 mm in diameter). Pomeroy's *Z* formula is

$$Z = \frac{\text{EBOD}}{S^{0.50}\,Q^{0.33}}\;\frac{P}{b} \tag{6-12}$$

in which the symbols are as defined in Chap. 6

The value of *Z* obtained by using the equation approximates sulfide generation probability as follows:

- When $Z < 5{,}000$, significant sulfide is rarely produced.
- When $5{,}000 \leq Z \geq 10{,}000$, marginal conditions for sulfide generation exist.
- When $Z > 10{,}000$, sulfide buildup is likely.

The *Z* formula is generally successful in predicting the incidence of sulfide generation in gravity sewers. However, it is not absolute. For example, sulfide has been found in a sewer with a *Z* value of 6,000, and not found in one with a *Z* value of 28,000 (1).

3. Flow-slope Relationship

Wastewater flow and wastewater characteristics in small sewers are quite variable. Even though the velocities are often low and solids settle out, the sewer may remain free of sulfide. This is due to the shallow depth of flow, the low oxygen uptake rate, and the contribution of fresh flow containing dissolved oxygen. These observations were responsible for studies which confirmed the relevance of the flow-slope relationship to the potential for sulfide generation.

In larger pipes with greater flow, sulfide build-up is attenuated only if adequate oxygen is present in the wastewater. This is particularly critical because the reaeration rate is usually less in large sewers because of flatter slopes. A crude parameter of oxygen supply is the elevation loss occurring over a given distance and given time. Thus, the rate of energy dissipation is a major factor associated with sulfide build-up. A typical assumption for gravity interceptors is that the oxygen content of interceptor atmosphere is close to that of standard air. If there is reason to believe that this is an incorrect assumption, the following analysis is not valid.

Figure 8-3 may be used as a qualitative guide to evaluate sulfide build-up potential for various flow-slope conditions in a sanitary sewer. The vertical scale is the elevation loss, or "effective slope," averaged over distances equal to 1 hour of flow time. For small collecting sewers, the actual slope of the pipe should be used. The horizontal scale is the daily 6-hr high flow.

Use of Fig. 8-3 is limited to sewers flowing less than two-thirds full. Further, the curves are particular to EBOD concentrations. In calculating the effective BOD (see Eq. 3-7), the average 6-hr high-flow BOD for the day and average temperature for the three warmest months of the year are typically used. Where diurnal BOD data are not available, it may be assumed that the 6-hr high-flow BOD is 1.25 times the

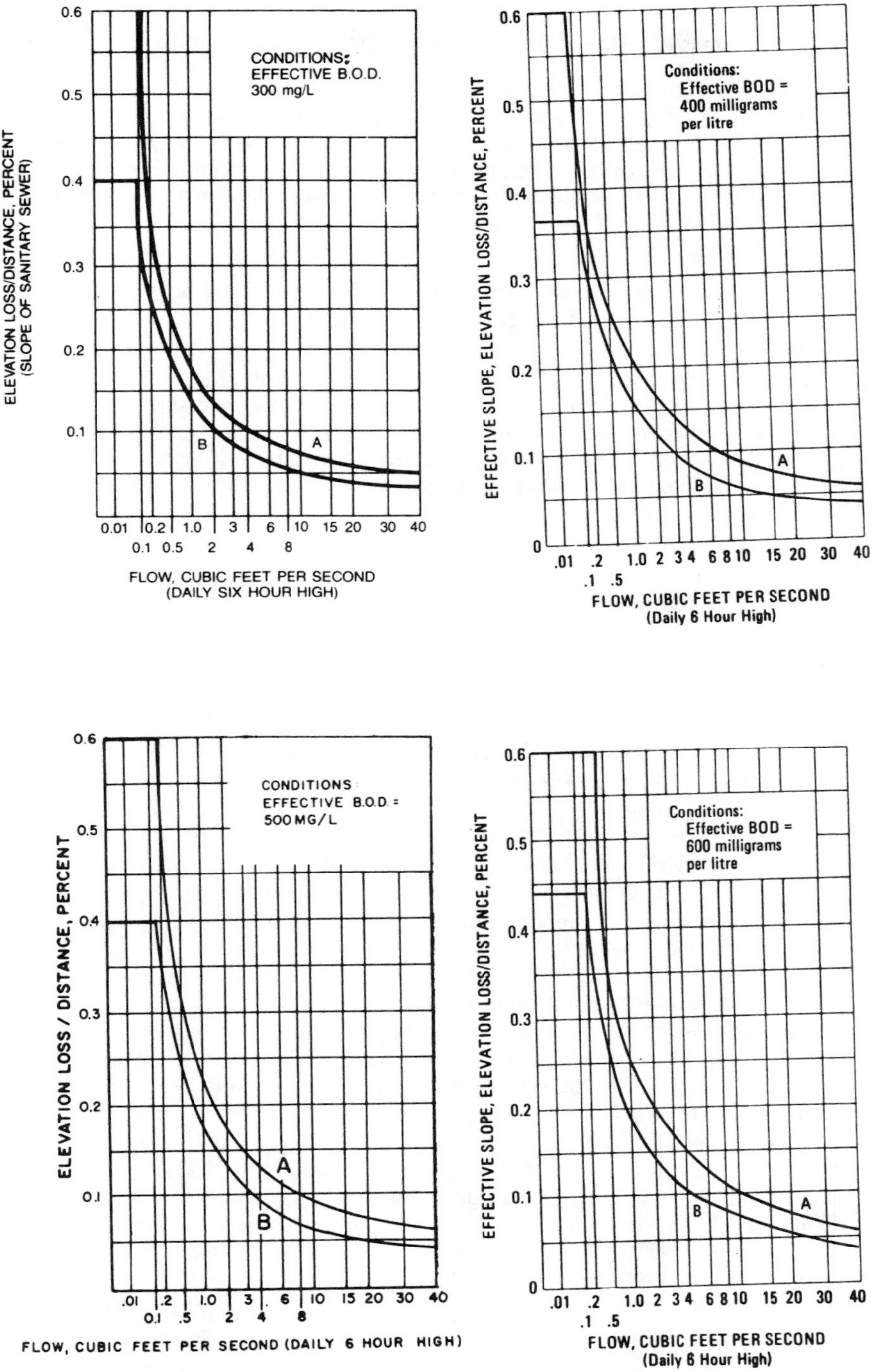

Figure 8-3. Flow-slope relationships.

BOD of a flow-proportioned 24-hr composite. Figure 8-3 may be interpreted as follows:

- A system functioning with a flow-slope relationship value on Curve A will produce very little sulfide, rarely more than 0.1 or 0.2 mg/L of dissolved sulfide. The annual average dissolved sulfide concentration under such conditions will be only a few hundredths of a milligram per liter. When the effective slope of the system is higher than Curve A, sulfide concentrations are negligible.
- A system with a flow-slope relationship value falling between Curves A and B is considered marginal for sulfide generation.

The positions of the curves in Fig. 8-3 change when the EBOD varies (curves for EBOD of 300, 400, 500, and 600 mg/L are shown). However, the cut-off effective slope of 0.6% for Curve A at small flows remains the same, because the determining factors are not associated with EBOD. Even though the relationship between the required slope and EBOD is complex, similar sulfide conditions result if the effective slopes are increased or decreased in proportion to the square root of the EBODs.

Color code methodology. —The color code methodology utilizes the A/B curve to locate the areas of possible sulfide problems and to forecast the probability of sulfide in a sewer. The methodology is as follows:

1. Obtain as-built plans of the existing system. These plans should contain the sewer size, slope, invert elevations, materials of construction, and manhole locations. If the plans are not available, acquisition of the data is necessary.
2. For all trunk sewers being studied, calculate the effective slope. The distance over which the effective slope is measured will be approximately one to two miles for low velocities.
3. Plot the flow-slope relationship for each trunk sewer on the appropriate flow-slope graph. When the flow rate is not known, calculate the flow rates with the sewer being 1/3 and 2/3 full, and plot these as a range.
4. Force mains, inverted siphons, and surcharged gravity sewers are the most probable sulfide generators. Therefore, the locations and details for all force mains, inverted siphons, and surcharged sewers must be identified. Additionally, any sulfide control measures in use must be defined.
5. Color the respective flow-slope graph (Fig. 8-3) being used. That is, color the area below Curve B red, the area between Curve A and Curve B yellow, and an area above Curve A equal to the area between Curve A and B green. Color the remainder of the area above the green area blue.
6. Color code the sewer system maps. That is, each sewer is colored the same as the area on the graph where the flow- slope relationship of the sewer is plotted. The colors graphically display the sulfide potential of each sewer reach in the system. When the flow rates are unknown and a range is assumed, the sulfide potential may fall into two or more categories. In such cases, the sewer should be dash-lined with appropriate alternating colors.

Once the potential problem areas are identified, consideration should be given to conducting field studies to more accurately evaluate the nature of the sulfide problem. Similarly, the color-coded map can be used to estimate the sulfide potential. For example, if all the tributary sewers are colored green or blue, it is reasonable that the wastewater entering the trunk would not exhibit significant sulfide concentration. Therefore, no field studies would seem warranted. However, if the tributary sewers are colored red or yellow, or both, field sulfide studies should be considered.

Force mains.—Force mains which flow full generally do not experience the corrosion problems often observed in gravity sewers. However, force mains are recognized as prime sources of sulfide production. Sulfide is generated at high rates by the slime layer which typically forms on walls of force mains. At the discharge point of the force main, dissolved sulfide can be released as hydrogen sulfide and cause serious corrosion, odor, and safety problems.

Pomeroy (9) published his equation for predicting sulfide build-up in force mains in 1959. The equation has been refined to the form identified in Chap. 6 of this manual.

Again, as with gravity sewers, low velocities can cause significant sulfide production, because of solids deposits that are created under these conditions. Equation 6-13 is valid only for the sulfide produced in the body of the sewage and in the slime layer. If a solids resuspension velocity is not attained in the force main, Eq. 6-13 can significantly underestimate the sulfide produced. (See discussion in Chap. 9 of minimum velocities in force mains.)

REFERENCES

1. American Concrete Pipe Association, *Concrete Pipe Handbook*, 1980.
2. Brater, E. F., and King, H. W., *Handbook of Hydraulics*, 6th ed., McGraw-Hill Book Co., New York, NY, 1976.
3. Brown and Caldwell, *Control of Corrosion and Odor-Phase I, Papillion Creek Wastewater System*, report for the City of Omaha, NE, Feb. 1985.
4. Chow, V. T., *Open-Channel Hydraulics*, McGraw-Hill Book Co., New York, NY, 1959.
5. Davy, W. J., "Influence of Velocity on Sulfide Generation in Sewers," *Sewage and Industrial Wastes*, Vol. 22, No. 9, Sept. 1950.
6. Foster, W. S., and Sullivan, R. H., *Sewer Infiltration and Inflow Control Product and Equipment Guide*, U.S. Environmental Protection Agency, Cincinnati, OH, July 1977.
7. J. B. Gilbert & Associates, *A Case Study - Prediction of Sulfide Generation and Corrosion in Sewers*, report for the American Concrete Pipe Association, April 1979.
8. Pomeroy, R. D., "Flow Velocities in Small Sewers," *Journal of the Water Pollution Control Federation*, Vol. 39, No. 9, Sept. 1967.
9. Pomeroy, R. D., "Generation and Control of Sulfide in Filled Pipes," *Sewage and Industrial Works*, Vol. 31, No. 9, Sept. 1959.
10. Pomeroy, R. D., and Bowlus, F. D., "Progress Report on Sulfide Control Research," *Sewage Works Journal*, Vol. 18, No. 4, 1946.
11. Pomeroy, R. D., "Sanitary Sewer Design for Hydrogen Sulfide Control," *Public Works*, Oct. 1970.
12. United States Environmental Protection Agency, *Process Design Manual for Sulfide Control in Sanitary Sewerage Systems*, EPA 625/1-74-005, Cincinnati, OH, Oct. 1974.

CHAPTER 9

DESIGNING SEWERAGE SYSTEMS TO MINIMIZE SULFIDE

A. INTRODUCTION

This chapter provides information on limiting or preventing the formation of sulfide in sewerage systems, or otherwise minimizing the presence of sulfide. Limiting the formation and presence of wastewater sulfide is the first line of defense in avoiding the problems caused by hydrogen sulfide. If wastewater sulfide cannot be eliminated or reduced to acceptable levels by good design and operating practices described in this chapter, then controls will be needed, as discussed in other portions of this manual.

The most significant factors to be taken into account in minimizing sulfide are summarized in Table 9-1. This summary is based on U.S. experience of the last 20 to 30 years. It does not include all factors which minimize sulfide, but concentrates on situations that have caused most problems. Also, these factors are within the control of wastewater authorities, engineers, and system operators.

B. SMALL SANITARY SEWERS

This section discusses sulfide aspects of small sanitary sewers, generally sewers of 18-in. diameter or less. These sewers typically serve local areas and portions of communities. For many communities, however, 18-in. (46-cm), 15-in.

TABLE 9-1. Summary of Major Factors in Minimizing Sulfide

1. Limit the use of closed conduit systems (force mains, siphons, and surcharged sewers). If a closed conduit system must be used, limit its length and provide adequate velocity. Limiting the use and length of force mains in raw wastewater conveyance systems is probably the single most important factor in minimizing sulfide in collection systems.
2. Make sure that wastewater velocities in both gravity and pressure pipes are adequate to prevent deposition and accumulation of solids, especially during periods of low flow.
3. Keep the velocity up in gravity trunk sewers and interceptors, to the point where wastewater surface reaeration is adequate to prevent sulfide build-up. To be successful in this, there are several parameters to consider, and sulfide modeling may be advisable to evaluate them.
4. Eliminate direct discharge of sulfide to the wastewater collection system from industrial and septic waste sources.
5. Minimize the accumulation of solids in the treatment plant at any location where they will become anaerobic and septic. Where anaerobic treatment processes are used, carefully assess the effects of sulfide produced and how either the sulfide or its effects will be controlled.

(38-cm), and even smaller pipes are used as trunk sewers, transporting all (or nearly all) the sanitary flow to the treatment plant. Little sulfide is normally generated in these sewers for the following reasons:

1. The wastewater flow is relatively fresh. The residence time of wastewater in sewers of these sizes is normally less than 2 to 3 hours. Since the oxygen utilization rate of fresh municipal wastewater is low, relatively high dissolved oxygen levels are maintained in the flow.
2. Oxygen transfer rates from the sewer atmosphere to the wastewater are high due to the hydraulics of smaller sewers, the variable flow rates, high oxygen content of sewer atmospheres, and frequent points of turbulence from changes in flow direction and sewer junctions.
3. High-strength industrial wastewater is normally not found in small sewers, because larger capacity pipes are needed to carry these flows.

Despite the factors listed, which limit the presence of sulfide in small sewers, there are situations in which sulfide can be generated.

1. Adequate Velocities and Slopes

For typical municipal wastewater with its associated grit load and solids content, 2 fps (0.61 m/s) at design conditions (full pipe flow) is commonly used as the minimum velocity for small sewers (1). This criterion implies that a velocity of less than 2 fps is acceptable for flows less than design conditions, and field experience supports it. Velocities of 1.5 fps (0.46 m/s) and even 1.0 fps (0.3 m/s) are acceptable for these pipes during low-flow periods and during the early years of development of an area, despite the fact that grit can build up, and some organic materials can become trapped in the deposits at such low velocities. The high oxygen content of the wastewater, and its low oxygen utilization rate, largely overcome sulfide generation occurring in these deposits. The designer of larger sanitary sewers, especially sewers larger than 36 in. (90 cm) in diameter, should recognize that the special conditions within small sewers, permitting velocities less than 2 fps, normally do not exist in larger sewers.

TABLE 9-2. Minimum Slopes for Small Sewers

Pipe diameter, (in.) (1)	Minimum slope (2)
6	0.0049
8	0.0033
10	0.0025
12	0.0019
15	0.0014
18	0.0011

Note: For a given flow and slope, velocity is little influenced by pipe diameter.

Table 9-2 shows minimum slopes for small sewers, assuming velocity = 2 fps at half-full and full flow, Manning's n = 0.013, and use of the Manning formula.

Preferably, slopes should be steeper than shown in Table 9-2, especially if:

1. There is a long trunkline transporting raw wastewater several miles, with little addition of fresh wastewater to the trunk along its route.
2. High-strength wastewater from industrial discharges is included in the flow being transported long distances.

2. Design and Maintenance Practices

Good design and maintenance practices will practically eliminate sulfide problems in small sewers. They are discussed in greater detail in sewer engineering texts (1, 9). In addition to the minimum velocity and slope criteria discussed above, recommended practices include:

1. Assuring good ventilation of sewer atmosphere; i.e., prevent air blockage down the pipe.
2. Preventing partial solids blockages of pipes by adequate inspection and cleaning.
3. If pump stations are absolutely necessary, limiting the length of force main, and designing the force main with proper attention to sulfide control.

3. Alternative Collection Systems

In special circumstances, wastewater collection systems other than conventional gravity sanitary sewers have been used. These systems include a wide variety of options and may involve septic tank pretreatment, or at least grinding, before the wastewater enters the collection system. One of the advantages of these systems is that smaller diameter collection pipes are used. The collection system is often either entirely pressurized or under vacuum. In some cases, a mixture of pressurized and gravity sewers is possible.

In most of these types of collection systems, the wastewater is anaerobic, and, therefore, subject to sulfide generation. Septic tank effluent often contains dissolved sulfide. Therefore, wastewater managers need to evaluate these systems carefully. Proposed systems using alternative sewers should receive considerable evaluation if they will serve large areas and the flow from them constitutes more than a small percentage of the total flow at the treatment plant, or of the flow in conventional downstream sewers.

C. GRAVITY TRUNK SEWERS AND INTERCEPTORS

Gravity sewers, especially those above about 36 in. in diameter, can be subject to significant sulfide generation. The designer of large gravity sewers needs to carefully evaluate the proposed system and the conditions under which it will operate. Table 9-3 lists the type of information needed to design interceptors for minimum sulfide generation.

1. Velocities to Avoid Deposits

Establishing appropriate minimum velocity criteria for trunk sewers and interceptors is one of the most important steps in successful interceptor design. Adequate velocity is necessary to assure that deposits do not form in interceptors. Once formed, such deposits can be difficult to scour with available wastewater flow, and they provide ideal conditions for sulfide production.

TABLE 9-3. Wastewater Information Often Required to Design Interceptors to Minimize Sulfide

1. Information about wastewater flow and variability in the flows, including seasonal, diurnal, or other significant interval of variability. Also, peak hourly and minimum hourly flows at start-up, and in the future, for wet and dry conditions.
2. Grit information. The largest expected size of grit material should be identified, so that velocities can be developed to move this material. Quantity of grit material can also be important in designing the system.
3. Wastewater characteristics. These include primarily BOD, soluble BOD, temperature, pH, sulfide content of flow coming into the interceptor or from industrial discharges, relative freshness of the wastewater, and seasonal changes in these parameters.

Interceptor designers can utilize an approach involving critical shear stress to develop minimum velocities. A certain shear stress is needed at the bottom of the pipe to transport the heaviest material (i.e., grit). Important characteristics of grit material include particle size, amount of material to be moved, and cohesive properties of deposited material.

Particle size.—Most velocity criteria are developed to move grit; grit is considered the material most likely because of its size and density, to form deposits in sewers. The specific size of the grit material to be moved is important. The largest material is usually found in combined wastewater systems. Gravels in the range of 8 to 10 mm are relatively common in combined sewers. In sanitary sewers, the largest grit material is usually in the 1-mm to 5-mm range (5). For transport sewers where degritting facilities are provided upstream, the largest size of grit will depend on the type of grit removal facilities and their performance, particularly during peak flow periods when much of the grit load moves through the system. Typically, grit removal facilities remove most particles greater than 0.2mm.

The designer should determine the largest size of inorganic material likely to be found in the proposed interceptor. If a concrete pipe is designed to corrode, through "sacrificial" concrete losses, or if it is very likely that the pipe will corrode over time, then the size of the granitic aggregate in the concrete needs to be considered in grit transport. Where combined sewers discharge to an otherwise separated sanitary sewer, the larger inorganic solids generated in the combined system need to be evaluated.

Oversized, heavy objects,such as bricks, large stones, and pieces of pipe or steel can make their way into sewers on occasion. These objects are usually not considered in sewer sedimentation analysis, since they are assumed to be insignificant to formation of deposits. However, for evaluations in combined sewers, consideration of such objects may be important.

Amount of sediment load.—Grit production rates vary considerably, depending on the characteristics of the area served, condition and type of sewers, and type of industrial discharges to the system. Typical values are 2 cubic feet per million gallons of flow (average) to over 20 cubic feet per million gallons of flow for peak situations (10). (Note: 1 ft^3 = 0.028 m^3; 1,000,000 gal = 3,785,000 L.)

The rate at which grit enters the trunks and interceptors can be expected to vary seasonally. High flows, often occurring in wet weather periods, usually bring grit loads into the trunk sewers and move them to the treatment plant. At other than high-flow periods, grit accumulates in the upper reaches of sanitary sewers and other locations where velocity is inadequate to transport it.

Cohesion.—Cohesion of deposited particles is perhaps the least known factor in determining scouring velocities. The silt-clay percentage of sewer deposits is typically low, and research in creek channels (4) shows that if the silt-clay content is below about 5%, critical shear stress is a function of individual particle sizes and is not significantly affected by cohesive properties. However, wastewater contains grease, oil, and other organic materials not found in significant amounts in creeks and surface channels. These materials increase the cohesion of sewer deposits. Furthermore, following deposition, biological growth (slimes) can begin to form within the deposits, thereby increasing their cohesive properties.

Shear stress.—The average boundary shear stress T_o, created in an open channel is defined (3) as

$$T_O = w\ R\ S \tag{9-1}$$

in which w = specific weight of water, lb/ft^3; R = hydraulic radius, ft; and S = invert slope. This average shear stress will be more than adequate to encompass the local shear stress near the bottom of a circular pipe if the flow depth is at least one- third of the sewer diameter. For most practical purposes, the shear stress is probably valid for situations where flow is less than 1/3 depth (21).

The boundary shear stress needs to be sufficient to initiate motion of particles, since it should be assumed that under low flow conditions the largest grit material has been deposited. The critical shear stress of a particle is defined as the minimum boundary shear stress necessary to initiate motion. Lane (7) and Paintal (12) provide a review of research work on critical shear stress needed to move sand and sediment particles of different size under various conditions.

TABLE 9-4. Recommended Critical Shear Stress To Move Deposits In Sanitary Sewers

Recommended shear stress, psf (1)	Reference (2)
0.08	8 (Lysne, 1969)
0.08	12 (Paintal, 1972)
0.03 to 0.04	14 (Schultz, 1960)
0.03 to 0.04 for <1-mm size	21 (Yao, 1974)
0.06 to 0.08 for> 1-mm size	

Note: 1KPa = 20.8855 psf.

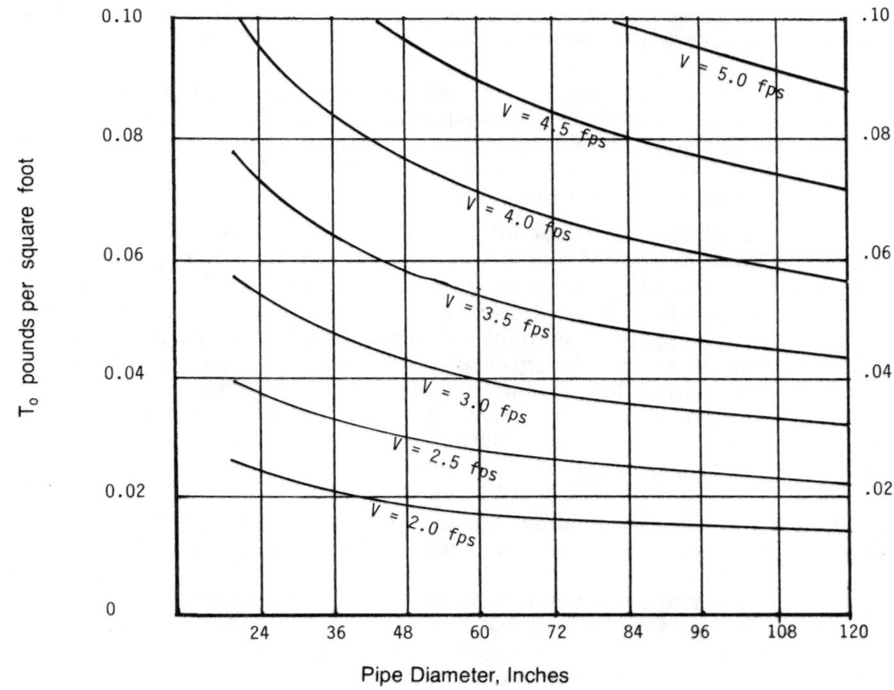

Figure 9-1. Minimum scouring velocities for gravity interceptors.

Researchers have offered various critical shear stress recommendations for sanitary sewer design. These recommendations range from about 0.03 to 0.08 pounds per square foot (psf), and are listed in Table 9-4.

The low end of the range would be appropriate when waste streams contain only small-sized grit particles (1 mm maximum) and limited grit load, and when frequent (perhaps daily) scouring-velocity events occur. The high end of the range would be appropriate for situations involving larger-sized grit, higher than normal grit production or cohesiveness of deposits, and less frequent scouring-velocity events.

Shear stresses of 0.10 psf and higher could be appropriate for larger-sized gravel materials found in combined sewers and some sanitary sewers. All these shear stresses are higher than would be necessary in clean water, because of the cohesive properties of wastewater and the effects of sediment particles in reducing turbulence in the boundary layer (8, 12). For transporting degritted wastewater with particles about 0.2 mm at the largest, critical shear stresses of about 0.01 to 0.02 psf have been suggested (21).

These shear stress recommendations can be used as guidelines in Eq. 9-1 to assess alternative sewer configurations. This type of analysis is helpful when assessing the low-flow or dry weather flow conditions in sewers during the early years of system operation.

Minimum velocities to achieve shear stresses required for full flow in pipes are shown in Fig. 9-1 for pipe sizes of 24 in. to 120 in. This graph is based on Eq. 9-1 and the Manning flow equation, and assumes circular pipe flowing half-full or full, with Manning's $n = 0.013$. As an example, if 0.07 psf shear stress is used in designing a large trunk, then a 60-in.-diameter pipe needs to flow at 4.0 fps when half-full to achieve self-cleansing velocities.

Frequency of cleansing flows.—The potential situation of low flows and formation of deposits occurs frequently in gravity interceptor design. The critical shear stresses can be analyzed and scouring velocities determined. The designer should then determine how frequently these scouring velocities should be achieved to provide assurance of minimal sulfide formation.

Some engineers have examined the amount of grit likely to be deposited during low flows, and placed a limit on the quantity of solids build-up which can be tolerated before scouring flows are required. The time needed for a mass of microorganisms to form in the deposits is in the range of 2 to 4 weeks under the typical summer climatic conditions of North America. If the growth of the microorganism mass within the grit deposits provides increased cohesiveness of the deposits, then scouring flows will need to provide even higher critical shear stress to dislodge the grit. If the critical shear stress is not provided naturally by daily or weekly peak flows, supplemental flushing flows may be required. The specific frequency will be subject to local conditions.

Flushing flows.—When minimum self-cleansing velocities cannot be achieved during the early years of operation of an interceptor or during certain seasons of the year, intermittent flushing flows are often used to scour the deposited material and move it downstream. For large interceptors, these flushing flows can be high. The source and storage of the flushing water may involve substantial costs. These costs must be balanced with increased costs of interceptor maintenance, interceptor corrosion, and loss of capacity if deposits are not removed.

Shape of conduit.—Most of the previous discussion assumes circular pipes. The circular cross section is generally preferable to other shapes for moving grit. If a cross section other than circular is to be considered, its ability to move grit and avoid deposits should be carefully evaluated.

2. Limiting Sulfide Buildup

A key element of interceptor design should be to provide sufficient reaeration from the water surface to offset sulfide generation. In this manner, sulfide control measures, such as chemical addition, become unnecessary. This approach may not be economically achievable, since it could mean frequent lift stations to provide increased slope and velocity for gravity sewers. However, the costs for additional lift stations could be attractive when compared to costs for chemical control of sulfide, corrosion protection, and other cost and non-cost factors associated with sulfide build-up.

Sulfide modeling.—The designer may need to use sulfide modeling to evaluate alternative interceptor design so that long-term economic comparisons between options can be made. The equations given in Chaps. 6 and 7 can be used, as long as the assumptions for the equations and models parallel the situation being evaluated.

Sewer ventilation.—Maximum oxygen supply to the flowing wastewater can be provided only if the oxygen content of the sewer atmosphere is close to the 21% found in air. It is not difficult to achieve this if there are frequent connecting pipes to the interceptor, and each of the pipes has sewer connections that are not far distant (13). Also, air may enter the interceptor at manholes, although solid sanitary-sewer manholes which allow very little air exchange are coming into wider use to limit inflow to the sewer from street or other runoff.

Suppression of oxygen content in sewer atmospheres can occur where there are few, if any, wastewater connections to the interceptor, manhole covers are solid, and there is a siphon, pumping station, or surcharged sewer upstream or downstream (or both), blocking the movement of air along the interceptor. It is not possible to achieve oxygen content close to zero, because the changing water level forces air out and pulls air in at cracks, holes, and leaks. However, it may be feasible for oxygen content to be depressed to the range of 10 to 15%. At this range, reaeration of the wastewater is reduced, and substantially more sulfide will be produced than in the case of 20% or 21% oxygen content in the sewer atmosphere.

Therefore, in reaches of interceptor which could incur reduced atmospheric oxygen concentrations, methods should be employed to assure that adequate fresh air is brought into the sewer. They could include ventilation pipes at key locations, or mechanical ventilation, usually located at the downstream end of a sewer (at a pumping station or treatment plant). The atmospheres of sewers contain toxic and odorous gases, and, therefore, care in venting sewers is required.

To maintain surface reaeration of wastewater, there must be adequate air space in the conduit and adequate wastewater-to-air interface. If circular pipes are more than 2/3 full, adequate reaeration becomes a problem. Therefore, when pipes will be flowing more than 2/3 full during a significant portion of the time (more than just peak infiltration/inflow events), this factor needs to be considered.

Inverted siphons.—Inverted siphons, sometimes called depressed sewers, are, in effect, short reaches of force mains. As such, they should be carefully designed to minimize sulfide formation. There is also concern for the potential blockage of air flow down the sewer. Inverted siphon design is described in Refs. 9 and 15. With respect to sulfide generation, it is best to:

- Minimize the length of siphon.
- Keep velocity up, preferably to 3 fps or higher, during most flow conditions (this requires multiple-barrel design).
- Provide good access for siphon cleaning and maintenance requirements.
- Provide air jumpers to maintain airflow.

Turbulent wastewater conditions often exist at siphon inlet and outlet structures. Therefore, if dissolved sulfide is present in the wastewater, high atmospheric hydrogen sulfide concentrations can be expected in the atmospheres of these structures.

To avoid venting at siphons, and concomitant odor problems, and to keep airflow moving down the gravity interceptor, air jumper pipes are often used connecting the siphon inlet structure to the siphon outlet structure. Sometimes these air pipes can be constructed above the sewer obstruction, and sometimes they must be buried along with the siphon barrels. If the air jumper pipe is buried, a sump is needed to remove condensate water which will collect in the pipe.

In some cases, air jumper pipes are not needed, and the foul air at the siphon inlet structure can be vented to the atmosphere in a standpipe. The siphon outlet structure then needs provision to allow fresh air to enter the sewer and be transported downstream. It should be recognized that under certain wind and meteorological conditions, the air movement in gravity interceptors can be countercurrent to the wastewater flow, thus causing foul air to be vented at whatever locations are available.

Junctions and drop structures.—Junctions and transitions are required on larger sewers when one or more branch sewers join or enter a main sewer. Care must be taken to avoid creating conditions which may contribute to the accumulation of solids in the sewer. Design aimed at streamlining the wastewater flows to reduce headlosses as they pass through these structures will also minimize the release of gases at these points. All junctions should have smooth channeled bottoms, and the invert of smaller sewers should be higher than the invert of larger sewers, if sewers of different sizes are joined. For large sewer junction structures, the hydraulic grade lines are usually matched as closely as possible. The objective should be to avoid hydraulic jumps, as these provide significant liberation of gases.

In some instances, a drop structure is cost effective and can be used to advantage to provide additional dissolved oxygen to the wastewater (15). This should only be considered when both of the following conditions are satisfied:

1. Dissolved sulfide concentration is low (less than about 0.2 mg/L). Otherwise, a large amount of hydrogen sulfide will be off-gassed and is likely to cause serious corrosion and odor problems.
2. Dissolved hydrocarbon gas levels are low. If drop structures create high hydrocarbon concentrations in the gas phase, explosive conditions can prevail in the sewer atmosphere at the drop structure.

Such drop or fall structures are rarely used on large sewers because of the potential sulfide concentrations and the chance for high hydrocarbon concentrations. Also good locations for a drop of several feet in the sewer grade line are seldom found.

Interceptor staging.—In some situations it may be appropriate to stage the construction of wastewater interceptors. Flows during the first several years or even the first decade or more will be so low that solids deposition and sulfide generation can become a long-term and costly problem for the agency. This issue needs to be evaluated as carefully as possible in the economic analysis of interceptor options during project planning and preliminary design work.

Maintenance and monitoring.—Good interceptor design needs to be coupled with adequate maintenance and monitoring to assure that the system is operating as intended. The design should complement long-term monitoring and maintenance needs by providing appropriate access and allowing maintenance work to proceed as safely as possible.

Since manholes and other structures normally provide the only points of entry into the interceptor, they must be designed to allow safe access by personnel. Maximum distance between manholes varies according to design standards and preference of maintenance staff, but 1,000 to 1,500 ft (300 to 450 m) is often the maximum distance used by agencies on pipe larger than 60 in. in diameter. Some agencies use smaller maximum distances between manholes.

A platform within the manhole at or near the top of the pipe is a useful addition, especially for large wastewater interceptors. Without some method for personnel to safely enter, stand, and work near the flow, without actually being in the wastewater, even routine inspections can be difficult and potentially dangerous. Any personnel entry into such interceptors must be effected in accordance with all safety precautions, including supplemental mechanical ventilation.

Inspection and monitoring can be conducted from a stable working platform near the flow. The walls of the pipe can be visually inspected, pH of the wall moisture and depth of flow determined, wastewater samples collected, and a check made for any solids deposits. Although these are relatively simple activities, they are important in tracking the conditions within the interceptor.

Interceptor cleaning and maintenance, and more definitive monitoring work, require a larger effort, typically with more specialized equipment. Access points to the interceptor must be designed to accommodate the type of equipment anticipated for use in cleaning and maintenance. Sometimes boats or rafts are used for interceptor inspection and maintenance work. Specialized monitoring that involves concrete coring, pipe liner inspections, and access to locations between manholes may result in unusual procedures with substantial safety needs. Gates have been constructed to block the flow in gravity interceptors for a period of a few hours at a time, to allow access to conduits downstream from the gate. This allows workers to do inspection and maintenance without encountering the problems inherent in deep, and sometimes high-velocity, wastewater flow.

D. PUMP STATIONS AND FORCE MAINS

Force mains are prime generators of sulfide, and their use should be limited as much as possible. Where force mains are required, however, the designer and operator can use methods and techniques to minimize sulfide.

For a lift station where the length of force main may be 100 to 200 ft or less (30 to 60 m), sulfide generation is likely to be minimal. Systems with the greatest potential for sulfide generation are long force mains (thousands of feet or several miles long) and those where velocity is inadequate to keep all solids moving. Potential low-velocity problems can almost always be solved by proper design.

1. Force Main Velocity

Minimum velocities in force mains are established so that the grit and associated heavy solids are moved through the pipe reliably to minimize, and preferably eliminate, build-up of deposits. Any such deposits are likely to be sources of sulfide generation.

Shear stress and scour velocity.—The previous subsection on critical shear stress applies to force mains as well as to gravity sewers. Critical shear stress must be created in force mains so that, essentially, the largest grit material is transported. In force main design, the Hazen-Williams formula is typically used in determining average velocity of flow (and hence flow rate) from a known pipe size and energy gradeline:

$$V = 1.318\, C\, (R)^{0.63}\, (Se)^{0.54} \qquad (9\text{-}2)$$

in which V = velocity, fps; C = roughness coefficient; R = hydraulic radius, ft; and Se = energy gradeline or energy slope, ft/ft.

For force mains, the roughness coefficient C depends on pipe material, velocity, size, and age of pipe. Common values of C are 120 for pipes of 36-in. diameter and greater, and 110 for pipes of 33-in. diameter or less. The energy slope and pipe size can be adjusted by the designer to achieve the desired velocity and flow rate. Peak velocities are frequently about 8 fps (2.5 m/s) due to significant energy loss at higher velocity.

Minimum scouring velocities for force mains can be developed from information on critical shear stress, and assessment of the pipeline and its associated pumping system. Force main design typically uses higher shear stresses than are used in gravity interceptor design for transporting the same wastewater. This is due to: (1) The variable profile of force mains (the force of gravity must be overcome when transporting grit up a rising leg of a force main); and (2) the greater chance for septic slime layers to form within force mains, thus increasing the cohesiveness of deposits.

Table 9-5 shows minimum velocity ranges for three different pipe sizes, and may be used to provide guidelines for design. The velocities shown are derived from Eqs. 9-1 and 9-2. These are minimum velocities, and the designer should consider higher values for greater assurance in transporting grit. A common grit resuspension velocity used in wastewater force main design is 3.5 fps (1.1 m/s). For some situations, and with large diameter force mains, this velocity is likely to be too low.

The scouring or self-cleansing velocities shown in Table 9-5 need to be achieved frequently enough to avoid sulfide production from the deposited material. The best situation is where scouring velocities are maintained continuously. However, this is not always possible. Daily scouring velocities are advisable. Less frequent scouring, such as weekly or even less often, increases the chance for sulfide generation. Specific situations may allow less frequent scouring without significant sulfide generation from the deposits.

There may be some cases in which lower critical shear stresses than those shown in Table 9-5 might be appropriate. They would include combinations of the following: (1) Grit is known to be composed of small-sized particles (such as 1 mm maximum); (2) grit load is limited; (3) frequent scouring velocities are achieved; and (4) cohesiveness of deposits is known to be low.

Pumping considerations.—Force mains fed by intermittent pumping systems (usually fixed speed pumps) are often designed to achieve scouring velocity every time the pump is activated. However, in many instances, intermittent pumping is undesirable and causes highly variable or slug flows to downstream interceptors and treatment plants. It can also cause highly variable (slug) loading of sulfide downstream. This is not desirable from an odor control and safety point of view.

Intermittent pumping also requires that wastewater be stored in the wet well. Wastewater storage results in solids deposition and sulfide generation. Because of these problems, intermittent pumping is usually employed only in small pumping stations.

TABLE 9-5. Minimum Scour Velocities for Force Mains

Force main size, in. (1)	Roughness coefficient, C (2)	Minimum Velocities, fps T_c=0.07 psf[a] (3)	T_c=0.10 psf[b] (4)
10	110	3.2	3.9
30	110	3.5	4.3
60	120	4.1	5.0

[a]Use for situations judged to be average or typical for the wastewater industry, and where scour velocity is achieved frequently.

[b]For situations involving transport of large-sized grit material (5 to 10 mm), heavy grit load, greater-than-normal cohesiveness of deposits, or substantial time between scouring- velocity events.

Continuous pumping is more desirable than intermittent pumping of wastewater, especially for larger pumping stations, since there is little storage time in wet wells and sudden changes in flows are largely eliminated. However, a common problem in designing continuously pumped systems is that during low-flow periods, and in the early years of operation, flows may not be sufficient to achieve scouring velocities.

Some designs have ignored the problem and, in many of these cases, when the system initiates operation, substantial sulfide is generated within the deposits which form. The sulfide causes problems downstream from the force-main discharge point.

The designer has several potential solutions to the problem, including the following:

1. Construct two pipelines instead of one. Then one pipe can handle the flow, up to a certain magnitude, and the second pipe comes into use at some point in the future.
2. Design a flushing system that can add enough flow to achieve scouring velocities in the force main.
3. Plan on conducting pipe cleaning operations frequently through the use of "pigging" or other equipment. A "pig" can be inserted in the pipeline and pushed through the pipe by the water pressure to scour the pipe surface.
4. Define a workable sulfide control technique that will minimize sulfide formation in solids deposits.

In some cases, combinations of solutions may be needed. These potential solutions should be evaluated at the time of project planning and preliminary design, so that all necessary costs are identified for the force main and pump station. Alternative analyses often need to be conducted to define which of several options provides the most cost-effective system. These analyses should take into account the additional costs incurred downstream, due to increased sulfide corrosion and odor, if the problem is not solved within the pump station/force main system.

2. Force Main Design Practices

Apart from force main length, the designer's primary tool to minimize sulfide formation is maintenance of scouring velocities. Once these velocity requirements have been satisfied, further assessment may be needed concerning the sulfide produced by the slime layer on the wall of the pipe, and other elements of force main design which can reduce or minimize sulfide and its effects.

Slime layer sulfide.—The sulfide produced within the slime layer on the wall of the pipe can be predicted by equations described in Chaps. 6 and 7. Increasing velocity cannot eliminate the slime layer, but will make it thinner. Note that adequate velocities and shear stresses drag the grit along the bottom of the pipe, thus effectively eliminating any slime layer and associated sulfide production in this zone.

Once the sulfide concentrations at the end of the force main are calculated for the various conditions expected, the concentrations can be evaluated to determine if they are acceptable. If they are not acceptable, sulfide generation within the pipe will need to be suppressed or controlled in some manner. A listing of common sulfide control methods for force mains is shown in Table 9-6. These methods are often applicable to control sulfide generated in solids deposits as well as in the slime layer; however, sulfide generated within deposits may be more difficult to control.

Discharge point.—The point at which the force main discharges to a gravity sewer,or where the discharged wastewater first encounters an atmosphere, are

TABLE 9-6. Common Sulfide Control Methods for Force Mains

1. Addition of air or oxygen to the force main, to reduce or eliminate anaerobic conditions in the wastewater and in part of the slime layer.
2. Addition of chemicals such as chlorine or hydrogen peroxide at the start of a force main, to suppress the slime layer biological activity.
3. Addition of oxidants at or near the end of a force main, to oxidize the sulfide which has formed within the pipe.
4. Addition of metal salts (primarily iron), to raise the oxidation-reduction potential to minimize reduction of sulfate to sulfide, and react with the sulfide generated to form insoluble metal-sulfide compounds.
5. Use of caustic wash on an intermittent basis to raise the pH to 12.5 or more for about 20 minutes, to kill the organisms producing sulfide.

locations that should be designed carefully to minimize the liberation of hydrogen sulfide. The wastewater discharged from the force main should not be dropped in a cascade or fall structure unless the hydrogen sulfide and other gases liberated by this extremely turbulent condition have been evaluated, and their effects fully explored and appropriately handled. Equilibrium or near-equilibrium may exist between sulfide in the liquid and hydrogen sulfide in the gas under this condition, which makes it possible to have a large quantity of hydrogen sulfide (perhaps over 1,000 ppm in specific circumstances) in the atmosphere of the structure containing the turbulent drop or fall. If the high gas concentrations can be tolerated, however, a drop or fall structure can be of considerable benefit in dissolving oxygen into the wastewater.

Force main effluent typically contains enough sulfide to make it advisable to bring the flow into the discharge structure with minimal turbulence. This usually involves discharging the force main at or below the water level.

If an oxidant is to be added to the end of the force main to oxidize sulfide formed, it is advisable to inject the chemical into the pipe upstream of the discharge point. The specific distance upstream depends on the chemical and its reaction time. Most chemicals should be dispersed quickly across the entire cross section of the waste stream, so that extraneous chemical reactions, having no benefit to sulfide control, are minimized.

Maintenance and monitoring.—A well designed force main has the appurtenances and features which allow maintenance and monitoring work to be accomplished with ease and safety. Such features relating to sulfide generation and control may include:

1. Stations to launch pigs to clean the force main, and pig recovery points downstream.
2. Special pressure manholes or structures to allow solids deposits within the pipe to be measured and monitored.
3. Provide flat slopes for up-gradient reaches so that minimum grit material will collect at these points.

3. Pump Stations

Sulfide generation can be minimized in pump stations through proper design of wet wells and pumps.

Wet wells.—The primary methods of minimizing sulfide production in wet wells are to reduce the amount of deposition of solids, and reduce the time during which wastewater and solids are stored in them. These considerations point out the need to limit the size of wet wells and to slope the sides near the bottom of a wet well (minimum slope of 1:1). Pump suction should generate enough velocity at its floor to move solids and eliminate dead spots. A typical maximum wet-well retention time is 10 minutes at average design flow conditions. Retention times during extended low-flow periods should be evaluated carefully to avoid excessively long retention times, such as 30 to 45 minutes. The wet well should be divided into two or more sections, and the pump suctions arranged to allow a section to be taken out of service for inspection and cleaning.

Medium to large pump stations are typically custom-built structures, commonly divided for corrosion and safety reasons into two isolated sides: the wet-well and dry-well sides. As the wastewater enters the station at the wet well, gases are volatilized and flammable gases are released. To minimize the chance for explosions, all equipment and electrical work on the wet-well side must be of

TABLE 9-7. Effects of Some Contaminants on Sewers

Waste characteristics (1)	Problems caused in sewer systems (2)	Remedial action (3)
Oil, grease and fat	coating and clogging; reduction in sewer capacity as a result of grease build-up, increased friction, and slower velocities	pretreatment, sewer cleaning
Acidity	corrosion and disintegration of concrete and steel pipe	neutralization, equalization
Alkalinity	corrosion of metal pipe, possible scaling	neutralization, equalization
Grit	excessive wear on pump stations; accumulation may clog piping	pretreatment, sewer cleaning
Coarse suspended matter	obstruction of piping	pretreatment utilizing screen and grit removal; cleaning of sewers
Heavy metals and toxic substances	in the presence of excess acidity or alkalinity, may produce poisonous gases	pretreatment
Excessive organic loadings	may create septic conditions in sewer lines, resulting in the production of hydrogen sulfide gas, which corrodes concrete structures and is also lethal	pretreatment; try to maintain aerobic conditions in sewer system; treat sulfide
Flammable and explosive substances	fires, explosions; noxious vapors create safety hazards for sewer maintenance personnel	prevent discharge into sewer; monitoring; use of safety equipment

TABLE 9-8. Some Characteristics for Evaluating Relative Intensity of Industrial Wastes

Waste parameters (1)	Relative Intensity Low (2)	Average (3)	High (4)	Extremely high (5)
BOD_5, mg/L	200	200-300	300-1,000	> 1,000
COD, mg/L	300	300-450	450-1,500	> 1,500
SS, mg/L	20	200-300	300-1,000	> 1,000
pH	6	6-9	9	> 9

explosion-proof and spark-proof construction. The wet-well and dry-well sides should be isolated by a vapor-tight dividing wall at both the substructure and superstructure levels. This requires that each have separate outside entrances, and that all conduit and pipe penetrations between them be caulked gas-tight. The isolation of the wet well from the dry well also serves to contain problems associated with sulfide to the wet-well side of the station.

Continuously pumped stations with variable-speed pump controls should be programmed to control the wet-well level to track the depth-versus-flow curve for the upstream sewer. Using this concept, cascading wastewater or excessive turbulence in the wet well will be avoided, thereby minimizing the release of hydrogen sulfide and other gases.

Ventilation.—Wet wells should be classified as "hazardous" in accordance with the National Electric Code, Article 500. Ventilation, if needed for employee access, should be designed to scavenge the wet well airspace frequently and efficiently. Both toxic and hazardous gases should be considered when selecting ventilation rates. The Ten-States Standards recommend that wet wells be ventilated at the rate of 12 air changes per hour. The National Fire Protection Association (NFPA) Code 820, Recommended Practice for Wastewater Facilities (guidelines currently available for public discussion), recommends continuous ventilation with 15 air changes per hour. While ventilation rates are most often expressed as air changes per unit of time, the designer should recognize that the air within the space is not being changed so much as it is being diluted by the operation of the ventilation system. For this reason, ventilation systems should operate continuously. To purge wet wells and all other access points in a pump station most effectively, fans should be provided for both supply and exhaust. High-energy diffusers are recommended for supply systems to distribute the incoming air throughout the ventilated space efficiently. Exhaust grilles should be located near the expected points of release of odorous and hazardous gases.

If the wet well does not need to be ventilated because it does not offer access for personnel, it should be considered a confined space. The National Institute for Occupational Safety and Health recommends labeling such areas as confined spaces, and warning staff of potential hazardous conditions within.

In some cases, the atmosphere of a wet well can be ventilated to the downstream sewer, thereby keeping the airflow moving downstream with the wastewater. More commonly, however, wet-well foul air needs to be vented to atmosphere at the pump station. Odor control may be needed on this foul air exhaust.

TABLE 9-9 Some Typical Industrial Pretreatment Schemes

Industrial waste (1)	Flow characteristics (2)	Normal contaminants (3)	Typical pretreatment (4)
Metal products	intermittent	BOD, COD, TSS, TDS, chlorine demand, color, coliforms, oil and grease, organic nitrogen	screening, oil and grease removal, equalization
Dairy products:			
milk handling, milk products	intermittent to continuous	BOD, TDS, COD, grit, chlorine demand, alkalinity, color, turbidity, detergents, foaming, coliform	oil and grease removal, equalization and neutralization
natural and cheese products	intermittent	BOD, COD, TSS, grit, alkalinity, color	oil and grease removal, equalization and neutralization
Beverages:			
malt beverage distilled spirits	intermittent to continuous	BOD, COD, TSS, grit, acidity, alkalinity	grit removal, coarse solids separation, equalization, neutralization
wine and brandy	intermittent to continuous	BOD, COD, TSS, grit, nutrient deficiency	grit removal, coarse solids separation, equalization, neutralization
soft drinks bottling	intermittent to continuous	grit, alkalinity	grit removal, neutralization
Textiles:			
wool	intermittent to continuous	BOD, COD, TSS, TDS, grit, chlorine demand, alkalinity, color, detergents, heavy metals, phosphorus, oil and grease, coliform	coarse solids separation, oil and grease removal, chemical precipitation, equalization, neutralization
cotton and synthetics	intermittent to continuous	BOD, COD, TSS, TDS, chlorine demand, alkalinity, color, detergents, heavy metals, phosphorus	coarse solids separation, chemical precipitation of heavy metals and color, equalization, neutralization
Leather tanning:			
chrome tanning and finishing	intermittent	BOD, COD, TSS, TDS, grit, acidity/alkalinity, heavy metals, oil and grease	grit removal, equalization, chemical precipitation, solids separation, neutralization

TABLE 9-9 continued

Industrial waste (1)	Flow characteristics (2)	Normal contaminants (3)	Typical pretreatment (4)
vegetable tanning	intermittent	BOD,COD, TSS,TDS, grit, oil and grease	coarse solids separation, grit removal, equalization, neutralization
Petroleum refining	continuous	BOD, COD, TDS, grit, heavy metals, oil and grease, phenol, sulfides	oil separation, equalization, chemical coagulation, dissolved air flotation
Metal finishing	intermittent to continuous	TDS, cyanide, ammonia, hexavalent chromium, heavy metals, oil and grease	equalization, neutralization, cyanide removal, chromium reduction, chemical precipitation
Fruit and vegetable	intermittent	BOD, COD, TSS, TDS, grit, color, detergents, acidity	grit removal, coarse solids separation, neutralization
Pulp and paper	mechanical pulping continuous	BOD, COD, TSS, grit, chlorine demand, heavy metals, acidity, coliform	grit removal, coarse solids separation, neutralization
Chemical pulping (bleached)	continuous	BOD, COD, TSS, TDS, grit, chlorine demand, acidity, heavy metals, color, coliform	grit removal, coarse solids separation, neutralization

E. INDUSTRIAL DISCHARGE CONTROLS

Many agencies control the direct discharge of sulfide to the sewerage system by limiting the operator of the industrial facility to a maximum allowable concentration, such as 0.1 mg/L. This obviously cuts the amount of sulfide introduced to the public sewers. In addition, many characteristics of industrial waste can contribute to or cause sulfide production in downstream sewers, and have other deleterious effects.

The characteristics of industrial discharges vary depending on the processes being used by a given industry at a given time. Consequently, it is often impractical to design the sewer system to handle concentrated industrial wastes that are incompatible with good design practices for municipal sewer systems. Objectionable characteristics are excess acidity or alkalinity; explosive and flammable substances; heavy metals and toxic substances; excessive organic wastes; oil, grease, and fats; and excessive amounts of coarse suspended matter and grit. The effects of these waste characteristics on the sewer system vary greatly, as summarized in Table 9-7 (modified Table 29-1 from Ref. 19).

Certain industrial wastes will disrupt wastewater treatment processes, harm sewer pipe and other structures, or add unduly to the cost of operating and maintaining municipal wastewater treatment facilities. Such wastes should either be excluded from the sewer system, with the responsibility for their disposal placed entirely on the industry, or they should be pretreated to reduce their effects on the treatment system (19).

Some industrial wastewater characteristics are summarized in Table 9-8. Typical industrial wastewater pretreatment schemes are listed in Table 9-9.

References 16, 17, and 18 discuss industrial waste pretreatment, sewer use regulation, and joint treatment of industrial wastes in municipal wastewater.

REFERENCES

1. ASCE and Water Pollution Control Federation, *Gravity Sanitary Sewer Design and Construction*, ASCE Manual No. 60 and WPCF Manual No. FD-5. Joint Task Force Preparation, 1982.
2. ASCE, *Sedimentation Engineering*, ASCE Manual No. 54, 1975.
3. Chow, V. T., *Open-Channel Hydraulics*, McGraw-Hill Book Co., New York, NY, 1959.
4. Dunn, I. S., "Tractive Resistance of Cohesive Channels," *Journal of the Soil Mechanics and Foundations Division*, ASCE, Vol. 89, No. SM3, Proc. Paper 2062, March 1963.
5. Finger, R., and Parrick, J., "Optimization of Grit Removal at a Wastewater Treatment Plant," *Journal of the Water Pollution Control Federation*, Vol. 52, No. 8, Aug. 1980.
6. Graf, W. H., *Hydraulics of Sediment Transport*, McGraw-Hill Book Co., New York, NY, 1971.
7. Lane, E. W., "Progress Report on Results of Studies on Design of Stable Channels," U. S. Bureau of Reclamation, Hydraulics Laboratory, Report No. Hyd-52, Denver, CO, June 1952.
8. Lysne, D. K., "Hydraulic Design of Self-Cleaning Sewage Tunnels," *Journal of the Sanitary Engineering Division*, ASCE, Vol. 95, No. SA 1 Proc. Paper 16404, Feb. 1969.
9. Metcalf & Eddy, Inc., *Wastewater Engineering: Collection and Pumping of Wastewater*, G. Tchobanoglous, ed., McGraw- Hill Book Co., New York, NY, 1981.
10. Metcalf & Eddy, Inc., *Wastewater Engineering: Treatment/ Disposal/Reuse,* 2nd ed., G. Tchobanoglous, ed., McGraw-Hill Book Co., New York, NY, 1979.
11. Paintal, A. S., "Design of Sewers to be Self-Cleaning," *Water & Wastes Engineering*, Jan. 1977.
12. Paintal, A. S., "Hydraulic Design of Self-Cleaning Sanitary Sewers," *Water and Sewage Works*, Scranton Gillette Publication, 1972.
13. Pescod, M. B., and Price, A.C., "Major Factors in Sewer Ventilation," *Journal of the Water Pollution Control Federation*, Vol. 54, No. 4, April 1982.
14. Schultz, H., "Uber die Berechnung der unteren Grenzegeschuindigkeiten in Kanalisationsnetzen," *Wasserwirtschaft und Wasserteknik,* Heft 7, 1960.
15. U. S. Environmental Protection Agency, *Design Manual-Odor and Corrosion Control in Sanitary Sewerage Systems and Treatment Plant*, EPA/625/1-85/018, Cincinnati, OH, Oct. 1985.
16. U. S. Environmental Protection Agency, "Federal Guidelines, Pretreatment of Pollutants Introduced into Publicly-Owned Treatment Works," Washington, DC, 1976.
17. Water Pollution Control Federation, "Joint Treatment of Industrial and Municipal Wastewater," Washington, DC, 1976.
18. Water Pollution Control Federation, *Manual of Practice No. 3-Regulation of Sewer Use*, Washington, DC, 1975.
19. Water Pollution Control Federation, *Manual of Practice No. 11-Operation of Wastewater Treatment Plants*, Washington, DC, 1976.
20. Yao, K. M., "Functional Design of Sanitary Sewers," *Journal of the Water Pollution Control Federation*, Vol. 48, No. 7, July 1976.
21. Yao, K. M., "Sewer Line Design Based on Critical Shear Stress," *Journal of the Environmental Engineering Division*, ASCE, Vol. 100, No. EE2, Proc. Paper No. 10480, April 1974.

CHAPTER 10

DESIGN OF CORROSION CONTROL SYSTEMS FOR NEW OR EXISTING FACILITIES

A. INTRODUCTION

The purpose of this chapter is to provide an overview of design options available for corrosion control in new or existing systems. Emphasis will be placed on "before-the-fact" solutions to potential corrosion in the design of new systems. Other important concepts pertinent to this chapter are covered elsewhere, including the Pomeroy-Parkhurst model for sulfide prediction (Chap. 6), the chemistry of sulfide (Chap. 2), and sulfide control by chemical and hydraulic means (Chap. 12). Covered within this chapter are prediction of design corrosion rates and a number of design options, such as pipe material, pipe lining and coating, concrete modification, sacrificial wall thickness, flooding, and protection of special structures.

B. PREDICTION OF DESIGN CORROSION RATES

Corrosion rates can be predicted for both concrete sewers and ferrous pipe.

1. Corrosion Rate for Concrete Sewers

Basic corrosion equation.—Pomeroy (10) has expressed the rate of concrete corrosion as a function of the flux of hydrogen sulfide gas to the pipe crown, the density and alkalinity of the concrete, and the fraction of product sulfuric acid that actually reacts with the pipe. This relationship is described by

$$C = 0.45 \frac{\phi_{sw}\, k\, CCF\ TCF}{A} \tag{10-1}$$

in which C = rate of concrete corrosion by sulfuric acid at the pipe crown, in./yr; ϕ_{sw} = flux of hydrogen sulfide gas to the pipe wall, gm/m^2-hr; A = alkalinity of the concrete; expressed as equivalent grams of calcium carbonate per gram of concrete; k = acid reaction factor; CCF = crown corrosion factor; and TCF = turbulence corrosion factor, equal to 1.0 for uniform flow conditions well removed from points of turbulence. The estimated corrosion rate in U.S. customary units C can be converted to metric units by

$$C^{*} = 25.34\, C \tag{10-2}$$

in which C^{*} = crown corrosion rate, mm/yr.

Evaluation of terms.—The acid reaction factor k is the fraction of sulfuric acid that reacts with cement at the pipe wall. This fraction is a function of the amount of H_2S released to the sewer atmosphere, the acid runoff, and failure of all the acid to react. When acid production is rapid, or if a significant amount of condensate is formed (i.e., warm wastewater and cold pipe wall), k may be as low as 0.30, because the acid runs down the pipe wall and back into the wastewater. For slow, uniform acid production, the reaction is more complete, and the value will approach unity. No actual measurement of the acid reaction factor has been made, but a range of values has been reported based upon correlation of measured corrosion with equation predictions, as shown in Table 10-1.

Unless the designer has specific system information available, a value of 0.95 should be used to estimate the maximum potential rate of corrosion in sewers larger than 60 in. (152 cm) diameter.

The alkalinity A, or acid neutralizing capacity of the pipe material, can be determined by analysis or can be estimated from the composition of the mix used in making the pipe. Anhydrous cement has an alkalinity, expressed as calcium

TABLE 10-1. Summary of Parameter Estimates for Calculation of Corrosion Rate

Date (1)	Publication (2)	Page reference (3)	Recommended[a] k (4)	Factor for mainline CCF (5)	Factor for structures [CCF X TCF] (6)	Design life, yr (7)
1974	EPA Manual	3-43, 3-45	0.3-1.0 (0.95)	1.5	N.R.[c]	N.R.
1978	Kienow/Pomeroy ASCE paper	9,14, 15, 18, 19, 21, 22	0.3- 1.0 (0.60,0.70)	2.0	N.R.	50
1980	ACPA Concrete Pipe Handbook	7-32 to 7-34	0.3-1.0 (0.70) (0.70)	1.5	N.R.	50
1980	Meyer WF paper	2674	0.3-1.0 (1.0) k = 0.70	N.R. 1.4	1.0-1.6 1.4-2.3	N.R. N.R.
1981	Hall/McLaren/Hyde	2,4	0.3-1.0 (1.0) k = 0.70	N.R. 1.4	1.5 2.1	N.R. N.R.
1982	ASCE Manual of Practice No. 60	64, 65	0.3 - 1.0 (0.70)	1.5	N.R.	50
1984	ACPA Sulfide Manual[b]	4-3	0.70	1.0	1.5	50 to 100
	ACPA Sulfide Manual	6-14	1.0	1.0	1.5	50
1984	Kienow SULF-5	2,7	(0.75)	2.0	4.0	50
1985	Kienow SULF.BAS	2,5	(0.75)	2.0	4.0	50
Recommended Design Value			0.70	2.0	4.0	50-100

[a] EPA used k = 0.95 to evaluate corrosion in Case Study pg. 3-45, and the Sacramento Case Study used k = 1.0 to evaluate corrosion in the Central Trunk. k in parentheses indicates value used in design example.

[b] k = 0.70 recommended for design, pg. 4-3, but k = 1.0 used for example calculation pgs. 6-14.

[c] N.R. = No recommendation.

Note: *CCF* = factor by which average perimeter corrosion is multiplied to obtain maximum crown corrosion, and *TCF* = factor by which crown corrosion in uniform flow pipeline is multiplied to obtain crown corrosion for pipe downstream of structures where moderate turbulence occurs.

carbonate equivalent, equal to about 1.18 times its weight. Thus, if the proportion of cement used, expressed as a percentage of the weight of the cured pipe, is multiplied by 1.18, the result is the alkalinity contributed by the cement. In concrete pipes made with granitic aggregate, the alkalinity is usually between 16% and 24%. In some areas, calcareous rocks, such as limestone or dolomite, are more available than any other hard rock and, therefore, are used as aggregate for concrete. Where calcareous rock is used in making concrete pipe, the equivalent alkalinity may be 100% or even higher.

The flux of H_2S to the pipe wall exposed above the water surface is expressed (10) as

$$\phi_{sw} = 0.45\,(sv)^{0.375}\,J\,[DS]\,\frac{b}{P'} \qquad (6\text{-}8)$$

in which

ϕ_{sw} = flux of H_2S to the pipe wall at 20°C (68°F), g/m^2-hr;

s = slope of the energy grade line of the liquid (i.e., pipe slope for uniform steady flow), dimensionless;

v = stream velocity, ft/s;

$[DS]$ = concentration of dissolved sulfide in wastestream, mg/L;

J = fraction of dissolved sulfide present as H_2S;

b = surface width of waste stream, ft; and

P' = exposed perimeter of pipe wall above waste surface,

The dissolved sulfide concentration DS can be measured directly, or approximated by using the Pomeroy-Parkhurst equation. The total sulfide concentration includes both dissolved and precipitated metal (iron, zinc, copper) sulfides. Dissolved sulfides consist of H_2S, HS^- and $S^=$ species. However, $S^=$ ion only becomes significant at pH levels above 12.0. The pKa for the H_2S/HS^- weak acid system is 7.0 at 20°C, given by

$$\frac{[S^=][HS^-]}{[H_2S]} = K_a = 1.0 \times 10^{-7} \qquad (10\text{-}3)$$

Assuming that dissolved sulfide DS comprises only H_2S and HS^-, one can calculate the J parameter for a given pH at 20°C by deriving it from the equilibrium expression as

$$J = \frac{[H_2S]}{[DS]} = \frac{1}{1 + \dfrac{10^{-7}}{10^{-pH}}} \qquad (10\text{-}4)$$

At 20°C, the J parameter is 0.99, 0.50, and 0.01 at pH levels of 5.0, 7.0, and 9.0, respectively. The hydrogen sulfide flux ϕ_{sw} can be estimated by the use of nomographs provided herein. The flux per 1 mg/L of dissolved, unionized hydrogen sulfide in pipes flowing half full can be determined by the use of Fig. 10-1.

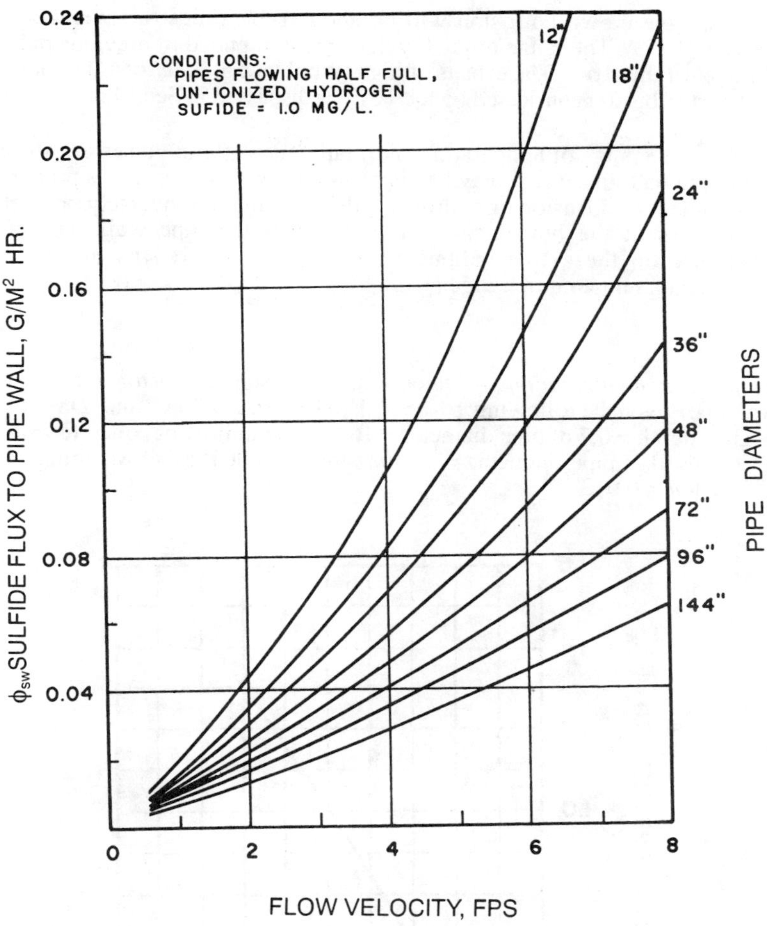

Figure 10-1. Effect of velocity and pipe size on sulfide flux to pipe wall under specified conditions.

The strong upward curvature of the nomograph lines is due to the increasing effect of turbulence, which increases the effective air-water interface area. The flow velocity for input to Fig. 10-1 is that velocity which prevails when the pipe is flowing half full. (The velocity corresponding to a specified design flow should not be used.)

The appropriate half-flow velocity can be determined from nomographs provided in most hydraulics reference books. To adjust the hydrogen sulfide flux at half flow to any other flow condition, Fig. 10-2 can be used. The use of both figures is demonstrated in the example problem at the end of this subsection.

The crown corrosion factor *CCF* is included in Eq. 10-1 to account for the increased rate of corrosion at the pipe crown. Relative to the average corrosion rate for the entire exposed pipe wall perimeter, corrosion will be greater at the flowline, less at 10 o'clock and 11 o'clock, and most pronounced at the crown (12 o'clock). In reaches that have uniform hydraulic conditions well removed from points of turbulence, the crown corrosion will be about 1-1/2 times the average perimeter corrosion (10, 3). The range of *CCF* values recommended in previous publications is shown in Table 10-1. The effects of high turbulence at junctions are not addressed by this factor, but are considered by the *TCF*, as discussed in Sec. H of this chapter.

Over the range of temperatures normally encountered in sewers, the effect of temperature on transfer rate may be disregarded. The transfer rate is proportional to the coefficient of diffusion of hydrogen sulfide in air, and inversely proportional to the thickness of the boundary layer of air near the pipe wall. The effects of temperature on these two parameters can be considered as offsetting in the environmental temperature range (13).

Example design calculations.—Determine the design corrosion rate for concrete sewer pipe, given the following: pKa = 7.0; pH = 6.0; $T = 20°C$; and $[DS] = 0.4$ mg/L. Design Q depth = 0.7 of pipe diameter = 16.8 in.; alkalinity of concrete with granitic aggregate = 0.2; pipe diameter = 24 in.; slope = 0.0015; and Manning roughness coefficient $n = 0.013$.

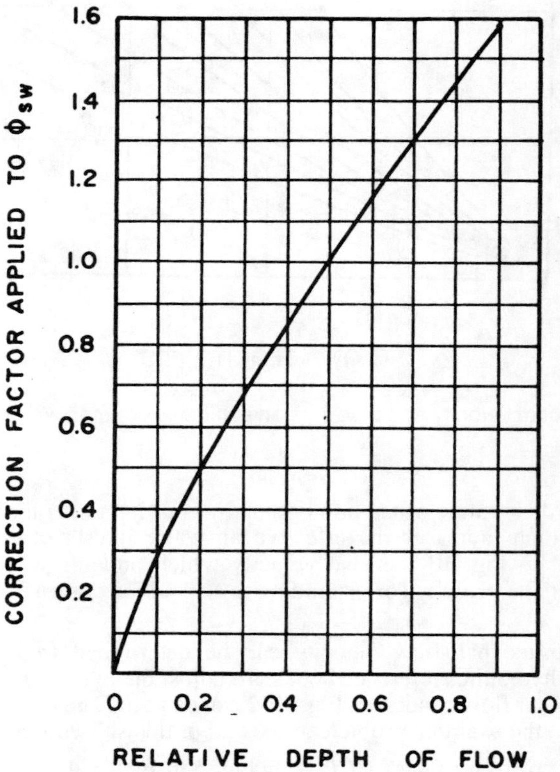

Figure 10-2. Factor to apply to ϕ_{SW} from Fig. 10-1 to calculate ϕ_{SW} for other than half-pipe depth

1. Determine the hydrogen sulfide flux at half-full flow: from Manning nomograph, v (full flow) = 2.80 ft/s and v (half-full) = 0.99 v (full) = 2.77 ft/s. From Eq. 10-4

$$J = \frac{[H_2S]}{[DS]} = \frac{1}{1 + \frac{10^{-7}}{10^{-6}}} = 0.91$$

Using Fig. 10-1 to determine ϕ_{sw} (half-full)

$$\phi_{SW} = 0.044 \text{ g/m}^2 - \text{hr per mg/L } [H_2S]$$

$$\phi_{SW} = (0.044) \{(0.91)(0.4)\} \text{ g/m}^2\text{-hr}$$

$$\phi_{SW} = 0.016 \text{ g/m}^2 - \text{hr}$$

2. Determine the hydrogen sulfide flux at design flow, using Fig. 10-2:

$$(\text{design flow}) = 1.3\ \phi_{SW}\ (\text{half} - \text{full})$$

$$\phi_{SW} = 0.0208 \text{ g/m}^2 - \text{hr}$$

3. Determine theoretical corrosion rate. Assuming an acid reaction factor of 0.95 and crown corrosion factor of 2.0, and a turbulence corrosion factor of 1.0, the solution of Eq. 10-1 is

$$C = 0.45 \frac{(0.0208)(0.95)(2.0)(1.0)}{(0.2)}$$

$$= 0.089 \text{ in./yr}$$

2. Corrosion Rate for Ferrous Pipe

Basic corrosion equation.—The corrosion of ferrous sewers by sulfuric acid involves the same processes as the corrosion of concrete pipe, and the ferrous corrosion rate also can be predicted by a slightly modified version of Eq. 10-1 (3).

Since the action of acid may stimulate the simultaneous reaction of the metal with oxygen (10), and direct reaction of H_2S and iron occurs, it has been suggested that, for design purposes, the total corrosion rate due to H_2S, H_2SO_4, and oxygen, can be assumed to be 10 times the rate for acid attack alone (15). It can be expresed as

$$C = 0.89\ \phi_{sw}\, k\ CCF\ TCF \qquad (10\text{-}5)$$

Example design calculation. —Using the same data as for the preceding example, the acid corrosion rate for a comparable ferrous pipe can be recalculated as

$$C = 0.89\ (0.0208)\ (0.95)\ (2.0)\ (1.0)$$

$$= 0.035 \text{ in./yr}$$

C. PIPE MATERIAL AS DESIGN OPTION

Various types of pipe material can be specified as part of the overall design strategy. Either corrosion-resistant or corrosion-sensitive materials can be employed, depending on predicted sulfide levels, required service life, and economic considerations. The designer has the option of simply specifying an acid-resistant material or specifying an acid-sensitive material together with other corrosion control strategies, such as:

- Providing O_2 to water.
- Providing chemical control of sulfide generation.
- Designing system hydraulics to avoid sulfide generation.
- Providing sacrificial concrete cover.
- Providing sacrificial metal thickness (steel or ductile iron).
- Using concrete modification, such as calcareous aggregate.
- Using a protective liner or coating.

Chemical and other strategies for control of sulfide are discussed in Chap. 12, while materials options are discussed in this chapter. The following materials, varying in sensitivity to acid corrosion, are available for sanitary sewer construction:

- Vitrified Clay Pipe (VCP) is virtually immune to acid attack. In older VCP lines, cement mortar joints expanded, and sometimes broke the bells. Other types of gasketed joints now in general use avoid this problem. However, VCP is brittle, and so needs special installation practice and care in handling and transport.
- Steel pipe is susceptible to direct corrosion by H_2SO_4, H_2S corrosion of the iron component, and the normal oxidation of iron.
- Cast Iron Pipe (CIP) is also susceptible to H_2SO_4 and H_2S, and normal oxidation corrosion. It generally lasts longer than steel, simply because a thicker pipe wall is used.
- Reinforced Concrete Pipe (RCP) is susceptible to acid corrosion. However, despite its vulnerability, concrete represents an important sewer pipe material, particularly for large trunk sewers. Concrete can be fortified against attack by using calcareous aggregate, increasing the cement content, or both, which provides additional alkalinity and acid neutralizing capacity.
- Asbestos Cement Pipe (ACP) is susceptible to acid corrosion. Due to its higher cement content, it corrodes at a slower rate than granitic aggregate concrete, although this attribute is offset by a generally thinner pipe wall.
- Thermoplastic pipes, such as Polyvinyl Chloride (PVC), Polyethylene (PE), and Acrylonitrile-Butadiene-Styrene (ABS), are flexible pipes. They are resistant to acid corrosion, although subject to strain corrosion in the presence of some materials, such as detergents, organic solvents, and fats and oils.
- Thermoset plastic pipes, such as Reinforced Plastic Mortar (RPM) and Reinforced Thermosetting Resins (RTR), are resistant to acid attack

providing the glass is not exposed to the acid due to cracking of the liner. Thermoset plastics are more sensitive to strain corrosion than thermoplastics, although not subject to attack by other chemicals.

D. PIPE LINING AND COATING AS DESIGN OPTION

Linings are differentiated from coatings in that the former are fixed to the interior wall of a pipe by projections embedded in the concrete at the time of manufacture, whereas the latter are fixed to the interior wall by adhesion. A typical thickness of a liner is on the order of 65 mils, while a coating thickness on the order of 40 to 125 mils is required.

TABLE 10-2. Los Angeles County Sanitation Districts Coating Study (11, updated June 1988)

			Performance[a]	
Coating (1)	Test Number (2)	Application (3)	Corrosion resistance (4)	Bond (5)
Senotex 3005	C-1 (Urethane)	2	1	3
Zebron	C-3 and C-39 (Urethane)	3	1	2
PR475	C-10 (Urethane)	2	2	1
Quantum	C-17 (Polyester)	1	1	1
Fosroc	C-22 (EpoxyMortar)	3	3	1
Aquata-Poxy	C-25 (Epoxy)	2	1	1
Vibrabond 500	C-26 (Urethane)	4	1	2
Concresive1305	C-28 (Epoxy)	1	2	1
PVC Panels	C-29 (liner)	3	1	
Acid Proof Cement No. 54	C-34 (liner)	2	2	
Allied Urethylene	C-35 (liner)	1	2	3
GS1490	C-36 (Urethane)	4		3
MainstayDS-4	C-37 (Epoxy)	1	1	1
120 Vinester	C-38 (Vinyl/Ester (Long cure time)	1	1	1
AlliedVinylthane	C-40 (liner)	1	1	

[a] Each coating was ranked for application, corrosion resistance, and bonding. 1 = good; 2 = some problems but not significant; 3 = significant problems; and 4 = failure.

1. Protection for New Systems

Liners for concrete pipe generally come in two forms: sheet forms, in which sheets of PVC or polyethylene are mechanically anchored to the pipe wall, or thin-walled pipe forms, such as smaller diameter PVC, HDPE, or glass reinforced polyester (GRP) pipes, which are inserted into a larger diameter conduit (2). Vitrified clay liners have been used, although generally not successfully, due to the porosity of the clay which permits slow diffusion of acid ultimately to attack the underlying pipe material.

Two principal groups of coatings have been employed: (1) Thermoplastic coatings, such as asphaltic/coal tar coatings and polyethylene, which are applied hot and cured by cooling; and (2) thermosetting coatings, such as polyurethane and epoxy, which "set" by chemical reaction caused by a curing agent which induces molecular cross-linking. Redner (11) reported the results of an extensive study of coatings by the Los Angeles County Sanitation Districts. Table 10-2 lists results for the more successful proprietary coatings studied.

2. Protection for Existing Systems

When unanticipated sulfide corrosion occurs in an existing collection system, the designer may specify in-place installation of a flexible liner. The process, currently available under the trademark Insituform, makes use of a resin-impregnated fiberfelt tube, which is inserted into a pipeline through an existing manhole. Cold water is used to inflate the liner against the existing pipe wall, and hot water is used to cure and harden the resin. The liner is resistant to sulfuric acid, and field evaluations have indicated effective performance (2, 4). According to a recent EPA study of a project in Northbrook, Illinois, the cost of the Insituform technique was about one-half of the conventional replacement cost, because traffic control, service reconnection, and surface restoration were important factors. Comparable savings were reported for a project in Dublin, Ireland.

3. Corrosion Resistance

The WRC Engineering Center (14) has summarized the chemical resistance of various polymeric materials in specific environments. These findings are shown in Table 10-3.

E. CONCRETE MODIFICATION AS DESIGN OPTION

Concrete modification can be considered as a design option using special cements or alkaline aggregates.

1. Special Cements

If modifications focus only on impregnating the pores of the concrete with wax, sulfur, tar, or resin, enhanced resistance is not produced (10). Sulfuric acid does not reach the interior of the concrete by way of the pores, but by attacking the solid phases.

The addition of a pozzolanic type of material has no effect. If a pozzolan is substituted for part of the cement, the rate of attack is increased, because the alkalinity is decreased (3, 10). Treating the pipe with a bactericide additive, such as SiF_4, produces a resistant surface. However, protection of the concrete in the presence of H_2S lasts only for a short time, because the bactericide must solubilize in order to be effective.

High-alumina cement (HAC) mortars and concretes have considerably greater resistance to very dilute acidity than similar specimens made with a Portland-type

Table 10-3 . Chemical resistance of polymeric materials to specific effluents and environments (14)*

Environment (1)		Material Suitability					
		Glass Reinforced Plastic (2)	Polyester Resin Concrete (3)	Polyethylene (4)	Polypropylene (5)	Insituform (6)	Unplasticised Polyvinyl Chloride (7)
Domestic Sewage		A	A	A	A	A	A
Effluent at Ambient Temperature	Acids	A (Note 1)	A	A	A	A	A
	Alkalis	B	B	A	A	B	A
	Organic Solvents	C	C	B	B	C	C
	Fats, Oil and Greases	B	B	B	B	B	B
	Sulfates	A	A	A	A	A	A
Effluent at Temperatures Significantly Exceeding Ambient		C	C	C	C	C	C
		Note 2	Note 2			Note 2	

Notes: A = Especially suitable; B= Normally suitable; and C = Check with manufacturer (may be unsuitable or require modification).
Note 1 = For acid strain corrosion tested GRP only; otherwise, Category C applies.
Note 2 = Resin formulation dependent.

* Table is provided courtesy of WRc Inc. and is taken from the latest design and specification guidance available to the UK water industry for sewer renovation.

cement. Among the hydraulic cements, only super-sulfated cement is usually credited with a higher acid resistance. To some extent, the behavior of HAC may be due to the absence of any calcium hydroxide in the set cement; but more probably it results from the presence of alumina gel which, by coating the hydrates, exerts a protective action. Alumina gel does not readily dissolve until the pH value of the surrounding liquid is below 4.0, and, in practice, the required resistance of HAC mortars and concretes to dilute acids is normally limited to pH values above 3.5 to 4.0. Attack will occur in stronger acid conditions, and HAC offers no protection against sewer crown corrosion where pH levels of 1.0, or lower, are typically observed. In fact, where pH is significantly below 2.0 or 3.0, HAC is frequently attacked more rapidly than are Portland cements.

The use of Type V sulfate resistant cement, while effective in protecting concrete against soil sulfates, has no influence on the sulfide attack which is observed in sewer pipe interiors.

2. Alkaline Aggregates

The rate of corrosion is influenced by the amount of reactive material in the pipe wall that will consume acid, most notably the concrete and aggregate alkalinity. Calcareous aggregate, as opposed to granitic aggregate, provides acid neutralizing capacity. The alkalinities of granitic versus calcareous aggregate can vary from less than 0.2 to as high as 1.0 equivalents $CaCO_3$. In many places where calcareous rock is not abundant, it has nevertheless been specified as aggregate for pipes used in sewer construction. In most of these cases, a calcium carbonate equivalent of 80% to 90% in the finished pipe has been specified.

Example calculation.—Considering the example presented in Sec. B1 of this chapter, once again solve part 3 and estimate the theoretical corrosion rate if an alkalinity of 0.9 were specified:

$$C = 0.45 \frac{(0.0208)\,(0.95)\,(2.0)\,(1.0)}{(0.9)}$$
$$= 0.020 \text{ in./yr}$$

It can be seen that the use of a calcareous aggregate can dramatically reduce the corrosion rate. In the example, the rate decreased from 0.089 in./yr to 0.020 in./yr.

F. SACRIFICIAL WALL THICKNESS AS DESIGN OPTION

When suitable slope-flow relationships to avoid sulfide build-up cannot be attained, a viable alternative approach is to design a sewer pipe for a designated service life. This design strategy acknowledges the inevitability of corrosion phenomena if unlined concrete or other corrosion-susceptible pipe is used, and sulfide control measures are not designed into the system. Thus, the designer can: (1) Provide additional concrete thickness to serve as a "sacrificial wall"; (2) specify the use of a calcareous aggregate to lower the corrosion rate; or (3) identify some combination of sacrificial wall thickness and calcareous aggregate use. Several equations have been developed by Kienow and Pomeroy (15) to aid in calculating the amount of additional wall thickness and the required alkalinity (i.e., the acid neutralizing capacity imparted by the calcareous aggregate). The rate of corrosion can be estimated from Eq. 10-1. Assuming a turbulence corrosion factor of 1.0, the general corrosion rate equation reduces to

$$C = \frac{0.45\, k\, \phi_{sw} \times CCF}{A} \qquad (10\text{-}6)$$

in which C = crown corrosion rate, in./yr; A = concrete alkalinity as equivalent $CaCO_3$, ϕ_{sw} = sulfide flux to pipe wall, g/m^2-hr; and k = acid reaction factor.

As indicated previously, the acid reaction factor k varies from 0.3 to 1.0 and represents the estimated fraction of the produced acid which reaches, remains on, and reacts with the exposed concrete wall. A value of 0.70 is commonly recommended for design purposes for sewers with moderate sulfide production rates. Higher values, up to 1.0, should be used for interceptors of very large diameter with slower sulfide build-up rates. The alkalinity A is expressed as a fraction and represents the amount of acid which a known weight of concrete, asbestos cement, or mortar lining can neutralize, as compared to the acid neutralizing capacity of pure calcium carbonate. Reported alkalinities for granitic aggregate, 50% calcareous aggregate, and 100% calcareous aggregate vary from 0.18 to 0.22, 0.4 to 0.6, and 0.8 to 1.0, respectively. Mortar lining alkalinity is on the order of 0.40, while asbestos cement is on the order of 0.50. The sulfide flux parameter, ϕ_{sw}, can be estimated using Eq. 10-2.

Equation 10-6 can be used to determine an anticipated corrosion rate and, based on a designated service life, to define specific pipe characteristics which will provide the required life span. This overall approach is known (11,15) as life factor design. The service life of a pipe can be estimated using $L = z/C$ (Eq. 6-10) in which L = service life, in years, and z = thickness of allowable material loss, in inches. The required sacrificial material thickness can be estimated by

$$z = LC \qquad (10\text{-}7)$$

One can use Eq. 10-7 to estimate the service life of a given pipe or, conversely, to define the required thickness of material which will provide a specified service life. Considering the latter approach, the parameter z is specifically defined as the thickness of the concrete cover over the inner reinforcement cage, in the case of reinforced concrete pipe. For mortar-lined steel or ductile iron, z represents the mortar lining thickness. In the case of asbestos cement pipe, or non-reinforced concrete pipe, z represents the thickness of allowable pipe wall loss, before structural properties are impaired. The specified service life generally ranges from 50 to 100 years. As indicated previously, the corrosion rate is a function of the sulfide flux and retention on the exposed pipe wall, as well as of pipe material characteristics, such as alkalinity. Thus, a designer can control the service life provided by a given pipe by specifying concrete cover thickness, mortar lining thickness, concrete or mortar alkalinity, or a combination of these characteristics.

Combining Eqs. 10-1 and 10-7 yields the life factor Eq. 6-27: $Az = 0.45\, k\, \phi_{sw}\, L(CCF)\,(TCF)$ .The life factor Az represents a multiplicative factor accounting for material thickness and acid neutralizing capacity. This suggests that there is some design flexibility in how the life factor requirement can be met. A manufacturer can vary the amount of calcareous aggregate and wall thickness to produce a pipe or lining with the required Az.

Several hundred miles of concrete pipe projects based on life factor design principles have been constructed in the United States, ranging from 24 in. to 108 in. in diameter. The approach was first employed in 1974 on the Herndon Avenue Interceptor Sewer project in Fresno, California. The project included 17 miles of pipe with diameters ranging from 39 to 84 in. (7). The design engineer specified Az

values ranging from 0.70 to 1.40. By comparison, a typical granitic aggregate pipe is characterized by an *Az* factor of about 0.12 in. Thus, for the Herndon Avenue project, the specified pipe has a life expectancy of 6 to 12 times that of conventional concrete pipe.

Example.—Concrete pipe is to be used for a trunk sewer line. The designated service life is 100 yr. Assuming a hydrogen sulfide flux ϕ_{sw} of 0.006 g/m^2-hr, a k value of 0.70, a crown corrosion factor *CCF* of 2.0, and a turbulence corrosion factor of 2.0, estimate the required *Az* value for this pipe.

From Eq. 6-27

$$Az = 0.45\, k\, \phi_{sw}\, L(CCF)\,(TCF)$$
$$Az = (0.45)\,(0.70)\,(.006)\,(100)\,(2.0)\,(2.0)$$
$$Az = 0.76\ in.$$

Estimate the required sacrificial cover thickness if a granitic ($A = 0.2$) or a calcareous ($A = 0.9$) aggregate is employed.

For a granitic aggregate

$$z = \frac{Az}{A} = \frac{0.76}{0.2} = 3.8 \text{ in.}$$

For a calcareous aggregate

$$z = \frac{Az}{A} = \frac{0.76}{0.9} = 0.84 \text{ in.}$$

Thus, the required pipe wall reinforcement cover would be 3.8 in. (96 mm) for the granitic aggregate example, and 0.84 in. (21 mm) if calcareous aggregates were used.

G. FLOODING AS DESIGN OPTION

Periodic flooding of pipeline segments can be used to dilute and remove sulfuric acid from the pipe wall. Such control has been reported effective in corrosion control at the Hampton Road Sanitation District, Virginia, and was considered but not installed in San Diego, California (5). The design for the latter site involved the use of a "Fabridam" inflatable structure of reinforced neoprene. The device would lie across the channel invert and could be inflated to back up flow in the main as desired. The frequency and timing of the flooding program would vary with specific site conditions.

H. PROTECTION OF SPECIAL STRUCTURES AS DESIGN OPTION

For pipes that may carry sulfides, reference manuals and texts recognize that the effects of sulfide corrosion are accelerated at points of turbulence, but a rational method of predicting the magnitude of such acceleration has not been developed. However, the designer should be aware of the likely trouble spots and the general protection schemes that are used. It should also be noted that if a wastewater stream does not contain H_2S, and if the other odors do not cause a problem, turbulence is advantageous. Much more oxygen is dissolved in a fall than during the same loss of

TABLE 10-4. Range of Turbulence Corrosion Factors From Sacramento County, California Study (8)

Pipe diameter, in. (1)	Predicted corrosion with k = 0.7, in. (2)	Observed[a] maximum corrosion at manholes in. (3)	*CCF* x *TCF* (4)
30	0.11	0.25	2.3
36	0.15	0.75	5.0
39	0.22	0.50	2.3
42	0.29	0.75	2.6
48	0.40	1.0	2.5
54	0.49	1.0	2.0
60	0.43	1.5	3.5

[a] The Sacramento Study (8) and the 1985 EPA Manual (3) reported the average observed corrosion in all the manholes in a given pipe diameter, rather than comparing calculated corrosion to the worst manhole in a given diameter,as shown above.

elevation in smooth flow. Elements considered in the following subsections are pipe junctions, manholes, and force mains.

1. Pipe Junctions

A major junction may be constructed within a vault or chamber entered by a manhole at its upstream end, and so designed that air which has been exposed to the high turbulence of the junction will be carried downstream and not exhaled from the manhole. If the wastewater will carry sulfide, the junction structure will probably need to be protected by a lining, and the pipe, too, will need to be either protected or made of noncorrodible material. The distance downstream that abnormal sulfide concentration will prevail will depend upon the turbulence of the air stream, its velocity, and the size of the pipe. The H_2S concentration of the air will probably be near a steady-state value at a distance downstream from the point of turbulence equal to 10 to 20 times the pipe diameter, or a minimum of 50 ft (15.2 m). Therefore, for that distance, protection should be provided against a corrosion rate much higher than the uniform flow corrosion rate. In Eq. 10-1, the turbulence corrosion factor *TCF* accounts for the proportional increase in corrosion rate due to the effects of turbulence. The range of estimates for *TCF* are shown in Table 10-1. The case study of the Central Trunk in Sacramento County, California (8) provided data from which (*TCF* x *CCF*) estimates of 2.0 to 5.0 were calculated, as shown in Table 10-4. Based on reported data, a design *TCF* of 2.0 is recommended. In addition, the value of sulfide flux ϕ_{sw} should be the higher of those predicted from either upstream or downstream pipe conditions.

2. Manholes

Manholes, as well as sewer pipes, may be affected by the acid produced from H_2S. If normal flow prevails through a manhole, the damage will be moderated, because the wall area is large relative to the water surface. Manholes are sometimes the site of high turbulence. Under these conditions the damage is likely to be severe if the sewage contains sulfide (9).

Turbulence in manholes can be minimized by proper construction of a smooth channel in the base section. Details on proper channel construction are included in ASCE Manual No. 60 (1) and other texts (13).

When either the sulfide concentration or the level of turbulence is high, manhole steps made of ductile iron, galvanized steel, or wrought iron will fail within a few years. Steps formed of 5/8-in. AISI Type 304 stainless steel rods have been used in a few places, and should have a long life under most conditions. In recent years, steps made of plastics or steel armored with plastic have been used. Aluminum has also been successfully used in many sanitary sewers. However, it is a very corrodible metal under the high sulfide conditions sometimes encountered in sewers. There is an increasing trend to omit steps entirely. This is sometimes done to reduce the danger that a corroded step might break under a person's weight.

3. Force Mains

The principal consideration in designing pressure mains and their discharge for sulfide control is to minimize the overall build-up of sulfides in the sewage during its passage through the system. Sulfide corrosion problems associated with pumping systems generally do not occur in the rising main, but if they arise or are aggravated by sulfide build-up in the pumping system, it is downstream in the receiving sewers or treatment works that problems will occur (12, 13). The interrelationship can be extended upstream to include the gravity pipes feeding into the pumping system. In those influent gravity pipes, the flow has been preconditioned, intentionally or unintentionally, to either minimize or encourage sulfide build-up and related corrosion problems. As discussed in previous sections, the designer has options to avoid sulfide build-up by preventive measures in the gravity lines leading to the pumping system. Also, several measures related to the force main may be considered.

If at all practicable, the selected route should permit a continuous, upward gradient from pump outlet to that of the rising main outlet. The overall length should be kept as short as practicable, thus restricting build-up to a minimum,whatever the final design arrangements. There is some disagreement about the effect of force main diameter. An empirical expression has been developed to define the force main diameter for a given pumping rate that results in the shearing of sulfide-producing slime growths (13). However, others have reported that selection of a small-diameter force main has no preventive effect on sulfide build-up (3, 10).

Finally, the discharge ends of rising mains should run full and remain full continuously. If this is impracticable, provision should be made for run-full conditions at least once a week. If this is not done, some corrosion of the crown is likely to occur. Detention times in wet wells are limited by most state regulatory agencies to 10 minutes at average flow (12).

REFERENCES

1. ASCE *Design and Construction of Gravity Sanitary Sewers*, ASCE Manual of Practice No. 60, ASCE, New York, NY, 1982.
2. EPA, *Sewer System Evaluation, Rehabilitation, and New Construction*, Manual of Practice, EPA 600 1/2-77-017d, Cincinnati, OH, 1977.
3. EPA, *Odor and Corrosion Control in Sanitary Sewerage Systems and Treatment Plant*, EPA 625/1-85/018, Cincinnati, OH, 1985.

4. Glasgow, D., *Existing Sewer Evaluation and Rehabilitation*, Manuals and Reports on Engineering Practice No. 62, ASCE, New York, NY, 1983.
5. Kienow, K., "The San Diego-West Point Loma 114 Interceptor Sewer Failure," *Bulletin*, California Water Pollution Control Association, Oct. 1980, and *Water and Sewage Works*, May 1980.
6. Kienow, K. K., and Pomeroy, R. D., "Corrosion Resistant Design of Sanitary Sewer Pipe," *Water and Sewage Works*, Reference Edition, April 1979.
7. Kienow, K. K., "Protecting Reinforced Concrete Sewers," *Water and Sewage Works*, Oct. 1975.
8. Meyer, W. J., and Hall, G. H., "Prediction of Sulfide Generation and Corrosion in Concrete Gravity Sewers, A Case Study," J. B. Gilbert & Associates, A Division of Brown & Caldwell, Sacramento, CA, April 1979.
9. Pomeroy, R. D., and Parkhurst, J. D., "The Forecasting of Sulfide Buildup Rates in Sewers," *Prog. Wat. Tech.*, Vol. 9, Pergamon Press, London, England, 1977, pp. 621-628.
10. Pomeroy, R. D., *Process Design Manual for Sulfide Control in Sanitary Sewerage Systems*, U. S. Environmental Protection Agency, EPA 625/1-7-005, Oct. 1974.
11. Redner, J. A., "Evaluation of Protective Coatings for Concrete," Los Angeles County Sanitation Districts, Compton, California; paper presented at the 59th Annual Water Pollution Control Federation Conference, Los Angeles, CA, Oct. 1986.
12. Tchobanoglous, G. T., *Collection and Pumping of Wastewater* , Metcalf and Eddy, Inc., McGraw Hill Book Co., New York, NY, 1981.
13. Thistlethwayte, D.K.B., *Control of Sulfides in Sewerage Systems*, Ann Arbor Science Publishers, Ann Arbor, MI, 1972.
14. WRC Engineering Center, "The Use of the Insituform Process for Sewer Renovation," *Report C2*, Philadelphia, PA, Sept. 1983.
15. Whipp, S., and Cooper, I., "Innovations in Sewer Rehabilitation Methods," *APWA Reporter*, Aug. 1984, pp. 12- 15.

CHAPTER 11

DESIGN OF ODOR ABATEMENT SYSTEMS FOR NEW AND EXISTING FACILITIES

A. INTRODUCTION

Odor has become an increasing problem at wastewater treatment plants in recent years; thus, better odor control strategies must be developed. Those planning and implementing a control strategy for a wastewater facility should consider many factors, including: quantity and source of foul air, specific odorous compounds and their concentrations, required degree of control, and system cost.

The emphasis in this chapter is on odorous air treatment processes, including description and typical design criteria. Despite the emphasis on foul air treatment processes, the engineer, manager, and wastewater system operator should examine the wastewater transport and treatment systems carefully to identify the sources of, and reasons for, odor generation. In many cases, an odor problem can be solved more easily and economically within the wastewater processing system than by foul air containment and treatment systems.

B. MINIMIZE PRODUCTION OF ODORANTS

Odorants are often a problem because they are produced or generated within the wastewater system, or are discharged in excessive quantities to the wastewater. The first approach to solving a problem should be to eliminate odorants at their sources. Some common sources are:

- Anaerobic wastewater transport systems which produce sulfide and organic odorous compounds.
- Odorants discharged directly to the wastewater by industrial or other sources. Source control can be effective in eliminating some of them.
- Odorants created by wastewater and sludge processing systems. Sometimes odorants are inadvertently produced through design or construction-related problems or operational difficulties.

Controlling odors at their source may also minimize hydrogen sulfide-induced corrosion and H_2S safety problems. Therefore, the evaluation of any specific odor problem should usually start with an examination of the sources of odor and the causes of odorant generation. Such an examination may lead directly to a process or related solution which avoids the need for foul air containment and treatment.

This manual centers on sulfide and H_2S issues. Hydrogen sulfide is not the only odorant of concern in wastewater systems, but it is usually one of the primary odorants, and is often the most significant. For information on odorants other than H_2S, see Refs. 18 and 19.

C. CONTAINMENT AND TREATMENT OF ODOROUS AIR

Containment and treatment of odorous air are required if other methods are inadequate in controlling the problem, or are determined to be less cost effective.

Containing odorants and transporting them to a foul air treatment system can be costly. Some sources do not lend themselves to a simple cover arrangement with exhaust fan and transport duct. Whenever a source of odors is covered, several factors need to be considered, including:

1. Movement of air and gases beneath the cover and the method and reliability of scavenging and containing odorants.
2. Potential humidity, condensation, and corrosion problems beneath covers.
3. Definition of employee access, and whether covered spaces are determined to be confined spaces.
4. Determination of ventilation rates if employee access is required. Both supply and exhaust fans are normally required for occupied spaces.

The costs to properly contain and transport foul air to a treatment system can easily match the cost of the odorous air treatment system. Therefore, in any economic cost comparisons of odor control options, containment and ventilation costs need to be developed.

The most common odorous air treatment processes currently in use in the United States are discussed in this chapter. They are:

- Odorants are absorbed onto either moistened media (soil, compost) or a wet slime layer, and biologically oxidized by microorganisms. This is termed biological treatment.
- Odorants are absorbed into the liquid phase,where a chemical oxidant treats the compounds. There are two major types of systems: recirculating liquid packed beds, and mist systems.
- Odorants are adsorbed onto the surface of a solid medium and held there. Sometimes odorants are partially oxidized while adsorbed on the medium. Eventually, the media are regenerated or replaced. These systems are usually called adsorbers.
- Odorants are oxidized through high temperature conversion in a furnace or similar system. This is termed combustion.

Other treatment processes are used less frequently, and are discussed briefly in this chapter. They include:

- Odorants are absorbed into the liquid phase and the liquid is eventually discharged (with the odorants) back to the wastewater system.
- Odorants are oxidized in the gas phase with ozone. These are called ozone contactors.
- Odorants are modified or masked, mostly by use of proprietary chemicals.

D. BIOLOGICAL ODOR CONTROL

Several different types of biological odor control systems have been used for municipal wastewater. Biological odor control is more common in Europe than in North America, but interest in biological odor control has increased in the United States in recent years.

1. Wastewater Biological Process

The simplest type of biological odor control system is to use the aerobic biological wastewater treatment process at the treatment plant to oxidize and treat foul airstreams generated in other portions of the plant. Both activated sludge and

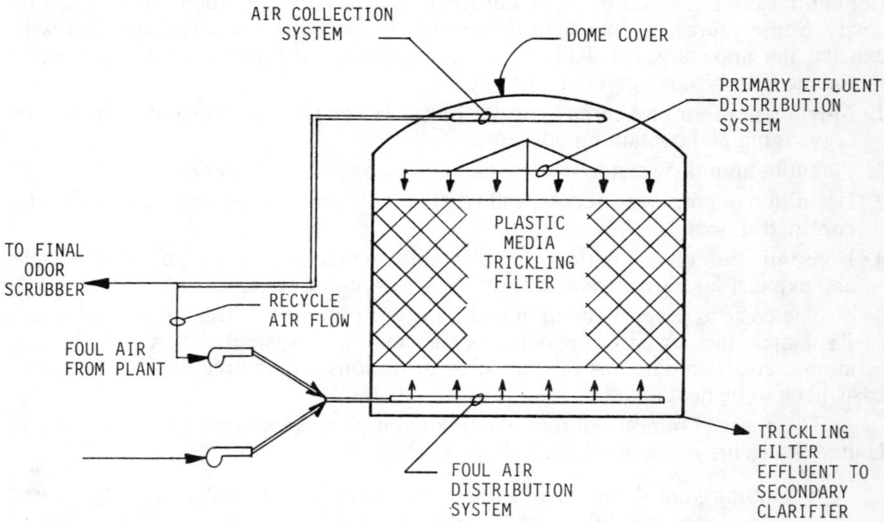

Figure 11-1. Trickling filter for odor control, Mesa, Arizona (2)

trickling filters have been used for this purpose with some success. With open activated sludge tanks and trickling filters, it is often difficult to measure the effectiveness of this type of odor control system, since the odorants become mixed with odorants from the biological wastewater process itself.

A trickling filter odor control system at Mesa, Arizona is shown in Fig. 11-1. Foul air from the covered headworks and primary clarifiers, and a pumping station, provide the air supply to the bottom of the plastic media filter. Air is recycled from top to bottom of the covered filters and a portion is withdrawn for final treatment in a mist scrubber. In this manner, the trickling filter acts as a first-stage scrubber which is intended to treat (oxidize) about 60% of the H_2S load applied to the filter. The H_2S, and perhaps other odorants, are absorbed into the biological slime layer on the plastic media and oxidized by the range of bacteria present. Absorbed sulfide appears to be oxidized to sulfate or elemental sulfur, or both, depending on the specific bacteria present.

2. Separate Trickling Filter

At several treatment plants, separate trickling filters, or certain portions of the trickling filter serving as the wastewater treatment process, have been used for odor control. Plant wastewater (usually secondary effluent, but sometimes primary effluent, or a combination of primary and secondary effluents) is used to load the filter from above, thus providing a source of food to create a biological slime layer on the media. Foul air is moved through the media so the odorants can be absorbed into the slime layer. A 10-second or greater foul air residence time is often used, and a nitrifying environment may be helpful (13).

3. Soil or Compost Filters

An increasingly common method of biological odor control is the soil or compost filter. In this system, a moistened bulk solid medium, such as soil or composted sludge, provides the contact surfaces for microbiological reactions to

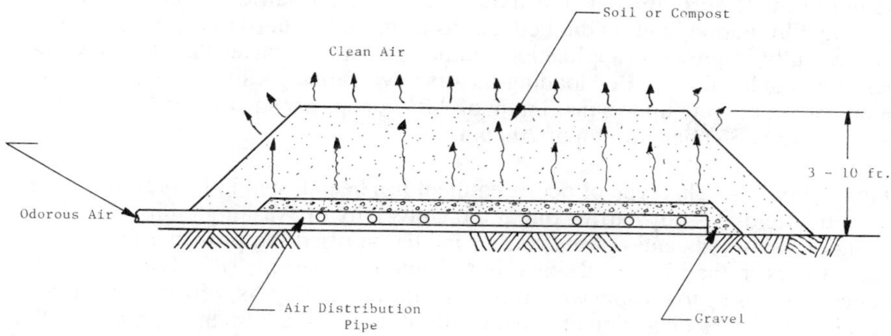

Figure 11-2. Soil/compost filter for odor control (18)

take place to oxidize odorants. Moisture content and temperature are important so that the organisms are properly cared for. Foul air residence times are often 15 to 30 seconds or longer in these systems.

Soil and compost filter beds can be used as a primary H_2S control system, or as a polishing step for exhausts from other odor control systems (1). The contaminated gas usually enters the soil or compost bed through perforated pipe located near the bottom of the bed. Sandy loam, compost, or mixtures of soil and peat moss have been used as soil filters.

Pilot studies on soil beds for H_2S control were conducted at the University of Washington (8). Major conclusions from the study were:

1. Moist loam soils were found to have excellent possibilities in efficient and low-cost removal of undesirable odorous gases from air streams. Sand and clay were less satisfactory soils.
2. The effects of ion exchange, chemical oxidation, and soil or water adsorption were negligible, the primary odor reduction mechanism being through the action of microorganisms.
3. Conditions should be maintained in an optimal range for bacterial growth. The soil should be kept moist.
4. Over a 3-month test period, H_2S gas concentrations of 15 ppm at a loading rate of 0.11 $m^3/min/m^2$ (0.35 cfm/ft^2) of soil surface were reduced to imperceptible levels in 81 cm (32 in.) of soil.
5. Effectiveness of the soil beds for removal of H_2S did not diminish during the 3-month test period.

Design of soil and compost filters. —Design of soil or compost filters is dependent on the nature and concentration of odorous compounds, air flow rates, target concentrations in exhaust gas, and characteristics of the soil or compost media (3). The design can be a simple filter bed or an elaborate system. Pilot tests should be conducted before designing full-scale facilities because of the many variables affecting design. A simple design is shown in Fig. 11-2.

A simple soil or compost filter system consists of a perforated pipe network surrounded by gravel, overlaid with sandy loam or compost. Air piping should be constructed of acid- resistant materials such as PVC (3). Valves may be used to direct odorous air. An underdrain system in moist climates may be desirable, and often a roof is needed to prevent excessive moisture. A manual or automatic watering system is almost a necessity. The bed area and depth is sized for the air volume required and to accommodate the anticipated hydrogen sulfide concentrations. As a practical limitation, soil media bed depths of up to 24 in. (61 cm) have been used successfully in previous applications, although depths greater than this have been tested in the laboratory. Bed loading factors have varied greatly from application to application, ranging up to a maximum of 2 ft^3/min air per square foot of bed surface area for an H_2S concentration of 20 ppm.

Performance. —This type of odor treatment has been tested in laboratories as well as in the field. A study carried out at the University of Washington in 1963, on the design of soil beds and optimum filter media, verified the proven effect of moist loam filters in the biological oxidation of hydrogen sulfide (8). Hydrogen sulfide concentrations up to 20 ppm were tested at varying bed depths, with the best removal efficiency of 95% occurring at a bed depth of 24 in. Headloss through the medium at this depth was approximately 2.5 in. (6.4 cm). Documented installations at Mercer Island in Seattle, Washington, at the Trans-Bay Pump Station in Coronado, California, and the Forest R. Gafner Wastewater Reclamation Facility in Leucadia, California, have verified the effectiveness of soil filter beds in the field. During the preliminary design stage, the designer should verify suitability of a soil bed or compost application, and establish site-specific design criteria for the bed sizing and choice of medium. In areas where soil freezing may occur, soil filter beds are not considered a practical application.

A composted sludge biofilter was built to replace a granulated activated carbon adsorber in the City of Jacksonville's Buckman Wastewater Treatment Plant (20). Odor removal efficiency increased from 68% to 88% with the compost system.

E. ABSORPTION AND OXIDATION

There are two basic types of absorption and oxidation treatment systems: the recirculating liquid packed bed, and the mist system. They are substantially different in practical usage and operation, and are discussed separately here. However, their common characteristic is that odorants are absorbed into a liquid medium containing a chemical oxidant which oxidizes the odorous compounds.

1. Recirculating Liquid Packed Bed Scrubbers

The recirculating liquid packed bed or packed tower scrubber is a common process used in many industrial applications to transfer gases into liquid streams. Gas transfer is often dependent on pH, and for odor control applications pH is quite important. Figure 11-3 shows some typical examples of recirculating liquid packed bed systems. Foul air residence time through the bed is usually limited to a few seconds (sometimes less than 2 s), so a strong driving force is needed for odorants to move from the gas phase into solution within the liquid stream.

Oxidation of odorants within the liquid.—The liquid solution contains an oxidant chemical so that treatment of the odorous compounds occurs within the liquid phase. Oxidants used include sodium hypochlorite, chlorine, potassium permanganate, and hydrogen peroxide. The chemical solution is sprayed over the packed bed. The solution falls through the packing to the sump beneath, and is repumped over the top of the packing. Fresh solution is added to retain the oxidizing strength, and a portion of the solution is wasted to remove treated products. If hydrogen sulfide is the primary odorant absorbed, then the soluble hydrosulfide ion (HS^-) is oxidized primarily to sulfate, although some elemental sulfur can be formed. The blowdown from the system is typically discharged to the wastewater treatment process. A mist eliminator is needed to recover liquid droplets from the treated airstream.

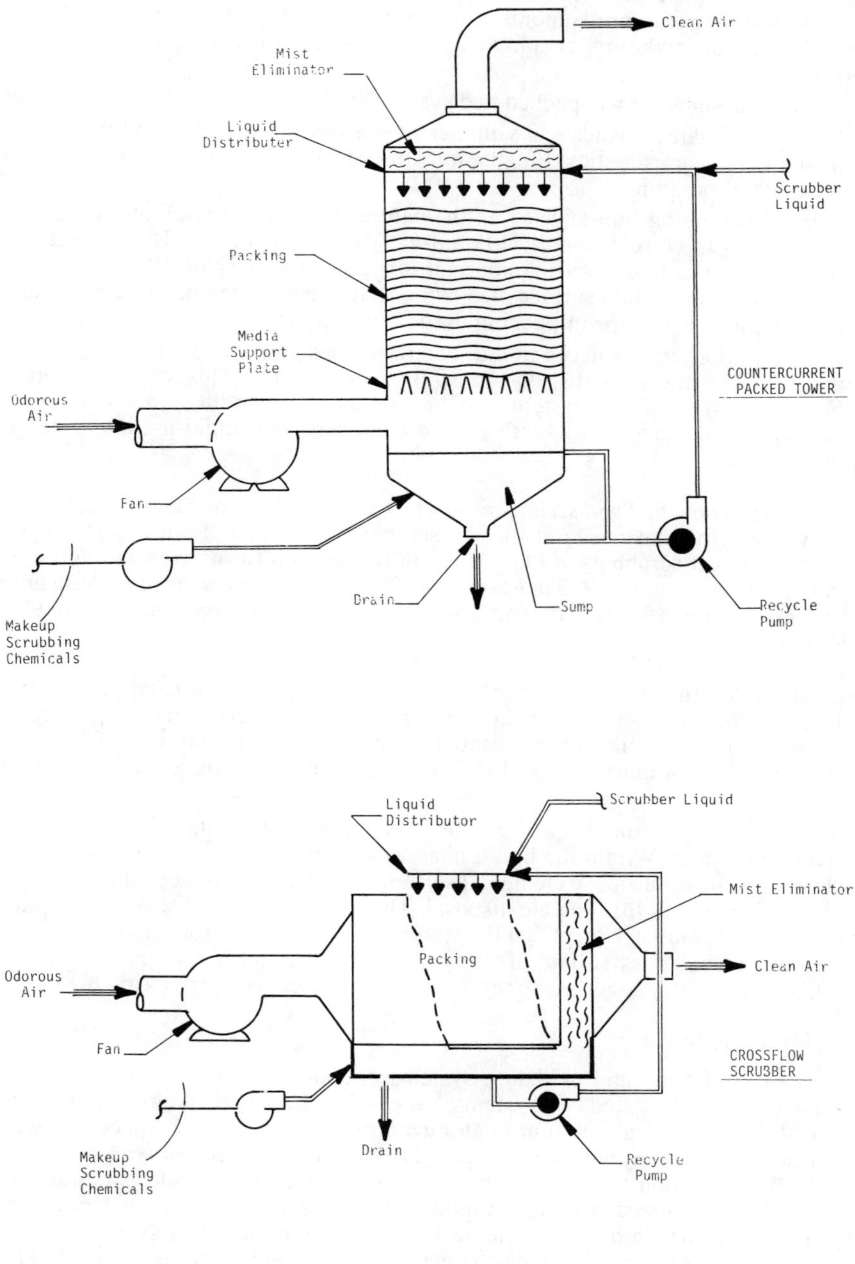

Figure 11-3. Typical wet packed bed systems (modified from Ref. 18)

The scrubbing solution feed loop includes a chemical solution tank, a metering pump and piping, a recirculation pump, and associated piping and valves. Instrumentation and controls monitor pressure drop through the packing and allow the addition of make-up chemicals. Materials of construction should resist corrosion.

Steps in the design of a wet packed bed system are:

1. Define the characteristics and volumes of the gas to be treated and the desired treated gas characteristics. Fluctuations in odor intensity of foul air should be recorded, if possible.
2. Select a scrubbing liquid based on the nature and concentration of the odorous compounds to be removed. The liquid-gas ratio is important. The contact area between gas and liquid is also important, since absorption is directly proportional to the amount of liquid surface area exposed to the gas stream. The contact area is a function of the type of packing used in the scrubber.
3. Conduct pilot tests, if necessary, to determine design criteria and performance, and based on these results, design a full-scale system. This step is important because design of the system can be complex, depending on the intended application, characteristics of the gas stream, scrubber liquid used, and design objectives.

Performance of packed bed scrubbers.—The County Sanitation Districts of Orange County, California have several foul air scrubbers in service. Performance of two wet packed tower scrubbers at Plant No. 2 in Huntington Beach was monitored for ten months from August, 1985 to May, 1986. The overall performance of these units is presented in Table 11-1. Efficiency, based on monthly average, varied from 93% to 98%.

Catalytic H_2S scrubbers.—Processes to oxidize hydrogen sulfide catalytically have evolved primarily from industrial applications. Such scrubbers in petroleum refineries normally operate on substantially higher input concentration of H_2S than are found in wastewater facilities. Limited performance data are available on these units.

A relatively standard wet packed bed is used to transfer H_2S into the recirculating liquid. Within the liquid phase, these processes often use proprietary solutions to oxidize sulfide to elemental sulfur. The sulfur is removed either as a bulk solid or a thick slurry for separate disposal. These systems operate as an absorption and oxidation scrubber, except that the sulfur may be removed for separate disposal rather than recycled as sulfate. This type of scrubber probably has its greatest potential use as a first-stage scrubber when input H_2S concentrations are high.

2. Mist Systems

Mist absorption and oxidation systems operate on similar principles to the recirculating liquid packed bed systems, except that there is no recirculation of the chemical. The chemical solution is atomized or sprayed through special nozzles which achieve small droplet or mist sizes. The foul air is mixed with the small mist particles in a reaction vessel, and the odorants are transferred into solution. The compounds are oxidized within the liquid particles by the chemical oxidant added. Any condensate which occurs in the vessel is wasted to the sewer system, and the fine mist particles are primarily discharged with the treated airstream. Figure 11-4 shows a diagram of a mist system.

Mist particle size.—A critical feature is the size of the mist particles. There are data to indicate that at least one manufacturer has produced mist particles which are in

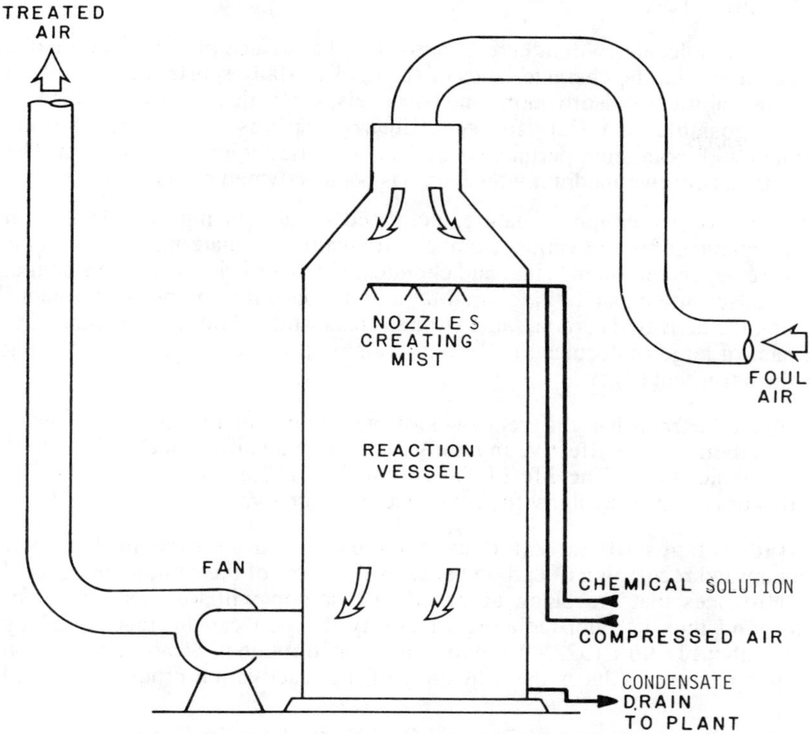

Figure 11-4. Typical mist scrubber

the 10-μ to 20-μ size range. The size of the mist particles required for a specified degree of odor control is not well defined, but it seems logical to produce the smallest particles possible to maximize the particle surface area in order to achieve better gas transfer.

Reaction time. —Since there is no chemical recirculation, the joint residence time of foul air and mist particles is another critical feature. Typical residence times for many wastewater applications are in the 10 to 15 second range, although systems with up to 30 seconds or more have been installed.

Chemicals used.—The primary chemical oxidant used in mist systems is sodium hypochlorite. Sodium hydroxide is used for pH adjustment. Since the chemicals are used in a once-through manner in mist systems, chemical dosing adjustment is important for proper performance and for chemical cost control.

Performance. —Performance of the mist absorption and oxidation systems has been good on many wastewater odor sources. Hydrogen sulfide and total odor reductions of 95% and greater are reported (15).

F. ADSORPTION

Odorous molecules will adhere or adsorb to the surface of bulk solid media, depending largely on the characteristics of the media. Many substances have been tested for their ability to adsorb odors, and silica gels, synthetic zeolites, and metallic oxides are possible materials for specific applications. Activated alumina impregnated with potassium permanganate has been used with some success. The most popular adsorptive medium, however, has been activated carbon.

Because of its nonpolar nature, activated carbon is highly selective in removing organic gases and vapors from the airstream. The materials and amounts adsorbed are dependent on physical and chemical characteristics of the compound. Maximum adsorbing capacity is favored by a high concentration of the substance surrounding the activated carbon, large surface areas, and freedom from competing contaminant in large molecules that fit and are strongly attached to the receiving sites on the adsorbent (9).

Activated carbon has different rates of adsorption for different substances. Activated carbon may be effective in removing hydrogen sulfide, and will work on reducing organic odors. The life of the carbon bed is limited; carbon must be regenerated or replaced regularly for continued odor removal.

Disadvantages of activated carbon adsorption include high power requirements and recharging of carbon beds. As the pores of the carbon are covered by the substances that are being adsorbed, the carbon will lose its ability for adsorption and thus its odor-reducing capability. Carbon can be reactivated by heating it to about 1,700°F (927°C) and forcing steam through it. Steam regeneration at lower temperatures reduces the efficiency of the reactivation process. Thermal

TABLE 11-1. Performance of Packed Bed Scrubbers in Orange County, California

Month and year (1)	H_2S in, ppm (2)	H_2S out, ppm (3)	Removal, percentage (4)
Aug. 85	160.8	6.3	96
Sept. 85	134.4	9.6	93
Oct. 85	96.8	3.3	97
Nov. 85	69.3	1.6	98
Dec. 85	60.8	2.3	96
Jan. 86	55.9	3.4	94
Feb. 86	53.4	2.6	95
Mar. 86	25.4	1.9	93
Apr. 86	46.6	2.4	95
May 86	63.6	1.6	97

Source: Courtesy of Orange County Sanitation District, California.

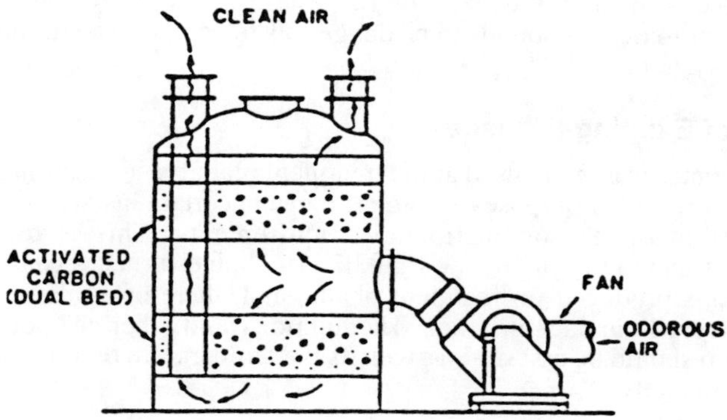

Figure 11-5. Typical activated carbon adsorber (18)

reactivation may be carried out in a furnace, and can be done either at the carbon supplier's plant or at the treatment plant.

Carbon can be impregnated with either sodium hydroxide or potassium hydroxide to increase its affinity for H_2S. Impregnated carbon can be regenerated by rinsing the carbon and then soaking it in a concentrated hydroxide chemical solution. Typically, chemical regeneration will remove primarily the sulfur compounds. Other compounds may not be removed and will accumulate. Impregnated carbon must be periodically replaced or regenerated with heat to control the accumulation of non-sulfur compounds.

Steps in the design of activated carbon adsorbers (18) are:

1. Characterize the contaminated air (volumes, concentrations of constituents) and the desired treated gas characteristics.
2. Select the adsorbent.
3. Conduct pilot studies, if necessary, to determine expected performance, useful life of adsorbent, design criteria, etc., and then apply pilot data to full-scale design.

A typical activated carbon H_2S adsorber system (see Fig. 11-5) consists of feed fans, air pretreatment units, activated carbon adsorber units, and associated piping and ductwork. Mechanical equipment required includes the vessel containing the adsorbent. Carbon adsorbers can be thin-bed and deep-bed. Thin-bed adsorbers are useful where odor concentrations are low and the effective mass transfer zone for adsorption is short. They have low resistance to air flow and thus require less power. Deep-bed adsorbers are useful when odor concentrations greater than 5 ppm are present, or when on-site carbon regeneration is used (18).

G. COMBUSTION OF ODORANTS

Combustion, or high-temperature oxidation, is an effective means of destroying odors, but it can be costly. Combustion should be seriously considered in the following instances: (1) Where foul odors are so strong that other, less costly, control methods are not adequate; or (2) where the foul air contains sufficient combustible or toxic compounds to be dangerous to handle in anything other than a combustion system.

1. Use of Existing Furnaces

If a furnace is already used at the treatment plant for incinerating sludge, grit, screenings, or for other purposes, odorous air can in certain instances be provided to meet a portion of its combustion air requirements. This is often the most cost-effective method to incinerate foul air. All sludge furnaces cannot accept the foul airstreams produced in treatment plants, and, therefore, this option must be evaluated carefully in each instance. The humidity and other characteristics of the foul airstream should be assessed, as well as the temperature required to oxidize the odorants completely.

2. Fume Burners

Two types of combustion process fume burners are presented in the literature for odor control: direct-flame oxidation and catalytic oxidation. In the direct-flame method, the key to complete combustion is maintenance of the proper temperature and contact time in the combustion chamber. Odorous gas is mixed with combustion air and exposed to temperatures ranging from 480°C to 815°C (900°F to 1,500°F) and contact times ranging from 0.3 to 3 s.

In the catalytic oxidation process, the presence of a catalyst (platinum-metallic screen) allows oxidation to occur at a lower temperature (315°C to 480°C) and in the absence of a flame. A burner preheats the odorous air to the temperature required for catalytic oxidation. The catalytic combustion mechanism, although not well understood, follows three steps: (1) Adsorption on the active surface; (2) chemical reaction (oxidation) on the surface; and (3) desorption of the reaction products (18). Fouling of the catalyst can occur due to sulfur compounds. The advantages of direct-flame oxidation over catalytic combustion include lower maintenance costs, less down time, and better odor control (19).

3. Design of Combustion Systems

The major considerations in the design of thermal combustion systems for odor control are temperature, residence time, and mixing. Temperature required for complete combustion is dependent on the compounds to be oxidized. Small temperature drops can result in substantial loss in oxidation efficiencies of specific compounds. Residence time is closely interrelated with temperature. Increased temperature results in expansion of the gas, thus lowering the residence time in the combustion chamber.

Mixing is often overlooked, but may be one of the most important design parameters (1). Achieving good mixing of the combustion air and foul airstream is critical for efficient operation. The mixture must be retained for a sufficient time to allow complete oxidation of the odorous components. Proper mixing can be accomplished through proper selection and sizing of burners, combustion chambers, and reactor configurations.

To maintain proper temperatures for complete combustion, temperature sensors should be located in the influent air ductwork, in the combustion chamber, and in the exhaust stack. To prevent human contact, high-temperature exhausts must be discharged through a stack. A suitable source of fuel must also be available.

Heat exchangers may help defray fuel costs for preheating influent air. Materials of construction consist of high temperature-resistant metals, insulating materials, and refractory materials that are used in the combustion chamber to withstand temperatures of 650°C to 800°C, and higher in some cases.

H. ABSORPTION AND RECYCLE

Absorption and recycle is a process to remove the odorants from the gas phase and solubilize them within the liquid phase (normally within a recirculating liquid packed bed system). The absorbed odorants are then typically recycled or wasted to the liquid stream. The gas transfer process is similar to the absorption and oxidation process, but there is no intentional oxidation or destruction of the odorants within the liquid phase.

The process is most applicable when one particular odorant is in high concentration, such as hydrogen sulfide or ammonia, and a first-stage scrubber is needed to reduce the concentration of that particular compound so the foul air can be treated effectively in a second-stage scrubber. The second-stage scrubber would then be a unit that could handle a wider variety of odorants.

An example of absorption and recycle would be a caustic scrubber working on a foul air with high H_2S concentrations (perhaps greater than 100 ppm of H_2S). The caustic scrubber could be used to reduce H_2S concentrations to about 5 or 10 ppm, or less, so the foul air could be treated in a second-stage unit. The H_2S removed from the air in a caustic scrubber is solubilized into the liquid, and the blowdown stream from the scrubber might be discharged to the sewer. The handling of the blowdown stream is critical, since a pH-drop to neutral will allow most of the sulfide to come back out of solution as hydrogen sulfide, and create an odor problem. If the sulfide in the blowdown stream is oxidized while the pH is relatively high, there may be little chance for an odor problem. Corrosion problems from various chemical reactions can be significant with these units if they are not designed and operated properly.

I. OZONE CONTACTORS

Ozone is a strong oxidant that eliminates odors by chemical oxidation. Ozone, an unstable gas, is generated on site. It can react rapidly with hydrogen sulfide, but is, itself, an irritating, odorous gas. Ozone treatment involves gas to gas molecular contact. Therefore, mixing and reaction time are critical, and humidity in the foul air may be a problem.

Design of an ozone odor control system needs to take into account the following factors: (1) Specific odorants; (2) concentration of odorants; (3) temperature of the exhaust gases; (4) humidity of exhaust gases; (5) retention time within the contact chamber; (6) the distribution and mixing of ozone within the contact chamber; and (7) the adjustment of ozone dosage rate to avoid discharging ozone in the exhaust from the contactor (10). Occupational and environmental health and safety aspects should also be considered.

Package ozone generators can be purchased for many odor control applications. Air-fed ozone systems require compressors, heat exchanger, air filters, and molecular sieve; an ozone generator and diffuser or injector; a baffled contact chamber; and piping, valves, appurtenances and controls. Safety features are usually included. Construction materials include polyvinyl chloride (PVC) pipe, fiberglass, concrete, and stainless steel (18); however, each part of the system should be carefully designed, because of the extremely corrosive nature of ozone.

J. MODIFICATION, COUNTERACTION, AND MASKING

Odor modification is a process that mixes two substances so that the mixture's odor is less intense than the odor of each single substance. Design for the use of a modification agent requires characterization of the odor and screening of various chemical agents to determine their effectiveness.

Odor counteraction is a process that reduces odor intensity by the interaction of two specific compounds in certain proportions. Several odor pairs have been well documented, such as: rubber and paraffin, cedarwood and paraffin, butyric acid and oil of juniper, pyridine and oil of wintergreen, ethyl mercaptan and eucalyptol, skatole and coumarin (6).

Odor masking attempts to disguise an unpleasant odor with a strong, pleasant odor. Typical masking agents include heliotropin, vanillin, eugenols, benzyl acetate, and phenylethyl alcohol (18).

The effectiveness of these methods is sometimes unpredictable in the long term, because of varying odor concentrations and ever-changing atmospheric conditions. Sometimes a more objectionable odor is produced. These methods may have more application for emergency situations or short-term attempts at controlling odors.

1. Design of Systems

These systems require characterization of the odor and screening of various chemical agents to determine their effectiveness. Subjective testing through an odor panel is often required. Important considerations when evaluating use of odor modification, counteraction, and masking agents are (19):

1. Any gas or vapor that may be toxic (such as H_2S) should not be disguised.
2. Odor modification should not be used as a substitute for good housekeeping.
3. If possible, odors should be controlled at the source.
4. Odor modification, counteraction, and masking agents must conform to local, state, and federal specifications and regulations.
5. Odor-modifying chemicals should always result in a lowering of the olfactory intensity.
6. Adding chemicals to the atmosphere could be a violation of local air pollution regulations.

2. Equipment Used

Equipment requirements for application of odor modification, counteraction, and masking agents are generally minimal and depend on the method of addition. Application techniques include the following (7):

1. Direct addition to the odorous material.
2. Spraying: hand spraying, automated spraying, stack spraying, or aerial spraying.
3. Metered feeding to the liquid stream.
4. Dispenser feeding to an airstream.

The odor-modifying agent may be sprayed in the air at the odor source or along the property line, added to the liquid, or incorporated into a floating chemical screen through which gaseous emissions from the liquid surface must pass (4).

K. ATMOSPHERIC DISCHARGE AND DISPERSION

The manner of discharge of a treated airstream to the environment can have a significant effect on the subsequent transport and dispersion of odorants and their

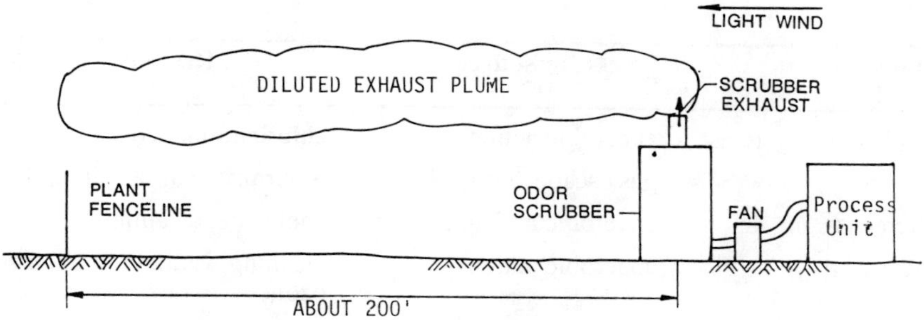

Figure 11-6. Allowable scrubber H_2S (16)

detection downwind. The specific atmospheric conditions at a site should be studied carefully, to make maximum use of any localized wind patterns, and to avoid conditions that allow only limited atmospheric dispersion of odorants. Effective dilution and dispersion of odorants can be a key factor in a successful odor control program.

In many cases, a relatively tall stack is the only effective way to achieve good atmospheric dilution. The stack discharge should be evaluated carefully, however, with respect to nearby structures, to determine whether the air discharged from the stack will be brought directly to the ground through downwash action caused by the wind.

Diffusing odorants over a large area can be another effective technique. For example, the odor benefits of discharging foul air to the air supply for an activated sludge system may result as much from the dilution of odorants over the entire tank area as from the biological oxidation of odorants within the wastewater. Also, a soil or compost filter having a large surface area in the open atmosphere can result in substantial diffusion of odorants over the bed surface, thus diluting the effects of remaining odorants in the treated air.

Diluting foul airstreams with odor-free air in an attempt to eliminate a problem should be critically evaluated. Two orders of magnitude, or more, are typically required to reduce wastewater odors to detection levels; therefore, dilution with odor-free air is normally not cost effective.

Atmospheric odor modeling can be a significant advantage in assessing the benefits of different foul air treatment alternatives and different atmospheric discharge alternatives. Such modeling work is only as accurate as the input data used. Therefore, reliable information on foul air strength and scrubber performance is critical, and valid local atmospheric data are needed.

The allowable hydrogen sulfide concentration in a scrubber exhaust can be determined from a relatively simple air quality dispersion model. Figure 11-6 shows a typical situation at a treatment plant or pumping station where a foul airstream on the order of 2 m^3/s (4,000 cfm) is discharged about 6 m (20 ft) above ground. With light, stable wind conditions, and in situations where the fence line is only 50 to 75 m (150 to 250 ft) from the scrubber, the scrubber exhaust will need to be limited to an H_2S concentration on the order of 0.1 ppm. Very efficient and reliable scrubbing is likely to be required to meet this concentration of hydrogen sulfide in treated exhaust air.

TABLE 11-2. Sources of Odors and Possible Remedies (20)

Location (1)	Odor source (2)	Remedy (3)
Collection systems	anaerobic action	chlorination, ozone
Weirs	gas escape, aerosols	minimize weir elevation drop
Screens	anaerobic action, puddles	enclosure, cleaning
Wet well	anaerobic action	cleaning, ventilation, chlorine, ozone
Grit tanks	evaporation, aerosols, washwater	housekeeping
Primary or final clarifiers	anaerobic action	increased sludge, scum removal, chlorination
Aeration tanks	anaerobic action, aerosols	proper mixing, clean walls
Disinfection	anaerobic action, chemical odor	clean walls, proper chemical dosing
Anaerobic digesters	gas escape	monitoring, proper operation, cleaning
Sludge handling and conditioning	anaerobic action, chemical odor	proper ventilation, proper operation, proper chemical dosing, cleaning
Incineration	gas escape, particulate escape	proper operation for combustion, gas scrubbing
Ponds	anaerobic action	aeration

L. PLANNING ODOR ABATEMENT STRATEGY

The development of an odor abatement strategy requires careful planning. Sources of odors and compounds in odorous air should be identified, causes of odors determined, if appropriate, and the abatement strategy and unit processes developed accordingly. After characterization of the foul air and development of treatment processes, pilot testing may be needed. Pilot systems can verify design parameters and develop information on expected system performance.

Most effective odor control systems utilize one or more of the control methods discussed in this chapter. Many of the odorants found in treatment plants have very low threshold concentrations so that it is necessary that the abatement systems remove 95% to more than 99% of the odorous substances. After design and construction of the odor abatement system, the operating staff should be trained in the proper operation of all systems. Many systems are complex and require regularly scheduled cleaning and proper maintenance if they are to operate effectively and reliably.

Table 11-2 summarizes various treatment plant odor sources and some potential control strategies.

REFERENCES

1. Bhatia, M.V., and Cheremisinoff, P.N., "Combustion Methods for Odor Control," in *Industrial Odor Technology Assessment*, Cheremisinoff and Young, eds., Ann Arbor Science, Ann Arbor, MI, 1975.
2. Brown and Caldwell, Design and construction documents for the Turner Ranches Wastewater Plant, Turner Ranches Water and Sanitation Co., Mesa, AZ, 1985 to 1987.
3. Carlson, D.A., and Leiser, C.P., "Soil Beds for the Control of Sewage Odors," *Journal of the Water Pollution Control Federation*, Vol. 38 (5), 1966, pp. 829-840.
4. Cheremisinoff, P., and Young, R., eds., *Air Pollution Design and Control Handbook*, Part 2, Marcel Dekker, Inc., New York, NY, 1977.
5. Dague, R.R., "Fundamentals of Odor Control," *Journal of the Water Pollution Control Federation*, Vol. 44, No. 4, April 1972.
6. Landrum, W. B., "Control of Sewage Plant Odors," CH_2M Hill Technical Memorandum, P.O. Box 428, Corvallis, OR 97339, 1987.
7. Lauren, O.B., "Odor Modification and Masking," *Proceedings*, Specialty Conference on the State-of-the-Art of Odor Control Technology, Air Pollution Control Association, Pittsburgh, PA, 1974.
8. Leiser, C.P., "Sewage Odor Reduction and Elimination by Use of a Soil Column," thesis presented to the University of Washington, Seattle, WA, in partial fulfillment of the requirements for the degree of Master of Science, 1964.
9. National Academy of Sciences, "Odors from Stationary and Mobile Sources," Committee on Odors from Stationary and Mobile Sources, National Research Council, Washington, DC, 1979.
10. Nebel, C.A., and Gottschling, R.D., "Industrial Odor Control by Oxidation with Ozone," in *Industrial Odor Technology Assessment*, P.N. Cheremisinoff and R.A. Young, eds., Ann Arbor Science, Ann Arbor, MI, 1975.
11. Neveril, R.B., Price, J.U., and Engdahl, K.L., "Capital and Operating Costs of Selected Air Pollution Control Systems," Air Pollution Control Association, Pittsburgh, PA, Jan. 1979.
12. Pollution Control Systems, *Odor Abatement*, Brochure No. 3710.
13. Pomeroy, R.D., "Biological Treatment of Odorous Air," *Journal of the Water Pollution Control Federation*, Vol. 54, No. 12, Dec. 1982.
14. Price, E.C., "Sewage Treatment Plants Combat Odor Pollution Problems," *Water & Sewage Works*, Oct. 1978.
15. Rafson, H.J., "Performance Results of Mist Odor Control Systems at Several Wastewater Applications," presented at the WPCF Annual Conference in Kansas City, MO, Oct. 8, 1985.
16. Schafer, P.L., "Upstream Sulfide Control at Kailua/Kaneohe," presented at the Hawaii Water Pollution Control Association Annual Conference, January 29-30, 1987.
17. Summer, W., *Odour Pollution of Air, Causes and Control*, Leonard Hill, London, England, 1971.
18. U.S. Environmental Protection Agency, *Design Manual-Odor and Corrosion Control In Sanitary Sewerage Systems and Treatment Plants*, EPA/625/1-85/018, Oct. 1985.
19. Water Pollution Control Federation, *Odor Control for Wastewater Facilities*, Manual of Practice No. 22, Washington, DC, 1979.
20. Water Pollution Control Federation, *Operations Forum*, Vol. 5, No. 2, Feb. 1988.
21. Yang, M.Y., and Cheremisinoff, P.N., "Wet Scrubbing for Odor Control," in *Industrial Odor Technology Assessment*, P.N. Cheremisinoff and R.A. Young, eds., Ann Arbor Science, Ann Arbor, MI, 1975.

CHAPTER 12

DESIGN OF SULFIDE CONTROL SYSTEMS

A. INTRODUCTION

Numerous methods are available to control the generation of sulfide in wastewater. Methods affect the oxygen balance in sewage, oxidize generated sulfide, react chemically with dissolved sulfide to form insoluble sulfide, or affect the sulfide generation capability of the sulfate or organic sulfur reducing organisms. The best choice for any specific application is dependent on the design of the sewerage system, the desired level of control, the operation and maintenance capability that can be devoted to the control method, and the local availability of chemicals or equipment. Direct treatment cost may not be the sole governing parameter.

Design information and case histories for several methods available to control the generation of sulfide in wastewater are presented in this chapter. The methods include: (1) Oxygen injection in force mains, inverted siphons, U-tubes, hydraulic falls, and sidestreams; (2) chlorination; (3) hydrogen peroxide; (4) iron and zinc salts; (5) shock dosing with sodium hydroxide; (6) potassium permanganate; (7) sodium nitrate; (8) ozone; and (9) bacterial cultures and enzymes.

Table 12-1 contains a list of contacts for the case history information presented in this chapter. The reader may obtain additional information from the specific agency referenced.

Much of the information assembled for this chapter comes from the EPA design manual (2). Additional reference to the design manual will only be made when necessary for clarity. Other references are included in this chapter where appropriate. The reader is referred to the EPA design manual for additional information and references on the design of sulfide control systems.

B. OXYGEN INJECTION

Maintenance of aerobic conditions, with at least 1 mg/L dissolved oxygen, is usually sufficient to prevent sulfide build–up in wastewater streams. Natural reaeration of the flowing stream is seldom enough to maintain 1 mg/L for any significant distance at most design velocities. Oxygen, either as air or pure oxygen, can be added to flowing wastewater by direct injection into force mains, lift stations, inverted siphons, U- tubes, pressurized sidestreams, or areas of hydraulic turbulence created by falls or jumps. The effectiveness of oxygen injection for controlling sulfide generation is dependent on the pressure at the point of injection, the energy expended in dissolving the gas, the temperature and pressure of the receiving stream, the method of injection, the profile of the receiving stream, the travel time for sewage from the point of injection to the point of control, the sulfide levels in the sewage entering the system, the sulfide generation rate, and the sewage reaeration rate.

The reader is cautioned that the inappropriate application of oxygen injection, particularly when using air, can exacerbate a sulfide problem. When inadequate oxygen transfer occurs, the excess gas can strip hydrogen sulfide from the

TABLE 12-1. Sulfide Control System Case History Facility Contacts

System (1)	Facility contact (2)
Oxygen injection into force main	Delta Diablo Sanitation District No. 7-A 2500 Pittsburg-Antioch Highway Antioch, California 94509 Steve Gittings, Operations Supervisor (415) 778-4040
Air injection into inverted siphon	County Sanitation Districts of Los Angeles County 920 South Alameda Street Compton, California 90221 Calvin Jin, Supervising Operations Engineer (213) 638-1161
Oxygen injection into fall structure	Sacramento Regional County Sanitation District 9660 Ecology Lane Sacramento, California 95827 Stanley Walton, Superintendent (916) 366-2241
Hydrogen peroxide	Palm Beach County Water Utilities Department P. O. Box 16097 West Palm Beach, Florida 33406 Jim Shamblin, Director of Operations and Maintenance (305) 641-3470
Ferrous chloride in anaerobic digesters	County Sanitation Districts of Los Angeles County 24501 South Figueroa Street Carson, California 90710 Blair Hansen, General Superintendent (213) 830-2400
Ferrous chloride injection in sewers	City of Mesa Utility Operations 55 North Center Street Mesa, Arizona 85201 Richard Bradford, Water Quality Engineer (602) 834-2011
Shock dosing with sodium hydroxide	County Sanitation Districts of Los Angeles County 920 South Alameda Street Compton, California 90221 Calvin Jin, Supervising Operations Engineer (213) 638-1161

TABLE 12-1. Sulfide Control System Case History Facility Contacts, (cont'd.)

System (1)	Facility contact (2)
Shock dosing with sodium hydroxide	City of Omaha, Nebraska 5600 South 10th Street Omaha, Nebraska 68107 William Moore, Dir. of Operations (402) 733-5465
Shock dosing with sodium hydroxide	Water Pollution Control Facilities 7525 Bertram Road Cedar Rapids, Iowa 52401 Dan Rockwell, Assistant Director (319) 398-5286

wastewater, and the corrosion potential can be increased. This can occur in gravity sewers, process tanks, and even low pressure areas in full-pipe sewers (force mains) where gas pockets develop.

1. Design Requirements

Estimating the air or oxygen requirements to achieve sulfide control has been based largely on the expected oxygen uptake rates in the wastewater. The oxygen balance in a wastewater stream can be described by (16):

$$\frac{dO}{dt} = R_f - R_r - R_e \qquad (12\text{-}1)$$

in which dO/dt = rate of change of oxygen concentration as the stream moves down the sewer, mg/L-hr; R_f = rate of gain of oxygen by surface reaeration, mg/L-hr; R_r = rate of loss of oxygen due to reaction in the stream (oxygen uptake rate of wastewater), mg/L-hr; and R_e = rate of loss of oxygen due to reaction on the slime layer, mg/L-hr.

In force mains, the surface reaeration rate R_f can be assumed to be zero. In gravity sewers R_f can be estimated using (15):

$$R_f = (2.10)\ (0.96)\ C_a\, T\, (Sv)^{3/8}\, D (d_m)^{-1} \qquad (12\text{-}2)$$

in which 2.10 = conversion factor from meters to feet = ($3.28^{5/8}$); 0.96 = empirical coefficient; $C_a = 1 + 0.17 (v)^2/g\, d_m$, where 0.17 = empirical coefficient and g = gravitation constant, 32.2 ft/s^2; T = temperature coefficient equal to unity at 68°F; S = slope of the sewage stream energy gradient, ft/ft; v = sewage velocity, fps; D = oxygen deficit, mg/L; and d_m = mean hydraulic depth, defined as the cross-sectional area of the stream divided by its width, ft.

Values for R_r, oxygen uptake rate in the wastewater, should be experimentally derived from sampling at several locations along the sewer reaches of interest. If this cannot be done, then Table 12-2 can be used as a general guide in selecting oxygen uptake rates.

Because its experimental determination is difficult and time consuming, R_e can be estimated using (15):

TABLE 12-2. Suggested Oxygen Reaction Rates (15)

Age of wastewater (1)	Suggested rate, R_r (conservative), mg/L-hr (2)
1 hr at 20°C or 0.5 hr at 30°C	5
2 hr at 20°C or 1 hr at 30°C	10
Over 3 hr at 20°C or 1.5 hr at 30°C	15

$$R_e = (1.81)\ (5.3)\ [O_2]\ (sv)^{1/2} r^{-1} \tag{12-3}$$

in which 1.81 = conversion factor from meters to feet = $(3.28)^{1/2}$; 5.3 = empirical coefficient; $[O_2]$ = oxygen concentration, mg/L; and, r = hydraulic radius of the sewage stream, ft.

Several approaches can be applied to estimate air or oxygen requirements for sulfide control. One approach involves field sampling and data analysis to determine R_r. Another procedure, one best utilized when insufficient sampling data are available for the wastewater, involves the following steps:

1. Calculate the gain of oxygen by surface reaeration R_f, using Eq. 12-2.
2. Select an oxygen uptake rate R_r that matches the calculated detention time from Table 12-2.
3. Calculate the oxygen uptake rate by the slime layer R_e, using Eq. 12-3.
4. Use Eq. 12-1 to determine the rate of change of oxygen, in mg/L-hr.
5. Multiply the result from Step 4 by the maximum detention time expected to yield the total oxygen required in mg/L.
6. Determine the mass flow rate of oxygen required, in lb/day, and calculate the gas requirements, in cfm.

An alternative for calculating air injection requirement in force mains is to use (15):

$$Q_a = \frac{(17 \times 10^{-4})\ VR_r}{P_1 \log\left[(P_1 + P_2)/P_1\right]} \cdot \frac{v_a + v_w}{v_a} \tag{12-4}$$

in which Q_a = required air flow at ambient force main pressure, cfm; V = volume of force main, cf; R_r = oxygen uptake rate, mg/L-hr; P_1 = ambient atmosphere pressure, atm; P_2 = gage pressure at injection point on force main, atm; $v_a = 3.0(d)^{1/2}$ the velocity of air relative to the wastewater, in fps; where d = pipe diameter, ft; and v_w = velocity of wastewater, fps.

Although Eq. 12-4 involves a number of approximations, it will generally indicate more than enough air for sulfide control. Eq. 12-4 is not useful at injection pressures greater than 40 psi, because of the large impact of pressure on both oxygen and nitrogen dissolution (15).

In practice, sulfide control is often accomplished with less oxygen than would be needed to fully supply the oxygen consuming capability of the wastewater.

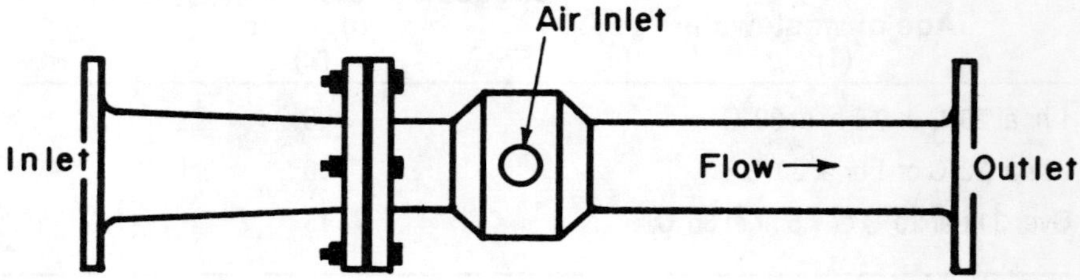

Figure 12-1. Typical Venturi aspirator (2)

2. Equipment

Once the amount of oxygen or air to control sulfide generation has been determined, it is necessary to decide upon a method of providing the oxygen. Oxygen can be provided from atmospheric air, either aspirated or compressed and injected into the wastewater system, or from a gaseous oxygen source.

Aspirated air. —Venturi-type aspirators have been employed for sulfide control in force mains and lift stations, but have gained little long-term use. The Venturi aspirator operates on the principle that liquid flowing at a high velocity through a nozzle of decreasing diameter creates a negative pressure at the discharge side of the restriction. Provision of an opening to the atmosphere at the discharge side of the restriction allows air to be drawn into the device. Turbulence created at the discharge provides for intimate mixing of the air and water. A typical Venturi aspirator is shown in Fig. 12-1.

Most Venturi aspirators require a minimum pressure at the inlet and a minimum wastewater flow rate to achieve the desired rate of air aspiration. Bypass valves and piping are recommended to allow for the diversion of flow so that maintenance can be performed on the aspirator.

Compressed air. —The hardware for the injection of compressed air usually consists of a compressor, a receiver or reservoir tank, air piping and valves, injection nozzles (if necessary), and a control system. A typical compressed air injection system is shown in Fig. 12-2.

The proper choice of a compressor is critical in developing long-term and minimum-maintenance operation. For small installations, energy consumption may not be a primary consideration, and the compressors are typically air-cooled, single-stage units. These are not heavy duty machines, and control systems are required that permit intermittent operation, by allowing the compressor to idle or to cycle through short off and on periods. For air requirements at larger installations, either minimum cylinder displacement or maximum speed should be specified. Speed for larger units should not exceed 1,150 rpm. Multiple compressors may be appropriate (15).

A rotameter or other visual flow indicator should be available so that the performance of the compressor can be observed. It should be installed in a bypass piping arrangement through which air does not normally flow. The air injection rate can be checked periodically by use of the bypass flow metering system.

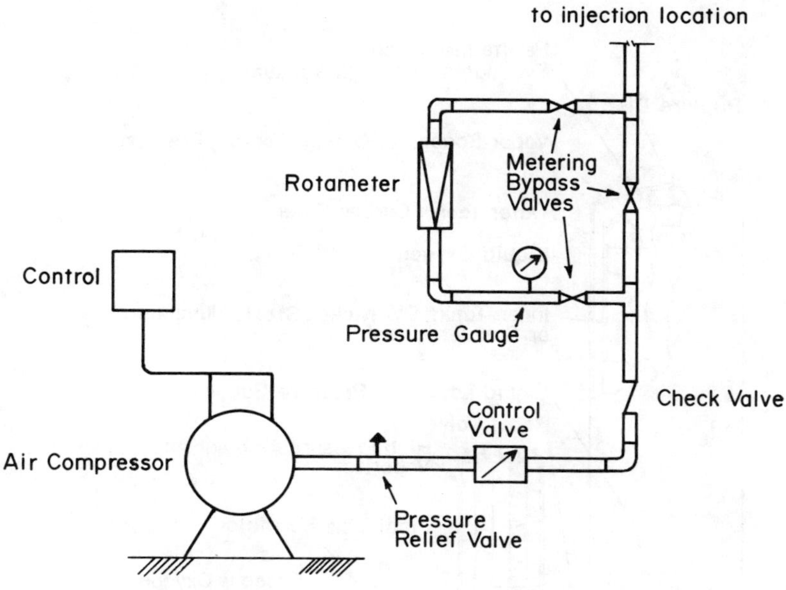

Figure 12-2. Typical compressed air injection system

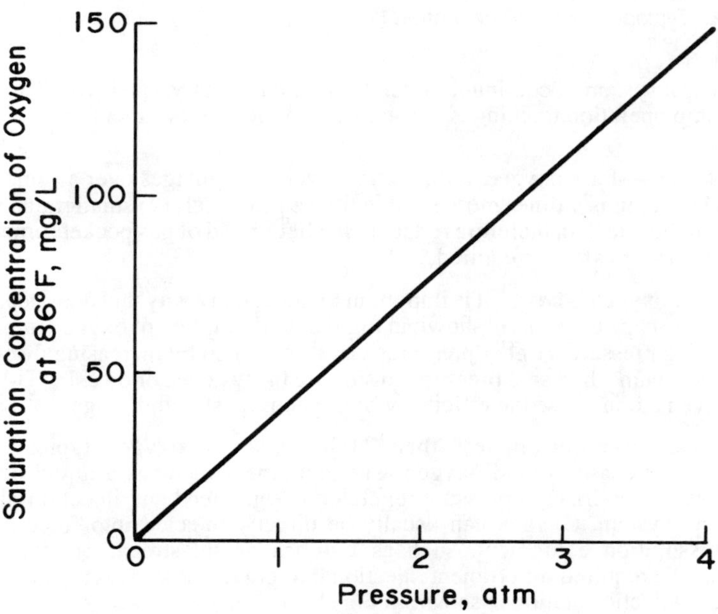

Figure 12-3. Saturation concentration of oxygen in water at different pressures (8)

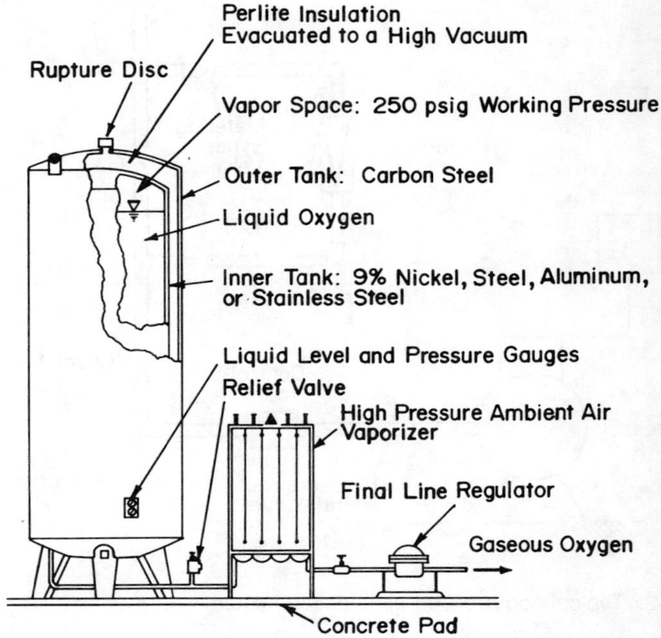

Figure 12-4. Typical liquid oxygen station (1)

A control system, if continuous operation is unnecessary, can involve pressure cycling, pump operation tracking, sewage flow monitoring, or timers.

Gaseous oxygen.—Pure oxygen addition has several advantages over air injection, such as: (1) Oxygen is 5 times more soluble than air in water; (2) smaller quantities of gas need to be injected, therefore reducing the likelihood of gas pocket formation; and (3) compressors are not required.

Because oxygen is costly, it is important to inject it in a way that does not allow a large part to escape unused. As shown in Fig. 12-3, dissolution of oxygen improves with increasing pressure. Higher pressures can be obtained by increasing discharge heads of force mains, by using pressurized sidestream systems, or by using U-tubes. The objective is to increase the efficiency of oxygen transfer and usage.

Applications requiring less than 2,000 lb/day of oxygen typically use commercially-purchased liquid oxygen. The equipment includes a liquid oxygen storage vessel, a vaporizer, a pressure regulator, oxygen feed and injection piping, and a control system. Oxygen can usually be directly injected into force mains. Separate dissolution equipment, such as U-tubes or sidestream pressurization systems, may be required for efficient injection into gravity sewers. A typical layout of an oxygen injection station is shown in Fig. 12-4.

The liquid oxygen injection tank, which can be purchased or leased from the supplier, is designed to regulate pressure automatically. The vaporizer allows for the vaporization of the liquid oxygen through the exchange of heat from the ambient air.

TABLE 12-3. Selected Studies of Direct Injection of Compressed Air Into Force Mains for Sulfide Control (2)

			Total Sulfide	
Diameter, in. (1)	Length, ft (2)	Air input, cfm (3)	Pump station, mg/L (4)	Discharge of force main, mg/L (5)
(a) Los Angeles, CA				
6	2,080	0	Avg.	0.32
		0.4	0.2	0.06
		0.8		0.06
6	1,120	0		0.20
		0.6	Avg.	0.31
		0.9	0.1	0.78
		1.3		trace
		1.3		0.05
		1.3		0.10
10	1,720	0		16.4
		1.3		12.4
		1.5		14.8
		1.5		12.9
		1.5	Avg.	12.3
		1.5	2.1	3.4
		1.6		3.2
		1.8		8.0
		1.9		7.5
		2.0		4.5
12	1,490	0		4.6
		3.4	Avg.	0.2
		4.2	0.7	0.2
12	1,970	0		5.7
		1.4		5.6
		2.2		3.8
		2.7	Avg.	1.9
		4.0	0.5	3.5
		10.5		0.25
		10.5		0.2
		10.5		0.25
(b) Port Arthur, TX				
16	3,840	0		2.3
		12		0.7
		24		2.1
		36	1.4	0.6
		48		0.3
		60		0.1
		66		1.5
		100		0.1
30	4,400	0	Avg.	5.6
		110	2.3	1.4

TABLE 12-3. Selected Studies of Direct Injection of Compressed Air Into Force Mains for Sulfide Control (2) (cont'd)

Diameter, in. (1)	Length, ft (2)	Air input, cfm (3)	Total Sulfide: Pump station, mg/L (4)	Total Sulfide: Discharge of force main, mg/L (5)
(c) Trinity River Authority, TX				
30	5,080	0	Range	5 to 8
		80	0 to 3	<0.3
(d) Sacramento, CA				
24	5,580	0		0.45
		10	Avg.	0.39
		14	0.3	0.26
		21 to 23		0.19

The oxygen piping is typically copper or stainless steel, although copper is more commonly used because of its lower cost and ease of installation.

Where oxygen requirements exceed 2,000 lb/day, generation on site may be more economical. The design of such a system is beyond the scope of this manual.

3. Force Mains

The most common method to increase the oxygen content of wastewater is through the injection of air or oxygen in the pressure piping from the discharge of sewage pumps. The increased pressure in a force main allows for greater dissolution

TABLE 12-4. Typical costs for Direct Compressed Air Injection into Force Main for Hydrogen Sulfide Control (2)

Conditions (1)	Capital cost [a], 1984 $ (2)
Flow = 1 mgd Diameter = 10 in. Length = 5,280 ft Pressure = 23 psig Air flow = 90 cfm	21,000
Flow = 10 mgd Diameter = 24 in. Length = 5,280 ft Pressure = 23 psig Air flow = 540 cfm	54,000

[a] Includes concrete pad, compressor, piping, valves, start-stop controls, pre-fab building, and installation labor.

TABLE 12-5. Performance of Direct Oxygen Injection into Force Mains (2)

Force main parameters, (1)	Oxygen rate, cfm (2)	Injection dose, mg/L (3)	Total Sulfide Concentration, mg/L Wet well (4)	Total Sulfide Concentration, mg/L End of force main (5)
(a) Port Arthur, TX				
Diameter = 16 in.	0	0	0.5	1.2
Length = 3,850 ft	2.0	16.8	1.5	0.1
Flow = 1.7 mgd	3.0	25.2	1.7	0.1
	4.0	33.6	1.0	0.2
	4.5	37.8	0.8	0.2
	5.0	42.0	1.9	0.1
	6.0	50.5	2.0	0.1
Diameter = 8 in.	0	0	2.3[a]	6.4
Length = 3,150 ft	5.0	79.4	2.3[a]	0.9
Flow = 0.9 mgd	6.0	95.3	2.3[a]	0.5
	7.0	111.2	2.3[a]	0.6
Diameter = 8 in.	0	0	0.9	10.7
Length = 5,420 ft	1.0	28.6	1.2	2.2
Flow = 0.5 mgd				
(b) Sacramento, CA				
Diameter = 21 in.				
and 27 in.	0	0	0.2	2.1[b]
Length = 9,610 ft	4.1	18.8	0.1	0.2[b]
Flow = 3.15 mgd	8.3	37.5	0.1	0.1[b]
	10.0	45.6	0.1	0.0[b]

[a] Weighted averages.

[b] Length and data are for the first sampling location, which is not the discharge point for the force main.

of oxygen into the wastewater. At 1 atm pressure and 70°F (21.1°C), water will dissolve approximately 2% air by volume. This increases to 4% at 2 atm, and 6% at 4 atm. The increased solubility of pure oxygen is illustrated in Fig. 12-3.

Aspirated air. —Venturi-type aspirators have been tested for sulfide control in force mains in Jefferson Parish, Louisiana, as part of a demonstration project on sulfide control (4). A Venturi aspirator was installed in an 8-in. force main. This device was capable of aspirating 5 to 6 cfm of air at a wastewater flow of 1,000 gpm, but proved inadequate for sulfide and odor control at the end of the force main. The Venturi aspirator was abandoned after the test program.

Compressed air injection.—Direct injection of compressed air in force mains has been reported in California (7, 16), Texas (20), Louisiana (4), and New Jersey (23). Data from selected studies are summarized in Table 12-3.

The air injection rates required for sulfide control vary depending on the wastewater characteristics, detention time in the force main, temperature and pressure of the system, hydraulic profile of the force main, and the degree of sulfide

TABLE 12-6. Typical Costs for Direct Oxygen Injection into Force Main for H_2S Control (2)

Condition (1)	Capital cost 1984 $ (2)	Annual chemical cost 1984 $ (3)
Flow = 1.0 mgd Diameter = 14 in. Length = 5,280 ft O_2 required = 180 lb/day	20,000	9,000
Flow = 10 mgd Diameter = 30 in. Length = 5,280 ft O_2 required = 680 lb/day	50,000	20,000

Note: Includes concrete pad, liquid oxygen storage vessel, vaporizer, piping, start-stop controls, and installation. Based on oxygen cost of $0.14/lb for 180 lb/day. Based on oxygen cost of $0.08/lb for 680 lb/day.

control desired. Design rules of thumb (19) have suggested providing air at a rate of 0.1 to 0.3 cf/gal of wastewater, or 1 to 2 cfm/in. of pipe diameter. In practice, these rules of thumb can either overestimate or underestimate the actual air requirements. Ratios calculated from the data in Table 12-3 range from 0.13 to 6.25 cfm/in. of pipe diameter. The procedures outlined under design requirements (Sec. B1) are recommended to estimate the air injection requirements more accurately.

The hardware necessary for the injection of compressed air into a force main has been previously discussed in this chapter. Typical installations do not utilize any diffuser. Air that is injected into a sloping force main coalesces into large bubbles within a very short distance of the point of injection regardless of the manner of air injection, so a simple flush nipple tap onto the sewage piping is often used. A flush nipple connection is preferably made at the springline of the pipe, to prevent any deposition of materials when the pipe is in a nonflow condition.

Estimates of the costs for the injection of compressed air into force mains are shown in Table 12-4. These costs include the concrete pad, appropriate size air compressor, piping, valves, control equipment, a prefabricated building enclosure, and the labor to install the system.

Pure oxygen injection.—Gaseous oxygen from a typical liquid oxygen station is simply piped directly to the force main downstream of the pump discharge check valve. The higher the force main pressure at the point of injection, the more oxygen transfer will occur.

Results from studies of direct oxygen injection into force mains have been reported at Port Arthur, Texas (20), and Sacramento, California (7). The data, which are presented in Table 12-5, show that direct injection of pure oxygen can successfully control sulfide generation.

Estmates of the costs for direct injection of pure oxygen into a force main are shown in Table 12-6.

The costs include the concrete pad, the liquid oxygen storage vessel and vaporizer, the necessary piping and controls, and the labor to install the equipment.

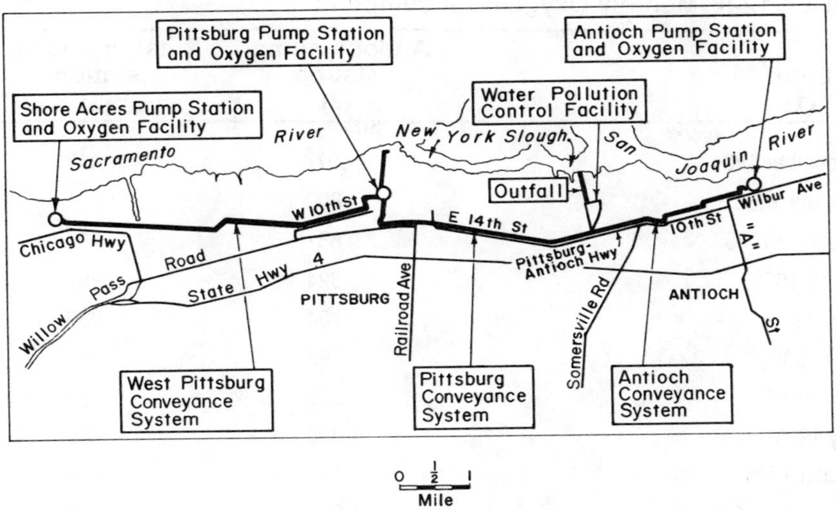

Figure 12-5. Location of pump stations, Delta Diablo Sanitation District 7A, Antioch, California

Oxygen suppliers often will lease the necessary liquid oxygen storage and vaporization equipment, thereby significantly reducing the capital expense.

Case history: Delta Diablo Sanitation District No. 7-A, Contra Costa County, California. —Regionalization of the sewage collection and treatment facilities for the cities of Pittsburg and Antioch and the unincorporated community of West Pittsburg, California, resulted in the construction of 5 miles of gravity sewer, 6 miles of force main, and three pump stations. Substantial sulfide generation was anticipated because of the length of the force mains involved. Pure oxygen injection was selected to control sulfide generation in the three pump stations. Figure 12-5 is a map of the service area showing the locations of the pump stations where pure

TABLE 12-7. Force Main Characteristics, Delta Diablo Sanitation Districts, California

Characteristic (1)	Antioch (2)	Pittsburg (3)	Shore Acres (4)
Design flow, mgd	5.5	8.5	3.8
Existing flow, mgd	4.0	2.0	0.5
Diameter, in.	24	24	16
Length, ft	13,690	13,090	4,465
Design detention time, hr.	1.4	0.9	0.3
Existing detention time, hr.	1.9	3.7	2.4
Existing discharge head, ft	65-125	81	

TABLE 12-8. Monthly Oxygen Consumption

Date (1)	Antioch pump station (2)	Pittsburg pump station (3)
December 1984	697[a]	1,113
January 1985	600	1,021
February 1985	631	597
March 1985	624	997
April 1985	734	969
May 1985	793	960
June 1985	972	1,003
July 1985	1,096	806
August 1985	1,141	1,153
September 1985	1,229	1,152
October 1985	1,083	1,328
November 1985	713	1,200
Total consumption	10,313	13,237
Average monthly consumption	859	1,103

[a] Consumption quantities are in pounds per day.

Yearly chemical cost, for Antioch = $13,686; for Pittsburg = $17,573.
Equipment cost, for Antioch= $8,400; for Pittsburg = $5,400.
Total yearly cost, for Antioch= $22,086; for Pittsburg = $22,973.

oxygen is injected into the force mains. Physical data for the force mains are presented in Table 12-7.

Wastewater flow entering the Delta Diablo Sanitation District No. 7-A Treatment Plant averaged 7.5 mgd in 1985. Wastewater is analyzed three times per week for sulfide concentration. Before oxygen injection was initiated, the total sulfide concentrations at the treatment plant ranged between 3 and 6 mg/L, and the dissolved sulfide concentrations varied between 1.5 and 5 mg/L. Since oxygen injection began, the total sulfide concentrations have varied between 0 and 1.5 mg/L, and the dissolved sulfide concentrations have varied between 0 and 0.7 mg/L.

Pure oxygen is injected just downstream of the sewage pumps in the force-main flow-metering structures. A flush nipple connection is used to connect the oxygen piping to the side of each discharge pipe. Oxygen injection rates are automatically varied in proportion to the flow. Two options are utilized if the sulfide level exceeds 0.1 mg/L. The first is to increase the oxygen dosage. The second is to clean the force mains. Pig launching connections were designed in the force mains to facilitate line cleaning. Periodic line cleaning during the summer months has been

Figure 12-6. Air compressor for the Western Avenue siphon, Los Angeles, California

found to be effective in helping the oxygen injection system control sulfide levels. Cleaning of each force main with the pig requires approximately four hours of operational staff time. Operating staff also change the oxygen injection location to the pig launch connection during the summer. They feel that increased turbulence at this location improves oxygen efficiency.

Daily oxygen consumption data were provided for both the Antioch and Pittsburg pump stations for the period December 1984 through November 1985. The average monthly consumption is listed in Table 12-8.

Sulfide data were provided for the end of the force mains from the Antioch and Pittsburg pump stations for the period July, August, October, and November 1985. July, August, and October represent maximum-use periods for the Antioch pump station, while August, October, and November represent maximum-use periods for the Pittsburg pump station. An analysis during this time period indicates that, at an oxygen consumption rate of 1,100 lb/day, the dissolved sulfide concentration was maintained below 0.3 mg/L at the Antioch pump station. At oxygen consumption rates above 1,200 lb/day, the dissolved sulfide concentration was maintained below 0.2 mg/L at the Pittsburg Pump Station. At doses under 1,200 lb/day, the concentration reached 0.8 mg/L at the Pittsburg Pump Station.

Calculations for the sulfide generation rates through each of the force main systems were made based upon Eq. 4-2. The sulfide flux coefficient M was 0.001 m/hr, and the effective BOD (EBOD) was 250 mg/L. The resulting sulfide generation rates were 2 mg/L-hr for Antioch and Pittsburg, and 2.8 mg/L-hr for Shore Acres. Estimates of the oxygen consumption needs for each pump station were determined to be 600 lb/day for Antioch, 800 lb/day for Pittsburgh, and 100 lb/day for Shore Acres. A comparison for the Antioch and Pittsburg pump stations of these predicted oxygen demands and the actual oxygen consumption indicates efficiencies between 55% and 67% for the oxygen injection systems.

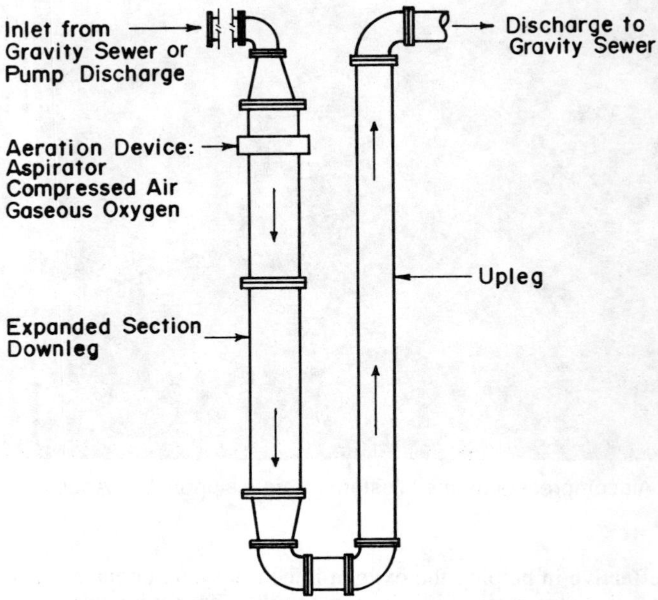

Figure 12-7. Typical U-Tube installation (20)

The installation costs for the three oxygen injection systems included costs for the concrete pad, necessary piping, valves, and controls as well as the labor for the installation. The installation costs were: Antioch $8,000 (1979), Pittsburg $5,000 (1981), and Shore Acres $5,000 (1981). The oxygen storage vessels and the evaporators are leased from the oxygen suppliers. Operating costs are not available for the Shore Acres pumping station. The 1985 operating costs for the Antioch and Pittsburg pump station systems total $45,059. Chemical costs were $13,636 for Antioch,and $17,573 for Pittsburg. Equipment lease costs were $8,400 for Antioch and $5,400 for Pittsburg. The operating costs averaged $60.00 per day at Antioch and $63.00 per day at Pittsburg.

4. Inverted Siphons

Inverted siphons can often be sources of significant sulfide generation, particularly if long detention times occur. If sufficient pressure exists at the low point of the inverted siphon and if access is possible, then air or pure oxygen injection can be used for effective sulfide control. The ideal location for air or oxygen injection is at the bottom of the descending leg of the siphon.

Equipment.—Equipment similar to force main injection (Sec. B3) is utilized.

Case history: County Sanitation Districts of Los Angeles County, Los Angeles, California. —A 7,400 ft (2,256 m), 24-in. (61-cm) diameter, rubber gasket RCP, inverted siphon was constructed to provide hydraulic relief for the Western Avenue pumping plant. The elevation rise from the base of the siphon to the discharge end

TABLE 12-9. Expected Oxygen Absorption in Wastewater Falls (15)

Fall, ft (1)	Oxygen deficit satisfied, percentage (2)
1	12
2	22
3	31
4	39
5	46
6	53
8	63
10	71
15	86.7
20	91.8
30	97.6

Note: Based upon K_H=0.4Lm in Eq.12-5.

is 115 ft (35 m). At an average flow of 3.6 mgd, the velocity is just under 2 fps (0.06 m/s), the travel time is 70 minutes, and the sulfide build-up through the inverted siphon varies between 2 and 4 mg/L. The siphon was constructed so that compressed air could be injected into the bottom of the siphon from the adjacent pump station, although no compressor was sized or installed.

Field determinations of R_r, the oxygen reaction rate of the sewage, produced a value of 10 mg/L-hr. Using Eq. 12-4 with R_r = 10 mg/L-hr, and the other parameters calculated from the siphon information, the air injection rate needed is 30 cfm.

The air requirement was also calculated by considering that the air injection point would be 2,700 ft (823 m) from the end of the siphon. Sulfide would be generated during the preceding 4,700 ft (1,433 m) of siphon, but the oxygen demand for that portion of the siphon would not have to be satisfied. A conservation figure of 4 mg/L was selected for sulfide generation through the siphon. Proportioning that level to the point of injection produced a sulfide level of 2.5 mg/L. Assuming that oxygen will react on a 1:1 basis with sulfide, then 2.5 mg/L would be needed to oxidize the existing sulfide at the point of injection.

Using Eq. 12-1, with the assumption that $R_f = 0$, R_r is the measured 10 mg/L-hr, and calculating R_e to be 6 mg/L-hr (at a dissolved oxygen level of 1 mg/L), produces an oxygen requirement of 16 mg/L-hr. Proportioning this 16 mg/L-hr requirement over the time period from the point of injection to the end of the siphon yields an oxygen requirement of 6.8 mg/L. Adding the 2.5 mg/L needed to oxidize the sulfide that is generated in the siphon upstream of the point of air injection yields 9.3 mg/L for an oxygen requirement. The corresponding compressed air injection rate is 15 cfm.

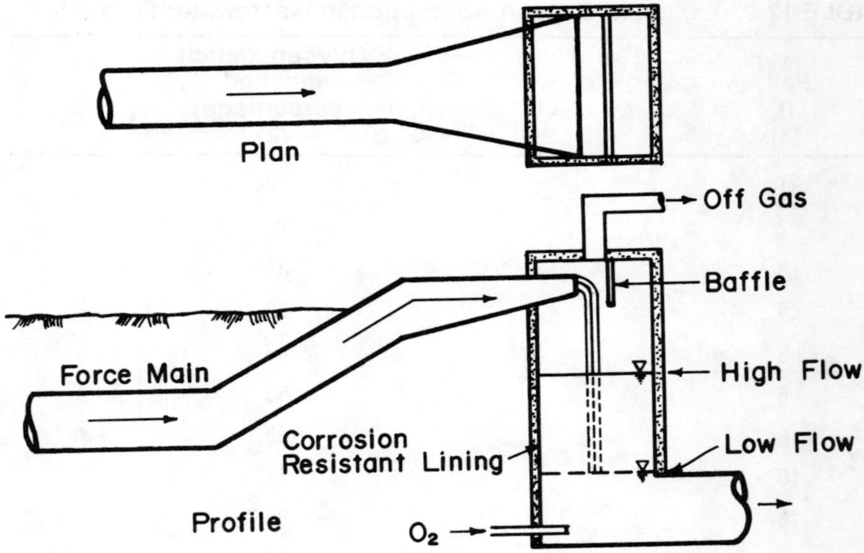

Figure 12-8. System for oxygen injection into a hydraulic fall (7)

A 30-cfm, 10-HP, single-stage, air-cooled compressor was selected. Compressed air was injected through a 2-in. (50 mm) air line to the low point of the siphon. A photograph of this installation is shown in Fig. 12-6. The compressor has a loadless start feature, a suction valve unloading feature when the reservoir tank is at full pressure, and a timer to operate the compressor on a 7.5 minutes on, 7.5 minutes off cycle. These features are all intended to extend the life of the compressor.

The air compressor was operated to discharge 13 to 15 cfm between 1980 and 1985. During this time period, the dissolved sulfide concentrations at the end of the inverted siphon averaged 0.1 mg/L, with a range between zero and 0.3 mg/L. The hydrogen sulfide concentration in the sewer atmosphere at the end of the siphon averaged 1.0 ppm, with a range between zero and 10.0 ppm.

The air injection system was temporarily shut off in 1986 because of concern about potential corrosion in the ascending leg of the unprotected RCP inverted siphon. A television inspection is scheduled. If no corrosion problems are observed, then air injection will be continued. If corrosion is found, then alternative methods of sulfide control and corrosion protection will be evaluated.

5. U-Tubes

A U-tube is a device that provides for the temporary pressurization of a gas liquid mixture by subjecting the mixture to a temporary hydrostatic head. This pressurization increases the gas transfer capabilities. A single U-tube can be visualized as a horizontally compressed inverted siphon. Single U-tubes can be constructed to any depth, but generally little advantage is gained by exceeding 100 to 150 ft, because the cost for construction increases significantly with very little additional benefit in oxygen transfer efficiency for the injection rates used for sulfide control.

A typical single U-tube installation is shown in Fig. 12-7. This is simply a pair of vertical pipes connected by a return bend. A U-tube configuration can take many other forms, including a pair of concentric pipes, a single pipe with a flat vertical partition, or a rectangular trough with a vertical partition to separate the down and up flow passages. Concerns with pipe plugging or fouling have led to the use of two pipes connected by a return bend for most test installations. Larger U-tubes that carry sufficient quantities of flow can be successfully constructed from two concentric pipes.

The cross-sectional area of the down leg is greater than that of the up leg, to allow greater detention time for oxygen transfer to occur in the down leg. The increase in velocity in the up leg prevents the depositing of solids. Typically, the velocities in the down leg are less than 2 fps (0.06 m/s), and are 4 fps (1.2 m/s) or greater in the up leg (13).

The location of the air injection point is usually near the top of the down leg, to provide for maximum contact time with the wastewater. Lowering the injection point reduces the headloss for the system, but decreases the dissolved oxygen concentrations at the outlet. Air injection can be accomplished by using an aspirator or an air compressor. Pure oxygen can also be used.

Testing with U-tubes has been conducted in Jefferson Parish, Louisiana (4); Port Arthur, Texas (20); and Sacramento, California (7). Most of this testing involved U-tubes that were 40 ft (12 m) in depth or shallower. In all cases, ineffective sulfide control was achieved. None of the U-tubes constructed for evaluation purposes remains in operation today. A significant amount of information has been developed for the design of U-tubes and the reader is referred to these sources for more information (13).

6. Hydraulic Falls or Jumps

Sudden changes in elevation, or other points of high turbulence, are notorious for the release of odors and hydrogen sulfide. At the same time, these locations can be points of significant oxygen transfer. The amount of oxygen absorbed in a fall is much larger than the amount absorbed by the smooth flow of water through the same loss in elevation (15). Testing has indicated that the oxygen concentration approaches saturation logarithmically with the height of the fall, and can be represented (15) by

$$2.303 \log \frac{D_1}{D_2} = 0.305\, K_H (h_1 - h_2) \tag{12-5}$$

in which 2.303 = conversion factor from ln to log; D_1 and D_2 = oxygen deficit upstream and downstream of the fall, mg/L; 0.305 = conversion factor from meters to feet; K_H = waterfall reaeration coefficient, L/m; and h_1 and h_2 = elevation of hydraulic energy lines upstream and downstream of the fall, ft.

The limited data available indicate that an average value of K_H = 0.4L/m is appropriate for wastewater (15). The percentage of the oxygen deficiency that can be expected to be satisfied as the result of falls of various heights is shown in Table 12-9.

Rather than allowing points of turbulence to be harmful, proper design can turn them into beneficial points of oxygen transfer. A diagram of an oxygen transfer system at a sudden change in elevation is shown in Fig. 12-8.

Equipment.—Equipment similar to force main injection is utilized to provide a source of oxygen.

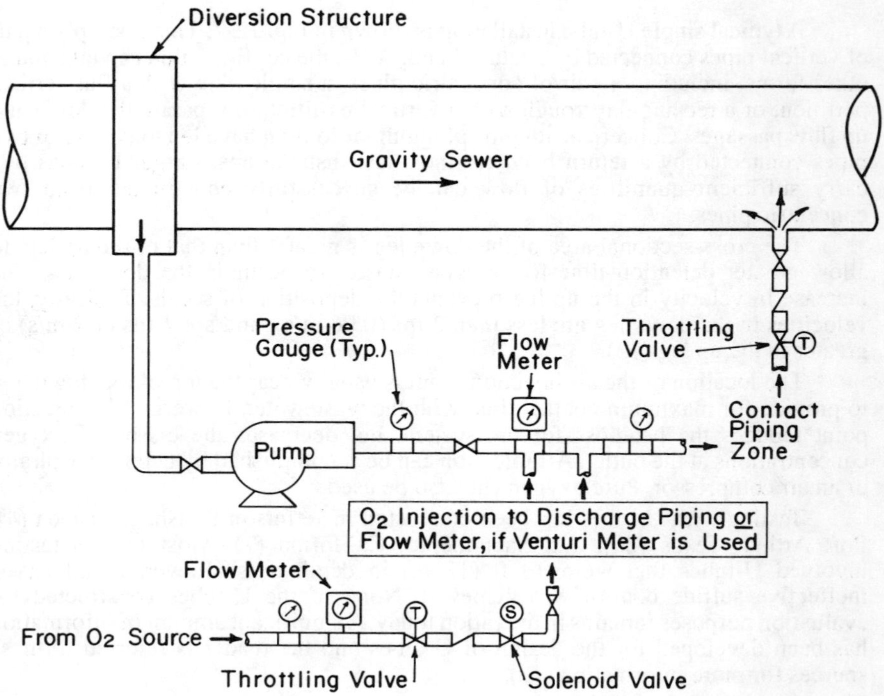

Figure 12-9. Components of a pressurized sidestream system

Case history: Sacramento Regional County Sanitation District, Sacramento, California. —Two oxygen injection fall structures have been in use in Sacramento, California, since 1983 to help control sulfide in the Sacramento Regional Interceptor System. Each structure is practically sealed, has a corrosion resistant interior, and is designed for a fall of about 15 ft (4.6 m) in an oxygen-enriched atmosphere. Pure oxygen is piped several miles, to two stations, from the Sacramento Regional Wastewater Treatment Plant, where high-purity oxygen is produced in a large cryogenic system for the 150-mgd activated sludge process. Both fall structures are at the ends of long force mains as described below:

> The Arden Oxygen Injection Station is located near the end of the six mile long, 60 in. diameter Arden Force Main. Average dry weather flow was 26 mgd in 1987, and the average velocity was about 2.0 fps. This flow is primarily municipal wastewater from the Arden and Natomas areas and other portions of northern Sacramento County.
>
> Uncontrolled sulfide was predicted to reach nearly 5 mg/L at the end of this force main (7). Wastewater from this force main joins other flows and proceeds about 10 miles through the large-diameter gravity Central Interceptor to the Sacramento Regional Wastewater Treatment Plant. Total sulfide level was predicted to be 5 mg/L in the Central Interceptor (7).
>
> The City Interceptor Oxygen Injection Station is located near the end of the almost two mile long, 72 in. diameter Sump 2 Force Main. Average dry weather flow was 24.5 mgd in 1987, and the average velocity was about 1.3 fps. This flow is mostly dry weather flow from the combined sewer area in the

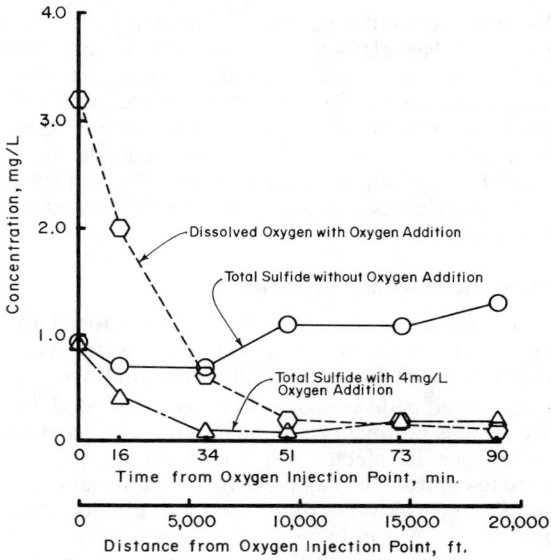

Figure 12-10. Effect of high pressure sidestream oxygen injection on total sulfide in 72 in. joint outfall "C," Los Angeles, California

central portion of the City of Sacramento. Uncontrolled total sulfide was predicted to be 1 mg/L at the end of this force main (7). Wastewater from this force main joins other flows and proceeds about seven miles through the large diameter gravity City Interceptor to the Sacramento Regional Wastewater Treatment Plant. Total sulfide was predicted to reach 4 mg/L in the City Interceptor (7).

When the Arden Station is operating on pure oxygen, about 52 scfm of pure oxygen is pumped into the structure. The structure atmosphere equalizes at an oxygen content of about 60%. Dissolved oxygen monitoring has shown 13 mg/L 3,900 ft (1,190 m) downstream. Similar parameters have been used at the City Interceptor Station. When air is used instead of oxygen, the structure atmosphere equalizes at 20% oxygen, and the dissolved oxygen reaches 5 mg/L 3,900 ft downstream.

A gas flow of about 50 cfm is pulled from each structure, mixed with ambient air, and scrubbed through a carbon filter at each site for odor control. These steps are successful in controlling odor release at both sites.

The dissolved oxygen added at both stations helps prevent sulfide build-up in the large-diameter gravity interceptors downstream of the fall structures. Dissolved sulfide concentration in these downstream interceptors is almost always less than 0.5 mg/L.

The primary problem with these stations has been high levels of hydrocarbons in the atmosphere of the fall structures. Explosimeters that monitor for the Lower Explosive Limit (LEL) are used in each station. When 25% LEL is reached, pure oxygen addition is shut down. This shutdown event occurs often. Investigation by the district into the source of the hydrocarbons has shown that it is primarily methane gas generated in the anaerobic conditions in the upstream force mains and gravity sewers.

The district has had only limited success in reducing these hydrocarbon levels. The Arden/Natomas system has relatively low flows (due to large capacity for future buildout), the velocities are low, and solids deposits occur. There has been a reduction in hydrocarbon levels in 1987 in portions of the Arden/Natomas system because of higher wastewater. In the City of Sacramento combined sewer area, there are probably many locations where solids deposits generate methane. Also, the velocity in the Sump 2 Force Main is low enough to allow solids deposition in the pipe. Reducing hydrocarbon levels in this system could be more difficult than in the Arden/Natomas service area.

7. High Pressure Sidestream Injection

One technique for providing efficient oxygen transfer for a large gravity sewer involves saturating a portion of the flow under pressure, with oxygen, then introducing the oxygen-rich sidestream back into the main flow.

The basic pressurized sidestream system is illustrated in Fig. 12-9. A sidestream, typically 10%, is removed from the gravity sewer and passes through a sewage pump; then oxygen is injected into the discharge piping. A Venturi or throttling valve is used to control the back pressure on the system. The high-velocity discharge of the oxygenated sidestream into the main gravity sewer provides for good mixing and dilution. A single point discharge, or possibly a manifold discharge, can be used for mixing with the main gravity sewer. Diffusers are not recommended because of the fouling potential with raw sewage. A grinder may be required upstream of the pump if ragging is a problem.

In a prototype installation in use at the County Sanitation Districts of Los Angeles County, California, 80% oxygen transfer has been obtained at an operating pressure of 70 psig. A 2-mgd sidestream is removed from 40 mgd of sewage that is tributary to a 72-in. diameter RCP sewer. The input of 4 mg/L of dissolved oxygen has maintained zero dissolved sulfide and 0.2 mg/L total sulfide in a three-mile reach of the sewer, with background total sulfide levels above 1.0 mg/L, as shown in Fig. 12-10. The operating costs for this system, including oxygen and power, have averaged $4.00/mgd, with the oxygen cost being one half of the total.

The capital cost for such a system would be high because it would involve the construction of a pump station, as well as a pure oxygen supply system (estimated in Sec. B3).

A proprietary "Vitox" system is a modification of the pressurized sidestream that reportedly requires only 30 psig pressure to provide oxygen transfer efficiencies between 40% and 60%. In this system, only part of the oxygen is dissolved in the sidestream. The remainder is dissolved when the stream is returned to the gravity sewer through a submerged nozzle or jet (12).

C. CHLORINATION

Chlorine, whether applied as chlorine gas into an aqueous solution or as sodium or calcium hypochlorite, will oxidize sulfide and depress the generation of sulfide as long as a residual exists. The effectiveness of chlorination depends upon the degree of sulfide control desired, the distance for which sulfide control is desired, the initial sulfide concentration, the magnitude of any competing reactions in the wastewater, and the degree of chlorine feed control desired. The amount of chlorine required for effective control can vary significantly from application to application.

Chlorine is added to wastewater as either hypochlorite or chlorine gas. Sodium hypochlorite is generally used as a matter of convenience for applications that are occasional, where dosages are small, or where, for safety reasons, chlorine

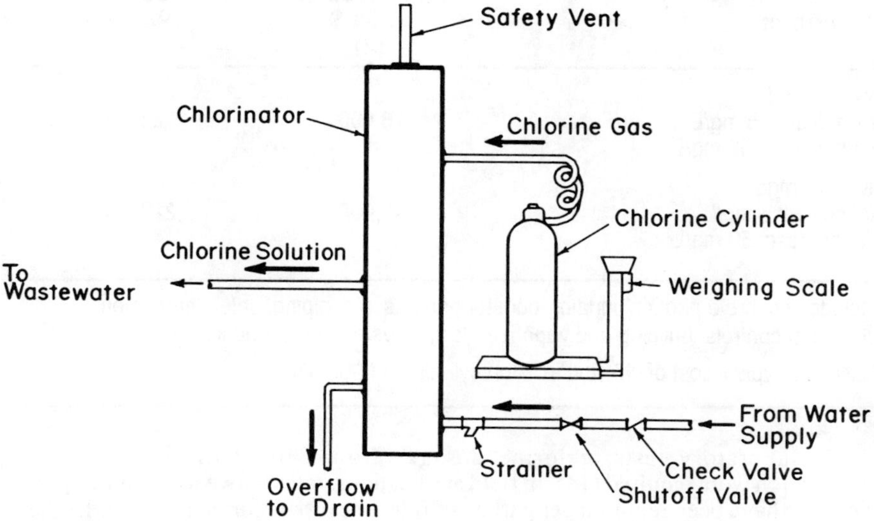

Figure 12-11. Typical chlorinator installation (4)

gas systems are excluded from consideration. The use of chlorine gas is usually chosen as a matter of economy when large quantities are needed, such as several hundred pounds per day.

Chlorine and hypochlorite can be used interchangeably. Chlorine in dilute solutions is, in fact, a mixture of hypochlorite and un-ionized hypochlorous acid. Concentrations and dosages of hypochlorite are expressed as the equivalent amount of chlorine.

Chlorine is a hazardous material. The design of any chlorination station must recognize the need to make appropriate provisions for safe handling and storage of chlorine. Poor mixing in sluggish wastewater flows can lead to localized fuming conditions as a result of chlorine overdose at the point of injection. These factors should be considered in evaluating the use of chlorine.

1. Design Requirements

If an excess of chlorine is added to wastewater containing sulfide, the sulfide is oxidized largely to sulfate. Theoretically, 8.87 parts by weight of chlorine are required to oxidize each part of sulfide to sulfate. If chlorine is added slowly, with vigorous mixing, to a pure sulfide solution, the sulfide may be oxidized to elemental sulfur. For this reaction, 2.22 parts by weight of chlorine are required per part of sulfide. Intermediate reactions between sulfur and sulfate can occur, including the creation of thiosulfate, trithionate, and sulfite, depending on a host of factors, including pH, temperature, organic matter, and the degree of mixing (15).

TABLE 12-10. Typical Costs for Chlorine Injection for Sulfide Control (2)

Conditions (1)	Capital cost[a] 1984 $ (2)	Annual chemical costs[b] 1984 $/yr (3)
Flow = 1 mgd Total sulfide = 5 mg/L Chlorine dose: 50 mg/L	18,000	23,000
Flow = 10 mgd Total sulfide: 5 mg/L Chlorine dose: 50 mg/L	40,000	228,000

[a] Includes concrete pad, chlorinator, booster pump, scale, piping, safety equipment, start-stop controls, building and vaporizer (larger system only), and installation.

[b] Based on typical cost of chlorine in 1-ton cylinders of $0.15/lb.

In laboratory tests, chlorine dosages from 3 to 9 times the sulfide concentration were required (14). In field applications, rates between 10 and 15 parts of chlorine have been required per part of sulfide, because chlorine reacts with other constituents in the wastewater, and because of less than ideal mixing conditions at the point of injection (14).

Flexibility should be designed in sizing chlorinators, storage tanks, and chemical feed pumps to be able to provide at least a 15:1 ratio of chlorine to sulfide. To minimize extraneous chlorine reactions, the chlorine injection point should be selected so that quick mixing across the entire cross section of flow is obtained. Full-scale testing should be performed to optimize the use of chlorine following the construction of the facility.

A field testing program in Sacramento, California, was conducted in 1974 at the Park Drive pump station (7). The average flow at the pump station was 2.5 mgd. The 5,000-ft, 24- in. diameter force main is tributary to a 27-in. gravity sewer. Normally, the chlorine feed is between 1,100 and 1,200 lb/day into the discharge side of the pumps for odor control in the downstream gravity sewer. Chlorination was terminated for one week, and dissolved sulfide levels were monitored at the end of the force main. The total sulfide reached 0.8 mg/L. Chlorination was then reinstituted at several different rates. At the lowest rate tested, which was 300 lb/day, or an equivalent chlorine dose of 14 mg/L, complete sulfide control was maintained throughout the test section. This rate equates to a dose ratio of 28:1. It should be noted that a fully generating slime layer probably had not developed in the force main during the one week that chlorination had ceased. Equation 4-2 would predict sulfide generation of 2.3 mg/L through the force main.

2. Equipment

Hypochlorite feed systems are generally used when the requirement for chlorine is less than 100 lb/day, when safety reasons exclude the use of chlorine gas, or when preliminary testing is being done prior to the design of a full-scale chlorine system. The equipment required includes a hypochlorite storage vessel, typically a fiberglass tank; a metering pump suitable for a hypochlorite service; hypochlorite solution feed piping, typically plastic; and a diffuser or injection system.

Most full-scale applications using chlorine gas utilize the typical chlorinators found in water and wastewater disinfection processes. A typical chlorinator installation is shown in Fig. 12-11. The equipment design requirements normally utilized for a wastewater disinfection should be followed (3).

Several control options are available for chlorine feed systems: manual adjustment of chlorine dose and a constant feed rate; varying the chlorine feed rate through the use of a preset cam controller; incremental adjustment of the chlorination rate based on the number of pumps operating; flow proportion control that adjusts the chlorine feed rate based upon an electronic signal from a flow meter; chlorine residual control that utilizes chlorine residual sensing equipment immediately downstream of the injection point to adjust the chlorine dose rate; and compound loop control that adjusts the chlorine feed based upon both flow and chlorine residual levels. The simplest and most inexpensive methods of control are generally utilized for those installations requiring small amounts of chlorine, because overdosing will not be a significant expense.

3. Costs

Typical costs for a chlorine injection system are presented in Table 12-10.

These cost estimates are based on the injection of chlorine into a pump station. The control scheme utilized is to interlock the chlorinator with the pump starting circuits so the chlorination occurs only as the pumps are operated. Cost includes a concrete pad, chlorinator, booster pump, chlorine weighing scale, necessary piping, safety equipment, start and stop controls, a separate building, and the labor to install the system.

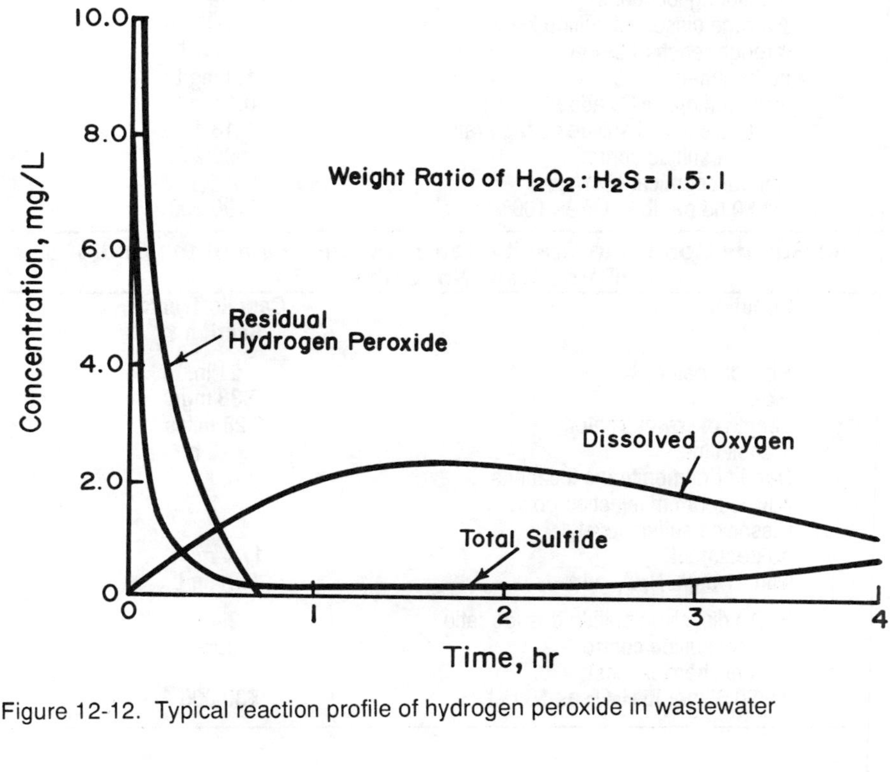

Figure 12-12. Typical reaction profile of hydrogen peroxide in wastewater

TABLE 12-11. Examples of Hydrogen Peroxide Use in Los Angeles, California (5)

(a) Sulfide and Odor Control at Treatment Plant-March 1973

Location	**Long Beach WRP Primary Sedimentation Tank**
Detention time	4.0 hr
Flow	3.0 mgd
Average dissolved sulfide level at primary effluent	
no treatment	3.1 mg/L
with 22 mg/L H_2O_2 added	0.2 mg/L
H_2O_2: dissolved sulfide dosage ratio	7:1
Percent sulfide control	94%
Annual chemical cost (@ \$0.65 per lb H_2O_2 as 100%)	\$130,000

(b) Sulfide Control in Gravity Sewer-September 1974

Location	**Joint Outfall C Units 7A & 7B**
Pipe diameter	48 in.
Flow	4.2 mgd
Length of sewer treated	2.5 miles
Transit time	1.7 hr
Number of downstream monitoring locations	4
Average dissolved sulfide level through reach of sewer	
no treatment	1.3 mg/L
with 23 mg/L H_2O_2 added	0.1 mg/L
H_2O_2: dissolved sulfide dosage ratio	18:1
Percent sulfide control	92%
Annual chemical cost (@ \$0.65 per lb H_2O_2 as 100%)	\$190,000

(c) Sulfide Control in Gravity Sewer Downstream of the End of a Force Main-November 1974

Location	**Castaic Trunk Section 1**
Pipe diameter	21 in.
Flow	0.38 mgd
Length of sewer treated	0.28 miles
Transit time	0.25 hr
Number of monitoring locations downstream of injection point	1
Dissolved sulfide level	
no treatment	17.4 mg/L
with 41 mg/L H_2O_2 added	0.2 mg/L
H_2O_2: dissolved sulfide dosage ratio	2.4:1
Percent sulfide control	99%
Annual chemical cost (@ \$0.65 per lb H_2O_2 as 100%)	\$30,000

TABLE 12-12. Large Gravity Sewers Tested by County Sanitation Districts of Los Angeles County, California (5)

Sewer (1)	Diameter, in. (2)	Test length, ft (3)	Peak flow, mgd (4)	Dissolved sulfide range, mg/L (5)
Joint Outfall A	114	14,500	55	2.0 to 4.0
Joint Outfall B				
Units 1B and 9A	78	23,000	75	0.1 to 0.5
Units 9B and 9C	90, 114, 120	23,000	185	0.6 to 1.0
Joint Outfall C	72	17,000	55	0.2 to 0.5

D. HYDROGEN PEROXIDE

Hydrogen peroxide will react directly with hydrogen sulfide at high concentrations in pure solutions. In acidic and neutral solutions, hydrogen peroxide will theoretically react on a 1:1 weight ratio with hydrogen sulfide to form elemental sulfur. In alkaline solutions, with a pH greater than 8.5, hydrogen peroxide will theoretically react on a 4:1 weight ratio with hydrogen sulfide to produce sulfate. In the pH region between neutral and alkaline solutions, neither reaction is considered dominant.

Hydrogen peroxide is not reaction-specific with sulfide in wastewater. The actual ratio of hydrogen peroxide to dissolved sulfide required for wastewater applications is dependent on the initial sulfide level, the degree of sulfide control required, and the concentration of other pollutants in the sewage.

In wastewater, hydrogen peroxide is also decomposed by the enzyme catalase, produced by a wide variety of living cells, to oxygen and water. The amount of oxygen is 47% of the weight of hydrogen peroxide added. In this way, hydrogen peroxide actually can act to supplement the oxygen supply in sewage. The biological oxidation of sulfide under these conditions would proceed at a ratio somewhere between 4 and 8 parts of hydrogen peroxide per part of sulfide (15).

Hydrogen peroxide can be effective in controlling sulfide generation, although it is generally not the least expensive alternative from a chemical standpoint alone. Its application, however, requires minimal capital expenditures, its reaction with wastewater produces harmless by-products, and its use requires very little maintenance.

1. Design Requirements

The chemical is commercially available in various concentrations: 35%, 50%, and 70% by weight. The concentration used is dependent on the application and the quantity of peroxide required. Bulk shipments of hydrogen peroxide are 50% or 70%. For safety reasons, 50% solutions are typically purchased for sulfide control applications. For small installations, drums can be purchased; larger installations rely on bulk shipments by tank truck or tank car. Storage tank capacities will vary depending on the anticipated hydrogen peroxide use and the available delivery schedules.

The rate of reaction between hydrogen peroxide and sulfide is not instantaneous, which is why most manufacturers recommend adding peroxide a minimum of 10 to 15 minutes upstream of where control hydrogen is desired. For a higher degree of control, it may be required to dose as far as 60 minutes upstream

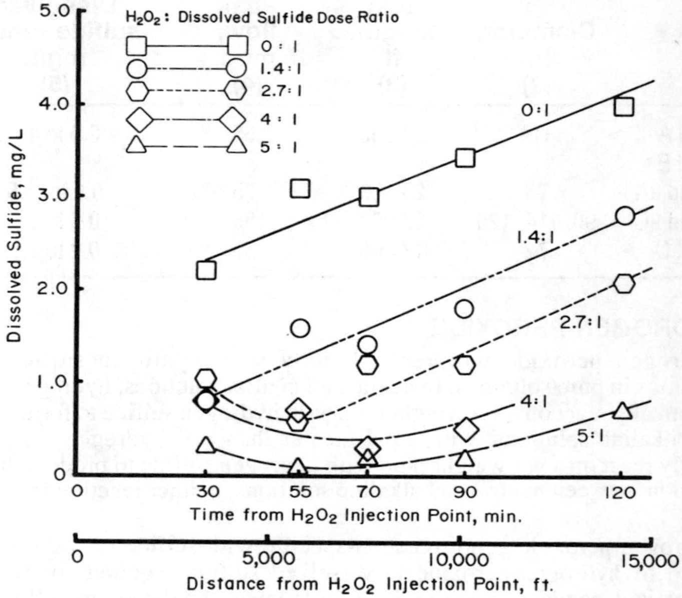

Figure 12-13. Effect of hydrogen peroxide doses on dissolved sulfide in 114-in. joint outfall A, Los Angeles, California (5)

before sulfide levels are reduced to near zero. The half life of hydrogen peroxide ranges from 20 to 40 minutes, the decomposition products being water and oxygen. A typical reaction profile for hydrogen peroxide in wastewater is shown in Fig. 12-12. The curve indicates that 90% of the hydrogen peroxide is reacted within 10 to 15 minutes. The reaction is completed in 20 to 30 additional minutes. Below a sulfide level of 1 mg/L, the reaction rate is fairly slow, and total sulfide control is not obtained.

The first step in designing a hydrogen peroxide addition system is to conduct a laboratory test to generate an initial estimate of the requirements of hydrogen peroxide to sulfide, and also to get an idea of the time required for the reaction to take place, so that a suitable injection point can be determined. Note that laboratory testing is conducted under ideal mixing conditions, with no slime layer present as would exist on the wall of sewers. The results obtained may not be duplicated under actual field conditions. This fact is recognized by the manufacturers, because they generally recommend adding two to three times the estimated amount.

The degree of sulfide control required will be a major deciding factor on the dose ratio designed. As a general observation, most municipalities use hydrogen peroxide for controlling odors through reduction of sulfide levels. This application typically requires a less stringent degree of control than that needed when the object is to prevent corrosion.

To properly determine the capacity of the metering pump, it is necessary to characterize the diurnal flow rates and fluctuations in the mass of sulfide in the sewer

Figure 12-14 Hydrogen peroxide dosing installation at ECR 241 pump station, Palm Beach County, Florida
(*Courtesy*: Los Angeles County Sanitation Districts)

system. It is good practice to specify a pump having a maximum discharge of twice the anticipated feed rate. The pump should be adjustable over a range of 10% to 100% of its maximum capacity. Multiple pumps that are integrated into a manifold system can be operated on a timer circuit to account for periods of high flow and high influent sulfide.

Three examples from the use of hydrogen peroxide in Los Angeles, California, are summarized in Table 12-11 (5).

The first use listed was at the Long Beach Water Reclamation Plant for sulfide and odor control through a primary sedimentation tank. The flow through the tank was 3 mgd, with the retention time of 4 hr. At a hydrogen peroxide to dissolved sulfide dose ratio of 7:1, the average dissolved sulfide at the primary effluent was reduced from 3.1 mg/L to 0.2 mg/L. The odor problem at the treatment plant was primarily associated with start-up operations. The use of hydrogen peroxide significantly diminished these odors, but no odor control was required under normal operating conditions.

The second example listed is a 48-in. diameter gravity sewer that carried a flow of 4.2 mgd. At a hydrogen peroxide to dissolved sulfide dose ratio of 18:1, the average dissolved sulfide level through 2-1/2 miles of sewer was reduced from 1.3 mg/L to 0.1 mg/L.

The third example listed involves sulfide control in a 21-in. diameter gravity sewer downstream of the end of a 16-in. diameter force main that is approximately one mile in length. Attempts to dose hydrogen peroxide at the pump discharge were ineffective, because of the long detention time in the force main. At an average flow of 0.3 mgd and a detention time of nearly 3-1/2 hr, the dissolved sulfide concentration at the end of the force main was 17.4 mg/L. At a hydrogen peroxide to dissolved sulfide dose ratio of 2.4:1, the dissolved sulfide levels were reduced to 0.2 mg/L at a point 0.28 miles (15 minutes) downstream of the end of the force main.

Figure 12-15. Hydrogen peroxide dosing equipment, Palm Beach County, Florida (*Courtesy*: Los Angeles County Sanitation Districts)

The County Sanitation Districts of Los Angeles County undertook an investigation of the use of hydrogen peroxide to control sulfide generation in large gravity sewers in 1985 (5). Testing was conducted in the main arteries of the Sanitation Districts' collection system: Joint Outfalls A, B, and C. The individual outfall flows and sulfide levels are listed in Table 12-12.

The effect of hydrogen peroxide doses on dissolved sulfide in the 114-in. semi-elliptical Joint Outfall A is plotted in Fig. 12-13. Based upon the testing program the following conclusions were developed:

1. Ninety percent sulfide control could be achieved for a 30 to 60 minute reach of sewer using a 5:1 hydrogen peroxide to dissolved sulfide dose ratio for dissolved sulfide levels greater than 1.0 mg/L. For dissolved sulfide levels less than 1.0 mg/L, a 6 mg/L dose of hydrogen peroxide was necessary.

2. Treatment of a long gravity sewer would require multiple hydrogen peroxide injection locations, spaced no more than 90 minutes apart.

3. Maintaining 100% sulfide control through an entire reach of sewer would be difficult, even if multiple hydrogen peroxide dosing facilities were closely spaced.

4. Hydrogen peroxide reduced odor and hydrogen sulfide concentrations in the sewer gas between 60% and 90%, when 90% or greater sulfide control was obtained, based on limited data.

5. Stopping the dosing of hydrogen peroxide resulted in the immediate return of sulfide to normal levels, indicating that hydrogen peroxide did not have any conditioning effect on the sulfide generation capabilities of the slime layer that produces sulfide in the collection system.

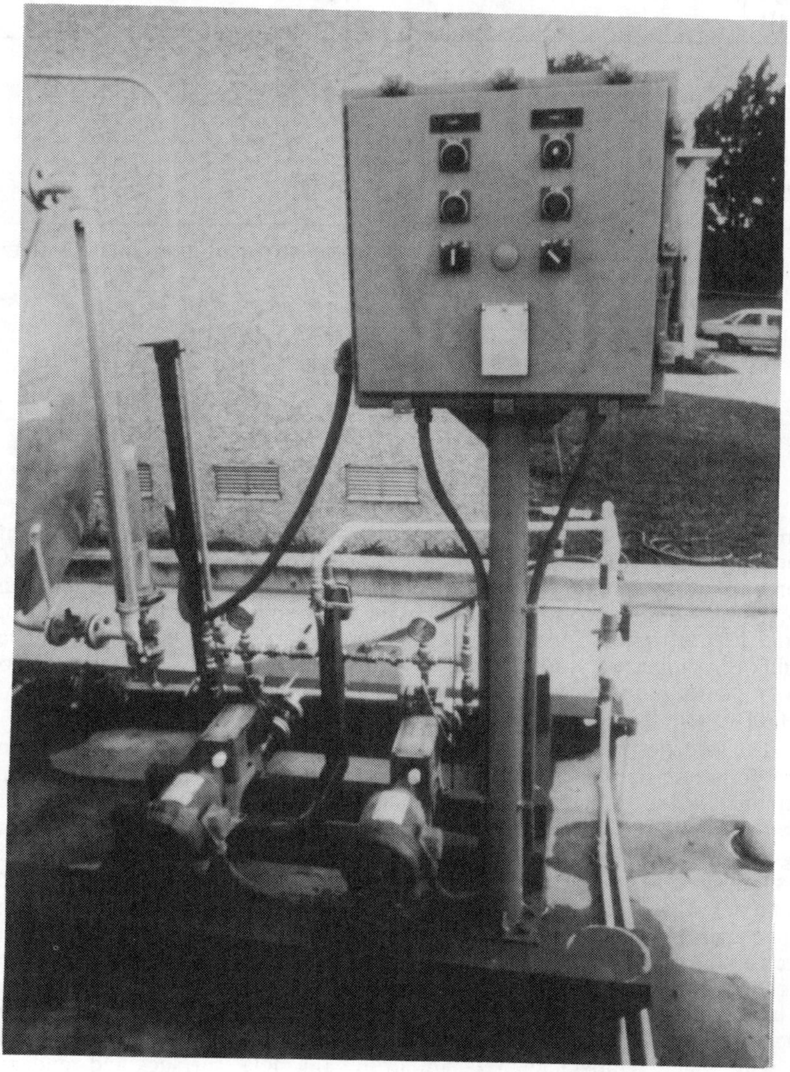

Figure 12-16. Hydrogen peroxide dosing pumps and controls, Palm Beach County, Florida.
(*Courtesy*: Los Angeles County Sanitation Districts)

TABLE 12-13. Typical Costs for Hydrogen Peroxide Injection for Sulfide Control (2)

Condition (1)	Capital cost, 1984 $ (2)	Annual chemical costs, 1984 $/yr (3)
Flow = 1.0 mgd Dissolved sulfide = 5 mg/L H_2O_2 dose = 25 mg/L	25,000	52,000
Flow = 10.0 mgd Dissolved sulfide = 5 mg/L	50,000	472,000

Note: Includes concrete pad, H_2O_2 storage tank, metering pumps, piping, valves, timer controls, safety equipment,and installation.
Dosage based on ratio 5:1.
Based on H_2O_2 cost of $0.68/lb as 100% H_2O_2 for 1 mgd flow.
Based on H_2O_2 cost of $0.62/lb as 100% H_2O_2 for 10 mgd flow.

2. Equipment

Equipment required for hydrogen peroxide addition is relatively simple. The two main components of a facility are a chemical storage tank and a chemical feed system. Chemical storage systems range in size from 55-gallon drums to 13,000-gallon bulk storage tanks. The feed system usually consists of chemical feed pumps and transfer and injection piping, along with pressure release valves. Occasionally, a water carrier system for the hydrogen peroxide may be necessary to enhance mixing of hydrogen peroxide and the sewage. Figures 12-14, 12- 15, and 12-16 are photographs of the components of a 1,000-gallon hydrogen peroxide storage and injection system in Palm Beach County, Florida.

The selection of construction materials in a hydrogen peroxide installation is crucial to assure product stability. Bulk storage tanks are typically of high purity aluminum alloy construction, while smaller drums are polyethylene lined steel. Aluminum, stainless steel, or PVC are used in piping. Diaphragm metering pumps are generally used and must be constructed of materials resistant to peroxide exposure, such as aluminum alloys and Teflon. All equipment used should be dedicated for hydrogen peroxide use exclusively, to avoid contamination. If hydrogen peroxide is contaminated, it will decompose and liberate large amounts of oxygen and steam. Contaminates include rust, dust, dirt, metals, and various organic materials. To minimize contamination, all equipment surfaces coming into contact with hydrogen peroxide must be passivated prior to use. Passivation, which consists of successive washings with a detergent, a solvent, water, nitric acid, followed by a rinse with water, removes catalytic impurities on metal surfaces and renders the surface inactive by the formation of an oxide coating.

Adequate safety precautions must be observed in handling hydrogen peroxide. Proper protective clothing and the availability of emergency eye wash and shower facilities are necessary for maintenance personnel.

TABLE 12-14. Hydrogen Peroxide Injection Scheme, Palm Beach County, Florida

Control pump station[a] (1)	Hydrogen peroxide storage site (2)	Hydrogen peroxide injection point[b] (3)
265	265	2,600 ft into 16-in. FM
EC 15	EC 195	3,600 ft into 10-in. FM
EC 204	System No. 2 Water Treatment Plant	3,600 ft into 30-in. FM
EC 227	EC 227	2,500 ft into 24-in. FM
EC 229	EC 56	4,900 ft into 30-in. FM
	EC 229	3,100 ft into 42-in. FM
EC 236	EC 236	2,500 ft into 42-in. FM
EC 241R	remote site at injection point	5,000 ft into a12 to 14-in. FM tributary to 30-in. gravity sewer
EC 241C	EC 241C	400 ft into 30-in. gravity sewer
ECR 241	ECR 241	2,300 ft into 18-in. FM tributary to 30-in. gravity sewer

[a] Sulfide and odor control is desired at the wet well.
[b] Distance upstream of the pump station in Col. 1.

3. Costs

Typical costs of hydrogen peroxide injection systems are shown in Table 12-13.

These costs include a concrete pad and containment berm, hydrogen peroxide storage tank, metering pumps, piping, valves, timer controls, safety equipment, and installation. Hydrogen peroxide suppliers will often provide a range of professional services including pilot studies, start-up services, and post construction monitoring. In addition, economically advantageous arrangements often can be made to lease the equipment.

4. Case History: Palm Beach County, Florida

In 1979, the Palm Beach County Water Utility Department began operating a regional sewage transmission system to serve the east central region of the county. The collection system consists of 42 miles (67.5 km) of force mains, ranging in diameter from 4 in. to 54 in. (10 to 137 cm), and 25 pump stations. The system is used to transport sewage to the East Central Regional Wastewater Treatment Plant operated by the City of West Palm Beach. This pump station-force main system is designed for an average flow of 50 mgd, but is currently handling only 15 mgd. Severe sulfide generation occurred in the transmission system, and numerous citizen complaints were generated over odors emitting from the pump station wet wells.

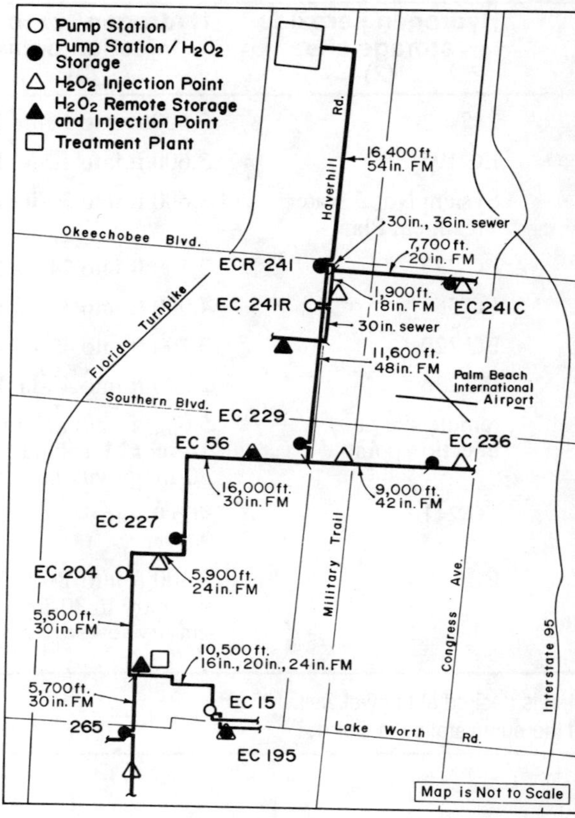

Figure 12-17. Hydrogen peroxide stations, Palm Beach County, Florida

Palm Beach County had 12 hydrogen peroxide dosing stations in operation in 1986 to combat the sulfide and odor problems. Each station injects hydrogen peroxide 20 to 40 minutes upstream of the pump station where sulfide and odor control is desired. The distance may vary from 400 ft to 5,100 ft (122 m to 1,555 m) upstream of the pump station.

A typical station consists of an aluminum or polyethylene storage tank and a dual pump system used to vary the dose rate depending on the time of day and time of year. Each pump is manually adjustable and operates at a constant dose rate. One pump is operated on a 24-hour timer to provide increased dose rates during selected periods each day. Each dosing system sits on its own concrete pad and is complete with automatic instrumentation panel and safety equipment. Figure 12-14 is a view of the west side of the ECR 241 Pump Station, showing the location of the hydrogen peroxide equipment. A closer view of the equipment is shown in Fig. 12-15. The discharge piping is visible leaving the concrete slab. The injection location is approximately 2,300 ft (700 m) upstream of the pump station into an 18-in. (46-cm)

TABLE 12-15. Performance of Hydrogen Peroxide in Palm Beach County, Florida

Control pump station (1)	Flow, mgd (2)	Uncontrolled total sulfide,[a] mg/L (3)	Hydrogen Peroxide Use lb/day[c] (4)	Hydrogen Peroxide Use mg/L[d] (5)	Controlled total sulfide,[e] mg/L (6)	Ratio, peroxide to total sulfide[f] (7)
265	1.0	2.0	350	21.0	0.5	14.0
EC 15	1.2	5.6	600	30.0	0.2	5.6
EC 204	2.3	5.5	450	11.7	0.5	2.3
EC 227	2.4	3.3[b]	240	6.0	0.2	1.9
EC 236	8.4	4.6[b]	900	6.4	0.3	1.5
EC 229	13.0	3.4[b]	1,040	4.8	0.4	1.6
EC 241R	0.8		230	17.2	0.2	
EC 241C	0.6		130	13.	<0.1	
ECR 241	14.2		340	1.4	0.4	

[a] Average total sulfide at the wet well of the pump station.

[b] Predicted total sulfide using Eq.4-2, with M = 1 x 10 M/hr, BOD = 150 mg/L, temperature = 82°F, and assuming dissolved sulfide is 90% of total sulfide.

[c] Average 50% hydrogen peroxide used from December 1985 through November 1986.

[d] 100% hydrogen peroxide.

[e] Average data from December 1985 through November 1986.

[f] Column 5 divided by the difference between Cols. 3 and 6.

tributary force main. The dual pump system and the control panel are shown in Fig. 12-16.

During the 12-month period from December 1985 through November 1986, the Water Utility Department used 1,812,000 lb (822,650 kg) of 50% hydrogen peroxide at a chemical cost of $530,000 ($0.29/lb). This is a daily use of 5,000 lb (2,270 kg)of 50% hydrogen peroxide, at a daily chemical cost of $1,450, to treat the 15 mgd of wastewater (330 lb/mgd and $100/mgd).

The layout of ten of the east county (EC) hydrogen peroxide stations is shown in Fig. 12-17. This portion of the system provides sulfide and odor control for the wet wells of nine pump stations listed in Table 12-14.

These ten hydrogen peroxide stations account for 86% of the hydrogen peroxide used by the Water Utility Department. The location of the hydrogen peroxide storage facility, the hydrogen peroxide injection points, and the pump stations are shown in Fig. 12-17 and detailed in Table 12-14.

Data were analyzed for the 12-month period from December 1985 through November 1986, and are summarized in Table 12-15.

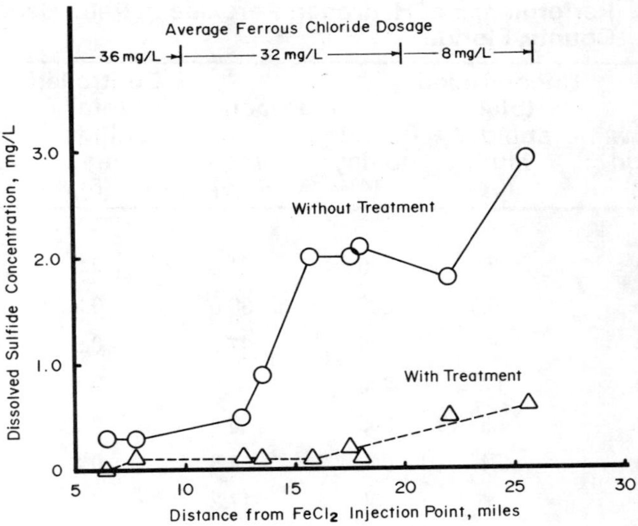

Figure 12-18. Effect of ferrous chloride on dissolved sulfide in joint outfall "B", Los Angeles, California

For each pump station at which sulfide control was desired, the amount of hydrogen peroxide used and the degree of control obtained were tabulated. Generally, the average total sulfide levels were maintained near 0.5 mg/L. Only limited data were available for sulfide levels prior to the initiation of the hydrogen peroxide dosing program. Where data existed, and where predicted levels could be calculated, estimates of the ratio of the hydrogen peroxide dose to the sulfide removed were made. The ratios, shown in Col. 7 of Table 12-15, varied from 1.5 to 14.0. This extreme differential reflects the difficulty in obtaining complete sulfide control, even with overdosing. The dissolved oxygen level at the wet wells of the nine pump stations varied between 1 and 4 mg/L, depending on the amount of turbulence in the wet well and the amount of hydrogen peroxide overdosing.

The first four dosing stations used were temporary, skid-mounted facilities leased from the chemical supplier. A monthly cost of $8,725 per pump station included 10,000 lb (4,540 kg) of 50% hydrogen peroxide, the equipment lease, start-up costs, and monthly monitoring efforts by the supplier. This generated a first year cost of $420,000. The Water Utility Department was able to move the portable equipment for testing throughout the collection system. Since then, permanent dosing equipment has been installed at 12 locations.

E. IRON SALTS

The salts of many metals will react with dissolved sulfide to form metallic sulfide precipitates, thus preventing hydrogen sulfide release to the atmosphere. For this method to be fully effective, the metal sulfide form must be highly insoluble. Copper, lead, silver, and zinc sulfide are extremely insoluble, although the three former salts are considered uneconomical. All four metals have been controlled for most industrial discharges to collection systems, and for treated wastewater discharges to the environment, and are not recommended.

TABLE 12-16. Performance of Ferrous Sulfate for Sulfide Control (2)

Location (1)	Average wastewater flow, mgd (2)	Average $FeSO_4$ dosage,[a] mg/L (3)	Dissolved sulfide, mg/L (4)
Clearwater, FL	5.0	0	6 to 8
		30	< 1
Naples, FL	5.5	0	5 to 20
		25	1 to 2
Boynton Beach, FL	12.0	0	10
		25	0.5 to 2.0

[a] $FeSO_4$ solution used was a proprietary product derived from the manufacturer of titanium dioxide, resulting in ferrous sulfate crystals.

Iron salts have been added to wastewater collection systems either as ferrous or ferric chloride or sulfate. Iron has generally been thought to become tightly bound to sulfide and other radicals in wastewater, thereby reducing its availability to form iron sulfide any distance downstream from the point of application. There have been applications, however, where the addition of iron has held dissolved sulfide concentrations to a few tenths of a milligram per liter for distances up to 20 miles (24). Ferrous and ferric chloride have also been used in wastewater treatment to control the hydrogen sulfide content of anaerobic digester gas. There can be concern about adding sulfate to the wastewater when using ferrous sulfate. If the generation of sulfide is depleting the available sulfate in the wastewater, then the addition of ferrous sulfate may allow more sulfide to be generated downstream. This may not be significant in most wastewater systems where sulfate concentrations are normally in excess of a reaction limiting level. However, in anaerobic digesters, the depletion of sulfate is more common, and the addition of sulfate is likely to result in increased hydrogen sulfide concentrations in the digester gas.

The literature reports that a mixture of iron salts containing one part of ferrous to two parts of ferric is superior for sulfide control compared to the reaction of either one alone (14). Full-scale testing on a large-diameter gravity sewer in Los Angeles, California, indicated that ferrous chloride alone was the most successful (24).

1. Design Requirements

Theoretically, the precipitation of one part by weight of hydrogen sulfide requires 4.5 parts by weight of ferrous sulfate and 3.8 parts by weight of ferrous chloride. The actual dosage requirements for a particular condition should be determined by field testing. The dosage required can be heavily dependent on the controlled sulfide level desired.

The County Sanitation Districts of Los Angeles County completed a full-scale evaluation of ferrous chloride to control sulfide levels in large gravity sewers in 1986 (24). The sewers that were tested are described in Table 12-12. The results of the study indicated: (1) Complete control of sulfide was difficult; (2) the dosage ratio necessary for 90% sulfide control was 20:1 if the dissolved sulfide level was greater than 1 mg/L, 20 to 60 mg/L if the dissolved sulfide level was less than 1 mg/L but

Figure 12-19. Ferrous chloride installation, Los Angeles, California. (Courtesy: Los Angeles County Sanitation Districts)

greater than 0.3 mg/L, and 200:1 if the dissolved sulfide level was less than 0.3 mg/L; (3) odor levels and hydrogen sulfide levels in the sewer gas space were reduced by 40% to 70%.

The effect of the addition of ferrous chloride in a 30-mile sewer in Los Angeles, California, is shown in Fig. 12-18. Uncontrolled dissolved sulfide reaches 3.0 mg/L at the end of the system. Currently, the addition of ferrous chloride is controlling the dissolved sulfide concentration in 20 miles of a gravity sewer carrying 200 mgd to a level of 0.5 mg/L or less, at a cost of $1,500 per day.

Results for ferrous sulfate control in Florida are summarized in Table 12-16.

The dose ratio of ferrous sulfate to dissolved sulfide can be calculated to range from 2:5 (which is lower than theoretical) to 5:1. Only partial control of sulfide occurs, and the residual sulfide levels are still sufficient to cause significant odor and corrosion problems.

2. Equipment

Ferrous sulfate and ferrous chloride are mildly corrosive acid solutions that require precautions for handling. The equipment required for an iron salt feed system includes a storage tank, chemical metering pumps, piping, valves, a control system, and a connection to the sewer system. Storage tanks are typically made from fiberglass because of the reduced cost. PVC piping is adequate. A concrete slab with a containment berm is necessary to provide emergency containment for any leakage. Figure 12-19 is a photograph of a typical installation; it is a dual storage tank installation. Ferrous solutions will stain most materials that they contact, so stain-resistant protective coatings should be specified if esthetics are of any concern.

TABLE 12-17. Typical Costs for Ferrous Sulfate and Ferrous Chloride Injection for Sulfide Control (2)

Condition (1)	Capital cost,[a] $ (2)	Annual $FeSO_4$ costs,[b] $/yr (3)	Annual $FeCl_2$ Costs,[c] $/yr (4)
Flow = 1.0 mgd Sulfide = 5 mg/L $FeSO_4$ dose = 30 mg/L $FeCl_2$ dose = 100 mg/L	10,000	170,000	40,000
Flow = 10 mgd Sulfide = 5 mg/L $FeSO_4$ dose = 30 mg/L $FeCl_2$ dose = 100 mg/L	21,000	1,700,000	400,000

[a] Includes concrete pad, storage tank, metering pumps, piping, valves, controls, and installation.

[b] Based on typical cost of $FeSO_4$ of $2.50/gal for solution containing 1.36 lb/gal $FeSO_4$.

[c] Based on typical cost of ferrous chloride of $0.13/lb anhydrous ferrous chloride in 32% solution containing 3.6 lb/gal anhydrous ferrous chloride.

3. Costs

Typical costs for ferrous sulfate and ferrous chloride injection systems, for the partial control of dissolved sulfide concentrations to a level between 0.5 and 1 mg/L, are shown in Table 12-17. The capital costs include a concrete pad, storage tank, metering pumps, piping valves, controls, and installation.

4. Case History: County Sanitation Districts of Los Angeles County, California

The Sanitation Districts' terminal treatment plant is the JWPCP (Joint Water Pollution Control Plant) located in Carson, California. The JWPCP is a 360-mgd advance primary treatment facility,with 200 mgd of high purity oxygen activated sludge secondary treatment. The JWPCP provides central solids processing for six satellite treatment plants and more than 1,100 miles of trunk sewers.

Anaerobic digestion of sludge is accomplished in 13 large circular digesters and 24 rectangular digesters. The circular digesters are approximately 500,000 cf each in volume. The rectangular digesters are approximately 100,000 cf each in volume. All the rectangular digesters, and one circular digester, process raw sludge from primary sedimentation. The 12 remaining circular digesters process a combination of raw sludge from primary sedimentation and thickened waste activated sludge from the high purity oxygen activated sludge secondary treatment system. Table 12-18 lists the operating parameters for the solids digestion system.

Beginning in 1982, air quality permits limited the hydrogen sulfide concentration in digester gas to an instantaneous level of 400 ppm. Until then, hydrogen sulfide concentrations in the digester gas had fluctuated between 200 ppm and 1,200 ppm. While many factors may have caused this fluctuation, the principal influence appeared to be the relationship between sulfur and heavy metals present

TABLE 12-18. Anaerobic Digestion Parameters, County Sanitation Districts of Los Angeles County Joint Water Pollution Control Plant [a]

Digestion parameter (1)	Primary sludge digesters (2)	Combined sludge digesters (3)
Number of digesters available		
rectangular	24	
circular	1	12
Normally in service		
rectangular	24	
circular	1	11
Being cleaned		
rectangular	3	
circular	1	
Active digester volume, cf	2,600,000	5,500,000
Primary sludge		
flow, gpd	1,200,000	2,100,000
solids, as percentage	3.9	3.9
dry solids loading, lb/day	390,000	690,000
volatile solids, percentage	75	75
Thickened waste activated sludge		
flow, gpd		290,000
solids, percentage		4.2
dry solids loading, lb/day		100,000
volatile solids, percentage		85
Daily volatile solids loading, lb/cf	0.11	0.11
Detention time, days	16.9	17.3
Volatile solids destruction, as percentage	54	51
Gas production		
daily volume, cf	2,300,000	5,300,000
volume per mass volatile solids destroyed, cf/lb	14.7	17.4
Average temperature, ^{o}F	94	94
Volatile acids, mg/L	30	45
Alkalinity, mg/L	2,400	3,000
Ferrous chloride usage	1,400	1,900

[a] Data for the week ending January 10, 1986.

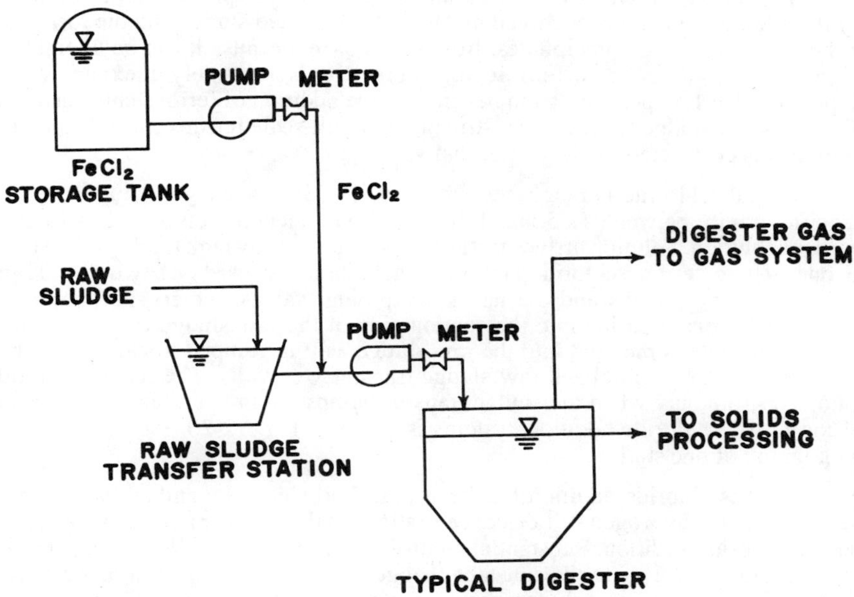

Figure 12-20 Schematic of ferrous chloride addition to anaerobic digesters, Los Angeles, California

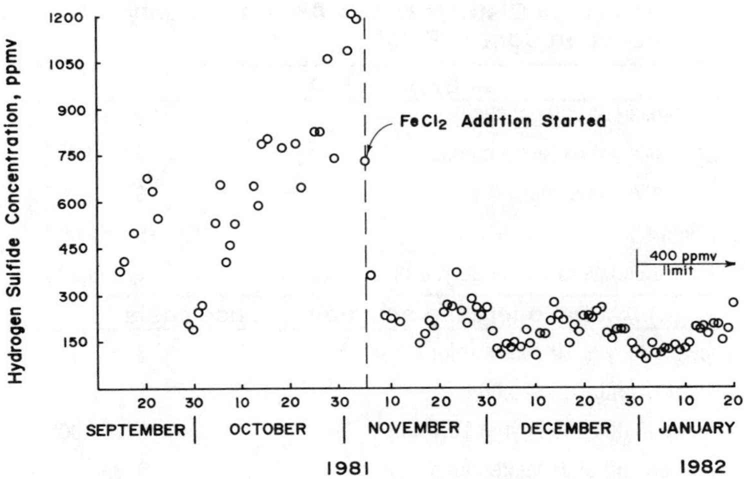

Figure 12-21. Digester gas hydrogen sulfide levels, Los Angeles,California

in the influent sewage. Increased industrial waste control had resulted in the general decrease in heavy metals influent to the JWPCP. As a result, control of the hydrogen sulfide concentrations in the digester gas was necessary.

The release of hydrogen sulfide into the digester gas phase can be controlled by the addition of certain dissolved metals to the digested sludge and the formation of heavy metal sulfide precipitates. Iron was chosen, because it can be tolerated in high concentrations in anaerobic digesters and is relatively inexpensive. In September 1981, experimentation began with the addition of ferrous chloride to the digester's raw sludge feed. The construction of a full-scale ferrous chloride addition system was completed by the end of that year.

Ferrous chloride is purchased as a 27% to 35% by weight solution, with a specific gravity between 1.35 and 1.40. The iron concentration averages 190,000 mg/L of solution. Solution arrives in 4,000-gal shipments by tank truck, and is stored at one of two ferrous chloride stations. Each station consists of two 10,000-gal fiberglass storage tanks and the necessary piping, valves, meters, and pumps to inject the ferrous chloride into the suction side of the raw sludge transfer pumps. Ferrous chloride is metered into the raw sludge as it is pumped to each individual digester from two centralized raw sludge transfer wet wells. The ferrous chloride pumps operate only when the sludge transfer pumps are in operation. A schematic of the ferrous chloride addition systems is given in Fig. 12-20. Figure 12-19 is a photograph of one station.

Ferrous chloride addition has been practiced since November 1981. Figure 12-21 shows the hydrogen sulfide concentrations in the digester gas, before and after ferrous chloride addition. Substantial control of the hydrogen sulfide concentration has been obtained. Due to the constant change in concentrations of iron and sulfate in the influent sewage and sludge, the dosage of ferrous chloride is adjusted daily if necessary. The ferrous chloride addition system is, unfortunately, slow to respond to sudden changes of hydrogen sulfide in the digester gas because of the detention time in the digestion system. This results in overdosing to guarantee meeting air permit requirements.

TABLE 12-19. Cost Of Ferrous Chloride Addition System, County Sanitation Districts of Los Angeles County Joint Water Pollution Control Plant

(a) Capital cost	
1. Four storage tanks (10,000 gal each)	$ 48,000
2. Two corrosion resistant metering pumps	$ 8,000
3. Piping and miscellaneous costs	$ 8,000
4. Total capital costs	$ 64,000
5. Annualized capital costs (8% over 20 years)	$ 6,500
(b) Annual operating and maintenance costs	
1. Electrical power (3 kw at $0.75/kw-hr for 4 hr/day)	$ 300
2. Operating labor (1 hr/day at $20.00/hr)	$ 7,300
3. Ferrous chloride (3,000,000 lb/yr at $0.13/lb)	$ 390,000
4. Total annual operating and maintenance costs	$ 397,600
Total Annual Costs	**$ 404,100**

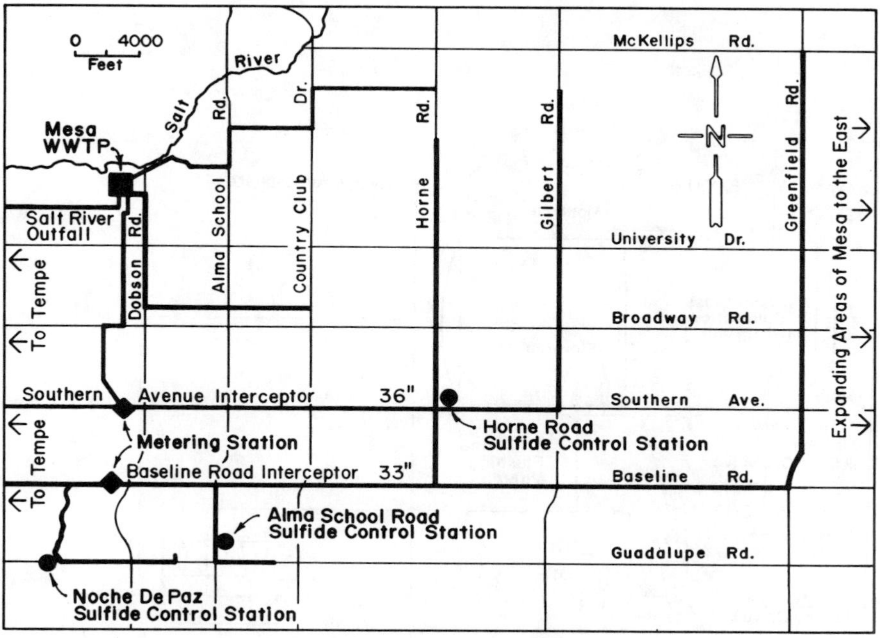

Figure 12-22. Ferrous chloride sulfide control stations and major sewers, Mesa, Arizona (9)

Usage has ranged from zero, during periods of low hydrogen sulfide concentration in the digester gas, to over 3,000 gpd. The average daily use, since 1984, has been 2,200 gal (8,525 L) of solution, which is 8,000 lb (3,632 kg) of anhydrous ferrous chloride, or 3,500 lb (1,590 kg) of iron. The yearly consumption is nearly 3,000,000 lb (1,362,000 kg) of anhydrous ferrous chloride. The price for ferrous chloride has been approximately $0.13 per pound of anhydrous ferrous chloride. The capital annual operating and maintenance costs for the ferrous chloride addition system are summarized in Table 12-19. The total annual cost is about $400,000.

The anaerobic digesters have shown no noticeable change in alkalinity, pH, gas production, or volatile acid concentration with the addition of ferrous chloride. The iron concentration in the digesters has averaged 600 mg/L. Laboratory-scale pilot digesters operated at ferrous chloride dosages 6 times the normal dose level showed no stress, with alkalinity holding at 2,000 mg/L. Testing at ferrous chloride doses 15 times the normal dose produced failure in the pilot digesters.

5. Case History: Mesa, Arizona

The City of Mesa, Arizona, is at the eastern upstream end of the regional wastewater system that serves the Phoenix metropolitan area. Long regional interceptors and high temperatures have made the Mesa collection system particularly vulnerable to sulfide generation. As part of the regional operating agreement, the City of Mesa has to control the dissolved sulfide concentration to 0.5 mg/L where the regional interceptors enter the City of Tempe. Figure 12-22 outlines the major sewer lines in the Mesa sewerage system. The Southern Avenue

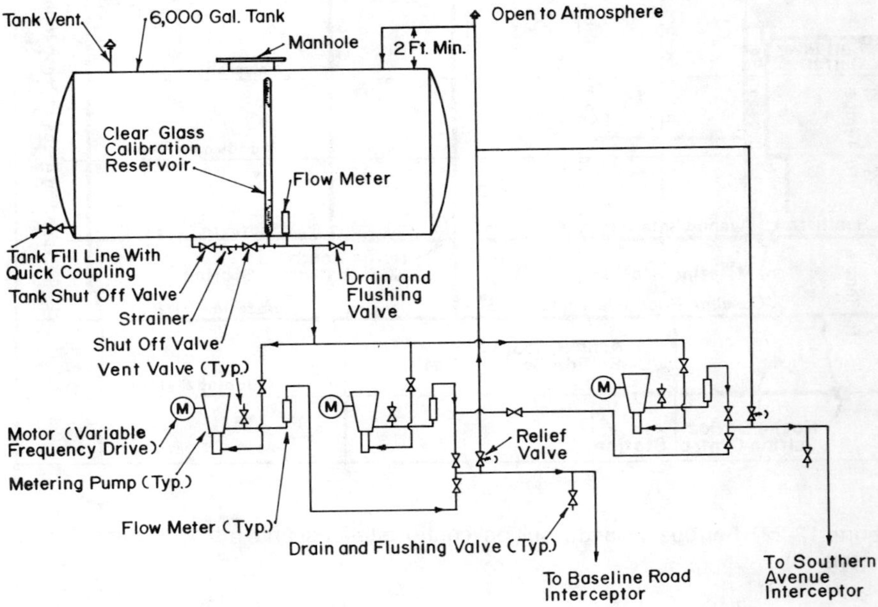

Figure 12-23. Schematic of ferrous chloride feed system, Mesa, Arizona (10)

Interceptor and the Baseline Road Interceptor are portions of the regional system that must receive sulfide control. Sulfide levels as high as 5 mg/L have been measured in both of these regional interceptors without sulfide control (9).

The City of Mesa placed the Horne Road Sulfide Control Station in operation in December 1985. Ferrous chloride is purchased as a 35% to 40% solution. It is shipped by 4,000-gal tank truck and stored in two, 3,000 gal each, cross-connected, fiberglass tanks. The station is situated so that ferrous chloride can be fed into both the Baseline Road Interceptor and the Southern Avenue Interceptor (10).

The ferrous chloride is injected with metering diaphragm pumps. The station has three pumps, one for each interceptor and one for standby. The chemical is fed directly to each sewer through a 1-in. plastic pipe. A schematic of this equipment arrangement is shown in Fig. 12-23. The pumps are designed to feed at twice the expected rates, and can also feed at rates below the minimum requirements (during the winter months).

TABLE 12-20. Ferrous Chloride Data, January Through August 1986, Mesa, Arizona (10)

Parameter (1)	Baseline Road Interceptor (2)	Southern Avenue Interceptor (3)
Flow, mgd		
minimum	3.9	10.0
average	8.3	12.1
peak	11.4	14.2
Metering station dissolved sulfide, mg/L:		
without treatment		
average	1.8	1.0
range	1.2 - 3.5	0.7 - 1.3
with treatment		
average	0.2	0.4
range	0.1 - 0.6	0.0 - 0.7
Ferrous chloride usage		
gpd	180.0	50.0
lb/day	840	240
mg/L	12	2.4
$/day	100	29

The ferrous chloride feed pumps have two modes for adjusting the feed rate: (1) The motor speed can be varied to change the number of plunger strokes per minute; and (2) the length of stroke can be manually adjusted. Timers automatically adjust the ferrous chloride feed rate by varying the speed of the motor. The timers are preset to adjust the speed of the motor at various times of the day as the sewage flow varies. Twice a year, to compensate for differences in summer and winter conditions, the length of stroke of each pump is manually adjusted. This adjustment reduces or increases the average feed rate; the automatic timer is then readjusted to closely approximate the diurnal fluctuation of the the sewage flow (10).

Dissolved sulfide measurements are made several times each week at the metering stations on the Baseline Road Interceptor and the Southern Avenue Interceptor. Data collected from the Horne Road Sulfide Control Station are presented in Table 12-20.

Since January 1, 1986, 95% control of dissolved sulfide has been provided.For the Baseline Road Interceptor, a dose of 180 gpd of ferrous chloride, or 12 mg/L, has controlled the dissolved sulfide concentration at the metering point to 0.2 mg/L. For the Southern Avenue Interceptor, a ferrous chloride dose of 50 gpd, or only 2.4 mg/L, has controlled the dissolved sulfide concentration to 0.4 mg/L at the metering station.

The Horne Road Sulfide Control Station cost $300,000 to construct. Its daily chemical cost is $130, and the power cost is $2.50 per day for the 1/2-horsepower pumps. The only continuing labor costs are to unload chemical shipments and to perform weekly monitoring of the feed rates and the sulfide concentrations. Two additional stations were constructed in March 1987 at a cost of $200,000: the Alma School Road Station and the Noche De Paz Station. These two new stations are located on the Baseline Road Interceptor. Operation of the three stations has been

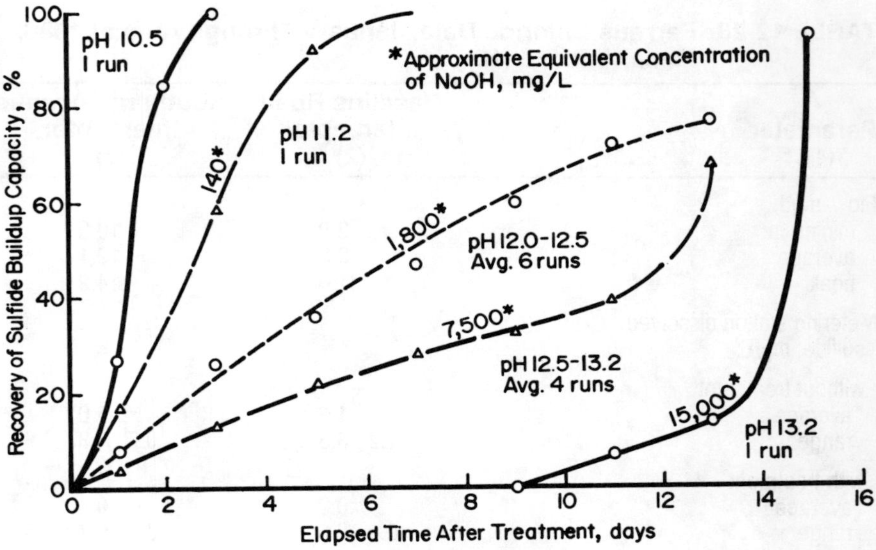

Figure 12-24. Impact of shock dosing with NaOH on recovery of sulfide build-up capacity, Los Angeles, California (16)

relatively trouble-free and has maintained dissolved sulfide below the 0.5 mg/L limit (10).

F. SHOCK DOSING WITH SODIUM HYDROXIDE

Sodium hydroxide, commonly known as caustic soda, is used in many applications including the chemical, pulp and paper, textiles, and petroleum industries. It is also used in the sanitary field to control sulfide in sewers. Caustic is typically made by the electrolysis of sodium chloride brine, which produces chlorine and hydrogen in addition to caustic. The caustic is usually concentrated to a 50% solution by removing water in multiple effect evaporators (11).

The proportion of dissolved sulfide in sewage that exists as hydrogen sulfide is dependent on pH of the wastewater as discussed in Chap. 2. At a pH of 7, equal concentrations of dissolved hydrogen sulfide and hydrosulfide ion exist at equilibrium, while at a pH of 8 only about 10% of the dissolved sulfide exists as hydrogen sulfide. The continuous addition of sodium hydroxide, or other strong alkalies, to maintain a high pH could reduce odors and corrosion, but this is generally not practical in most collection systems. Tributary flows and the production of carbon dioxide and organic acids from biological action tend to lower the pH. A decrease of only 0.5 pH units could result in a substantial increase in hydrogen sulfide in the sewer atmosphere (15).

The purpose of shock dosing with sodium hydroxide is to raise the pH of a slug of sewage to a high enough level to inactivate the bacteria in the slime layer which reduce sulfate to sulfide. The caustic does not oxidize or otherwise destroy the sulfide that is already in the sewage. The caustic is not added continuously, but for only a short period of time, usually 30 minutes. While raising the pH does shift almost all of the dissolved sulfide to the hydrosulfide ion fraction in the slug of sewage, that is not the purpose of the shock dosing. The sole purpose of shock dosing

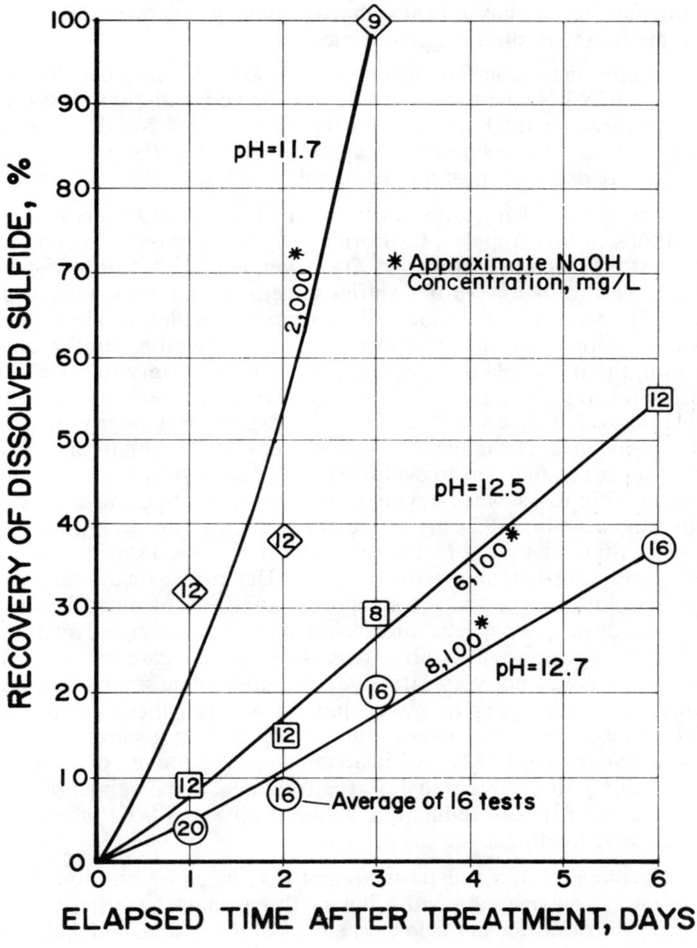

Figure 12-25. Impact of shock dosing with NaOH on recovery of sulfide in large sewers, Los Angeles, California

with caustic is to stop the reduction of sulfate to sulfide by the bacteria in the slime layer.

1. Design Requirements

Shock dosing is accomplished by discharging the caustic into the sewer at a constant rate for approximately 30 minutes. The amount of caustic discharged over this period is determined by calculating the volume of sodium hydroxide necessary to raise the pH of the 30-minute slug of wastewater at the farthest downstream portion of the sewer to be treated to a pH between 12.5 and 12.8. This amount of caustic can be discharged anywhere in the collection system upstream of the point where treatment is desired. The caustic slug will spread and dilute as it moves downstream. Field testing has indicated that overdosing may be necessary to achieve the desired pH for the desired length of time.

The County Sanitation Districts of Los Angeles County use 200 to 250 gal (750 to 945 L) of 25% NaOH per mgd of flow for sewers with flows less than 5 mgd. Larger sewage flows are treated with 150 gal (568 L) of 50% NaOH per mgd of flow, and the resulting NaOH concentration is approximately 6,000 mg/L. Results of field pH monitoring are used to adjust the treatment dose.

Testing of shock dosing with caustic in small-diameter sewers was conducted in the late 1960s in Los Angeles, California (16). The testing was conducted on a 4,600-ft (1,400-m) reach of a 24-in. gravity sewer, and a 1,400-ft (425-m) reach of a 10-in. diameter force main. Normal sulfide generation in both test reaches was less than 1 mg/L. The recovery of sulfide build-up capacity with time after shock dosing with caustic is graphically indicated in Fig. 12-24. Depending on the caustic dose and, therefore, the initial pH achieved, the times for recovery to 100% of normal sulfide build-up varied from several days to two weeks. Experiments in 1985 in Los Angeles (11) revealed that shock dosing with caustic was nearly as effective in controlling sulfide in large-diameter sewers as it has been in small- diameter sewers. The objective of the testing was to evaluate the degree of sulfide control in a 126-in. (320-cm) semi- elliptical sewer carrying peak flows of 60 mgd and dissolved sulfide of 2.5 mg/L, and a 72-in. (183-cm) diameter sewer carrying peak flows of 45 mgd and dissolved sulfide of 1.0 mg/L. The recovery of dissolved sulfide with time after shock dosing with caustic is shown in Fig. 12-25. Depending on the caustic dose and the initial pH achieved, the times for recovery to 100% of normal sulfide levels varied from two days to two weeks. Increasing the pH of the caustic treatment above 12.6 to 12.7 did not improve its effectiveness. Caustic treatment with pH below approximately 12.2 was not very effective. The effectiveness of caustic treatment varied somewhat from sewer to sewer, but, in general, these conclusions were reached: (1) Caustic treatment is not able to stop sulfide generation or eliminate dissolved sulfide completely; (2) caustic treatment at and above a pH of 12.6 reduces the sulfide levels by about 95% one day after treatment and by about 90% two days after treatment; and (3) odor reductions between 80% and 90% lasted for several days after a caustic treatment at a pH of 12.6.

The effectiveness of a caustic treatment is reduced by dilution as the shock dose moves farther downstream and tributary flows enter the system. Designing a shock dose to control sulfide in the lower reaches of the system and then performing the treatment in the upstream reaches of the collection system can extend the treatment effects throughout the collection system. In the Los Angeles testing, a total of 14 miles (22.5 km) of 33-in. diameter to 126-in. semi-elliptical sewer was treated with each shock dose treatment.

When tributary reaches of a collection system are treated, sufficient dilution of the high pH slug of sewage is normally achieved before entrance to any sewage treatment plant. Treatment of the major trunks and interceptor sewers tributary to a treatment facility results in a high pH slug moving through the facility. The Long Beach Water Reclamation Plant in California has a regular program of caustic treatment in its collection system. Although the entire collection system is not treated simultaneously, the influent pH of the treatment plant does reach a level of 11.8. Monitoring the pH through the activated sludge portion of this tertiary treatment plant has shown pH levels as high as 9 at the feed points in the aeration tanks, and 8 in the secondary effluent. The only noticeable effect upon treatment is a temporary increase in the final effluent turbidity of approximately two turbidity units (from 1.5 to 3.4 turbidity units) for a period not exceeding 6 hours after the caustic treatment. If the pH in any biological process cannot be raised above 9, or if the discharge pH limitations to the environment could be affected, then special handling of the shock dose may be necessary. The high pH slug of sewage could be stored in an out- of-service process unit for slow feeding back into the process, or pH adjustment may be necessary. The addition of chemicals to adjust pH could be economically prohibitive.

2. Equipment

Caustic is not flammable or combustible, but contact with water or acids may generate sufficient heat to ignite nearby combustible materials. Contact with certain metals, such as aluminum, tin, or zinc, will evolve flammable and combustible hydrogen gas. Caustic is corrosive to any tissue and can cause severe burns. Proper procedures, equipment, and clothing should be used to prevent ingestion, inhalation, or contact with skin or eyes. Emergency showers and eye wash facilities are essential.

Iron or steel equipment is generally satisfactory for the handling and storage of 50% caustic at moderate temperatures because of the very low corrosion rate (approximately 0.01 in./yr). At temperatures above 100°F, (37.7°C), more resistant materials, such as nickel, or plastic lined steel, should be used. Aluminum, tin, zinc, brass, bronze, and copper are not suitable for use with caustic.

If storage of 50% caustic is necessary, steps must be taken to guard against low temperature crystallization, which can occur at temperatures below 50°F (10°C) in unheated storage tanks. If tanks cannot be heated, then water can be used to dilute the solution to a concentration that will not crystallize. The manufacturer should be consulted for information on how to prevent crystallization.

The supplier of caustic would normally use vacuum trucks to fill the storage tank sites in the collection system or to directly perform the caustic treatment at the designated locations. Rental of vacuum trucks to perform treatment may be required. Some agencies purchase vacuum trucks to perform treatment themselves.

If storage tanks are constructed, steel or fiberglass materials are usually adequate. Containment is necessary because sodium hydroxide is considered a

TABLE 12-21. Typical Costs for Shock Dosing with Sodium Hydroxide

Flow, mgd (1)	Capital cost,[a] $ (2)	Annual chemical cost,[b] 1986$ (3)	Annual truck rental cost,[c] $ (4)
1.0	14,000	4,000	11,400
10.0	36,000	28,000	11,400

[a] Includes concrete pad, containment, fiberglass storage tank, piping, valves, meter, and the labor for construction. Storage tank sized for four treatments, or a minimum of 2,000 gal (7,570 L).

[b] Based on weekly usage of 150 gal (568 L) of 50% of NaOH per mgd at $0.48/gal.

[c] Rental of vacuum truck and driver to perform weekly treatment instead of constructing a storage site. $55 per hr; minimum 4-hr charge.

corrosive material. Pneumatic systems can be used for unloading and loading trucks. If chemical feed pumps are used, they must be suitable for caustic service. Gravity treatment can be performed at the treatment location, but the flow rate should be regulated with a rotameter and a control valve. Stainless steel is often used for pumping, metering, and control.

3. Costs

An estimate of the typical cost for the installation of a single caustic storage tank system is shown in Table 12-21.

The capital costs include the concrete pad and containment, the storage tank, piping, valves, meter, and the installation. Column 4 of Table 12-21 shows costs for the use of rental vacuum trucks instead of the construction of tank storage facilities. Arrangements can usually be made with the chemical supplier to deliver and inject the sodium hydroxide at a price per gallon that would provide substantial savings over the cost to rent trucks to haul and inject the solution.

4. Case History: County Sanitation Districts of Los Angeles County, California

Shock dosing with sodium hydroxide (caustic) has been the principal method of sulfide control used by the County Sanitation Districts of Los Angeles County for the last 25 years. Initially, the program utilized a dilute, waste caustic solution from industries to treat local trouble spots in the small- diameter portions of the collection system (sewers less than 36 in. in diameter, with flows generally less than 2 mgd). The loss of waste caustic solutions, uncontaminated by heavy metal concentrations, made it necessary to purchase concentrated commercial caustic.

Caustic is purchased as a 50% by weight solution of sodium hydroxide (NaOH) in water. There are 12.9 lb/gal in a 50% NaOH solution. Payment is based

Figure 12-26. Shock dosing with NaOH, Los Angeles, California

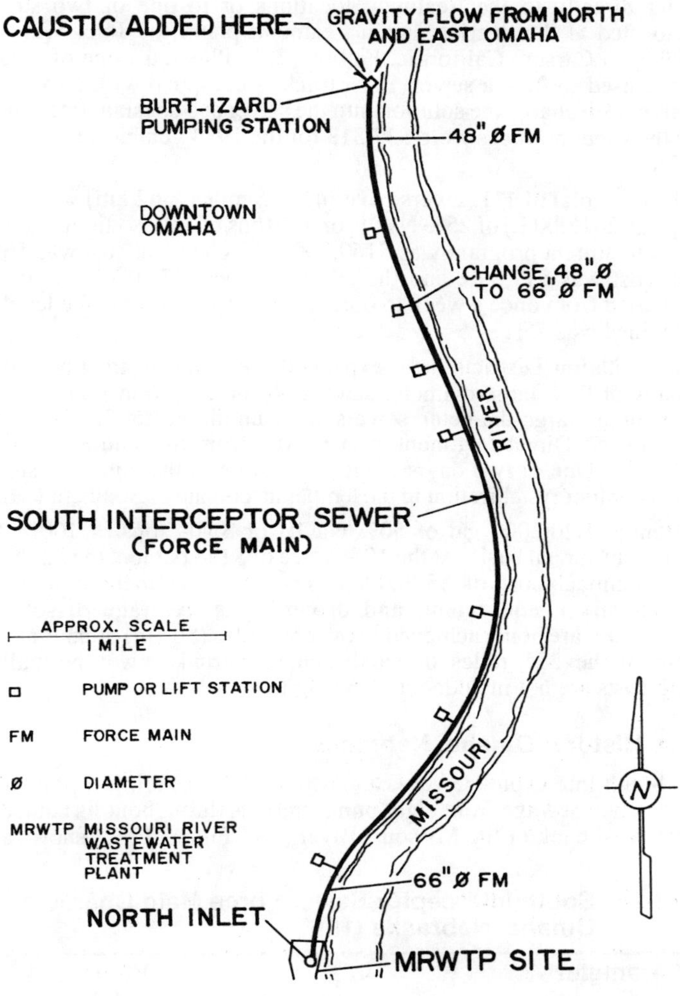

Figure 12-27. Shock dosing with NaOH, South Interceptor Sewer, Omaha, Nebraska (18)

on the tons of 76% sodium oxide used to manufacture the solution (same as tons of 100% NaOH). Prices have varied between \$100 and \$270 per ton during the last ten years. The 1986 price was \$148 per ton, or \$0.48 per gallon of 50% NaOH.

To guard against low temperature crystallization of the 50% solution, which can occur at temperatures below 50°F (10°C) in the unheated storage tanks, water is used to dilute the solution to 25% by weight NaOH. There are 10.5 lb/gal in a 25% NaOH solution. At the 1986 price of \$148 per ton of 76% sodium oxide, the cost for the 25% NaOH solution is \$0.19 per gallon.

Two of the districts' vacuum trucks are used to transport the 50% NaOH from the supplier directly to the treatment locations or to one of two storage tanks centrally located at the districts' main treatment plant, the Joint Water Pollution Control Plant in Carson, California. Figure 12-26 illustrates one of these vacuum trucks being used to treat a sewer. Each truck is equipped with a rotameter and a control valve to discharge the solution into the sewer at a constant rate. The purchase prices for the vacuum trucks were $37,313 for the 1974 vehicle and $81,281 for the 1983 vehicle.

In 1984, a total of 111 sewers covering 225 miles (362 km) were treated with 328,000 gal (1,241,000 L) of 25% NaOH or 430 tons of 76% sodium oxide. The total cost for this treatment program was $160,000. The chemical cost was $69,000; the equipment cost was $17,000; and the labor cost was $74,000. The frequency of treatment varied from once a week to only once a year. The average level of sulfide control obtained was 75%.

The Sanitation Districts have expanded their caustic treatment program to include many of their large-diameter sewers (48 in. to 126 in.) on a weekly basis. The treatment of large-diameter sewers uses undiluted, 50% NaOH to reduce hauling volumes. Direct treatment after pickup from the solution supplier is the normal practice. One or two day storage is possible in the unheated storage tanks without concern for crystallization in the temperate climate of Southern California.

Currently, 1,200,000 gal of 50% NaOH or 3,900 tons of 100% NaOH are being used on an annual basis. At the 1986 price of $148 per ton, the 3,900 tons make an annual chemical cost of $570,000. The total treatment cost is $700,000, including chemical, equipment, and driver costs. Average dissolved sulfide reductions of 70% are being achieved in over 50 miles (80 km) of major trunk sewer, in addition to the 225 miles of small-diameter trunk sewer normally treated. Monitoring costs are not included in these figures.

5. Case History: Omaha, Nebraska

The South Interceptor sewer is a 4.5-mile (7.2-km) reinforced concrete force main that receives sewage from eight pump stations throughout its course along the levee of the west bank of the Missouri River (18). Figure 12-27 shows a layout of

TABLE 12-22. South Interceptor Sewer Force Main Characteristics, Omaha, Nebraska (18)

Parameter (1)	Value (2)
Length, miles	4.5
Diameter, in.	48 and 66
Average flow, mgd	
daily	18
low	11
peak	26
Average velocity of flow, fps	1.2
Average wastewater 5-day BOD, mg/L	250
Typical wastewater pH	7.0 to 7.4
Typical wastewater temperature, ^{o}F	72 to 77

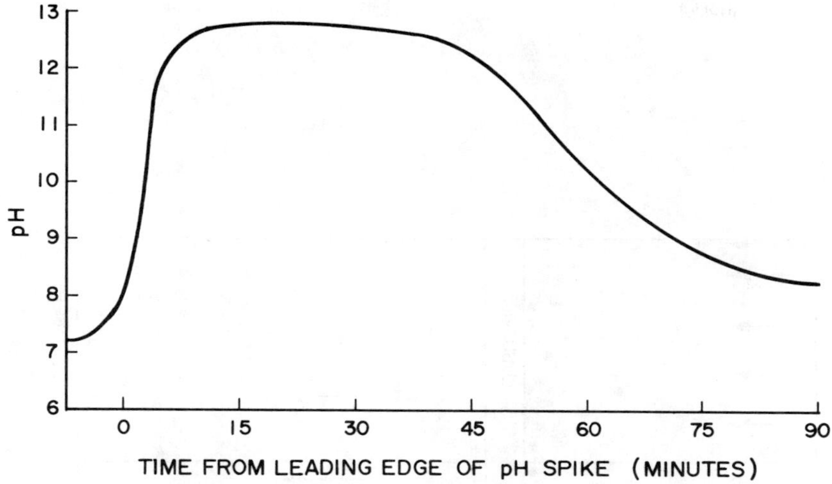

Figure 12-28. Typical influent pH profile during shock dosing with NaOH, Omaha, Nebraska (18)

the South Interceptor sewer from the Burt-Izard pump station to the North Inlet at the Missouri River Wastewater Treatment Plant. This long, flat force main, which varies in diameter between 48 and 66 in., receives wastewater from combined sewers in the central business district of Omaha, as well as from several industries. Sewer and wastewater characteristics are summarized in Table 12-22.

Predicted sulfide generation can be as high as 1.5 mg/L per hour, or more than 7 mg/L during the 5-hr transit time. The input of aerated sewage from the tributary

TABLE 12-23. Caustic Treatments in 1985 in South Interceptor Sewer, Omaha, Nebraska (18)

Month (1)	Number of caustic treatments (2)
March	0
April	0
May	1
June	2
July	5
August	5
September	2
October	1
November	1
Total	17

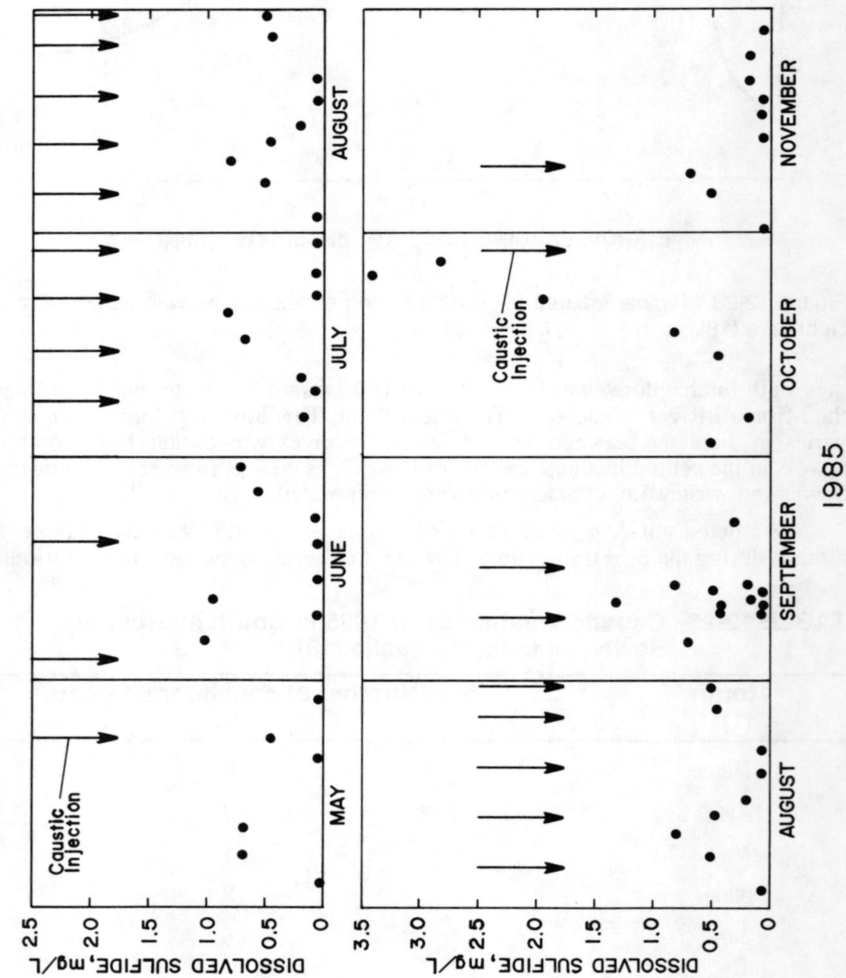

Figure 12-29. Dissolved sulfide levels, south interceptor sewer, Omaha, Nebraska (18)

pump stations is probably responsible for maintaining sulfide below 5 mg/L. Nevertheless, the resulting release of hydrogen sulfide at the headworks and the primary clarifiers has caused significant corrosion, odor, and safety problems at the treatment plant.

Shock dosing with sodium hydroxide was tested in 1984 and found to be beneficial. Full-scale caustic treatment was conducted during the 1985 warm weather season. Caustic was purchased in tank truck loads from commercial suppliers. A typical load is 3,500 gal (13,250 L) of 50% NaOH solution. The entire load is discharged to the wet well of the Burt-Izard pump station over a 20-minute interval. Shock dosing is usually performed during low-flow conditions, approximately midnight, so that the resulting high pH slug arrives at the treatment plant at approximately 5:00 a.m. A typical pH profile at the North Inlet of the treatment plant is shown in Fig. 12-28. This high pH slug, which exceeds pH 12.5, is diverted into two spare primary clarifiers to avoid raising the pH above 9 in the biological portions of the plant. The stored high pH wastewater is recycled for treatment over the following one to two days.

Caustic treatment was conducted 17 times during 1985. Table 12-23 summarizes the number of monthly caustic treatments during 1985.

One objective of the caustic treatment program was to maintain dissolved sulfide concentrations below 1 mg/L at the North Inlet. The variation in dissolved sulfide concentration between May and November 1985 is shown in Fig. 12-29, and each caustic treatment is also indicated. During periods when caustic treatment was conducted every week or two, the dissolved sulfide concentration at the North Inlet averaged 0.4 mg/L, with peaks of almost 1 mg/L. During a six-week period from mid-September to the end of October, when no caustic treatment was conducted, dissolved sulfide concentrations rose to about 3 mg/L. The low dissolved sulfide concentrations at the North Inlet in late September and early October are the result of rain storms.

Caustic purchases are the primary cost for this sulfide control technique. For 1985, the City of Omaha spent $44,000 for approximately 60,000 gal of 50% NaOH solution or $0.74 per gallon delivered and injected. Weekly treatment cost was $2,600. Other costs, which have not been estimated, included the city's staff labor and monitoring expenses.

The City of Omaha continued its caustic treatment program on this interceptor in 1986 and 1987,and intends to compare caustic treatment to chlorination when a proposed chlorination system is constructed.

6. Case History: Cedar Rapids, Iowa

The Main Interceptor in Cedar Rapids transports essentially all the city's sanitary wastewater to the Cedar Rapids Water Pollution Control Facilities (18). The layout of this 5.9-mile, 84- and 96-in. diameter gravity sewer is shown in Fig. 12-30. Sewer and wastewater characteristics are summarized in Table 12-24.

Several large corn processing industries discharge wastewater to this system, so most of the BOD and suspended solids loads are industrial in origin. Much of the BOD is highly soluble. The wastewater pH is consistently below 7, and the presence of high sulfide concentrations, along with high dissolved oxygen levels, indicates inhibition of oxidation in the sewage. The release of hydrogen sulfide at the headworks and primary clarifiers has caused significant odor and safety problems at the treatment facility. Corrosion in the unprotected reinforced concrete sewer is being investigated.

Shock dosing with NaOH was tested in early 1985 and implemented full scale shortly thereafter. Caustic is purchased in tank truck loads from commercial

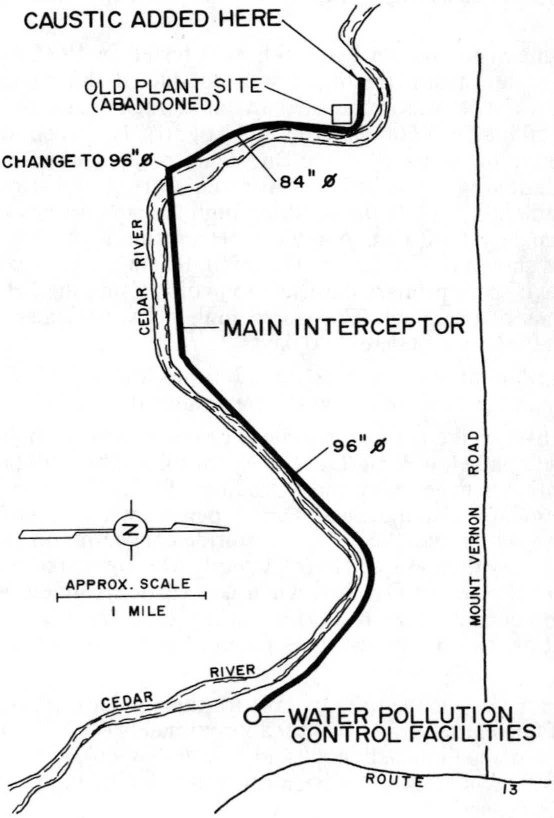

Figure 12-30. Shock dosing with NaOH, main interceptor, Cedar Rapids, Iowa (18)

suppliers. A typical load is 3,500 gal of 50% NaOH solution. The entire load is injected into the Main Interceptor over a 20-minute time period at 8:00 a.m. A typical pH profile of the sewage when it reaches the treatment facilities, by midday, is illustrated in Fig. 12-31. The pH generally reaches a peak of approximately 12.5. This high pH slug is diverted into a spare primary clarifier, where it is stored and recycled slowly to the treatment facilities over the following one to two days.

Caustic treatment was conducted 11 times in the Main Interceptor in 1985. Table 12-25 summarizes the number of caustic treatments per month.

One objective was to keep dissolved sulfide concentrations below 1 mg/L at the influent to the treatment facilities. Figure 12-32 shows the variation of dissolved sulfide concentration at the treatment facility, as well as the caustic treatments between May and November 1985. Dissolved sulfide concentrations increased to approximately 2 mg/L about two weeks following each treatment. Caustic treatment

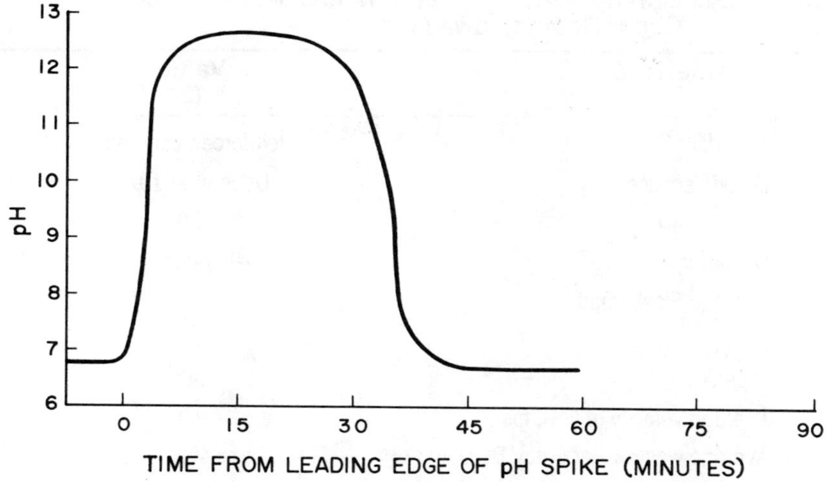

Figure 12-31. Typical influent pH profile during shock dosing with NaOH, Cedar Rapids, Iowa (18)

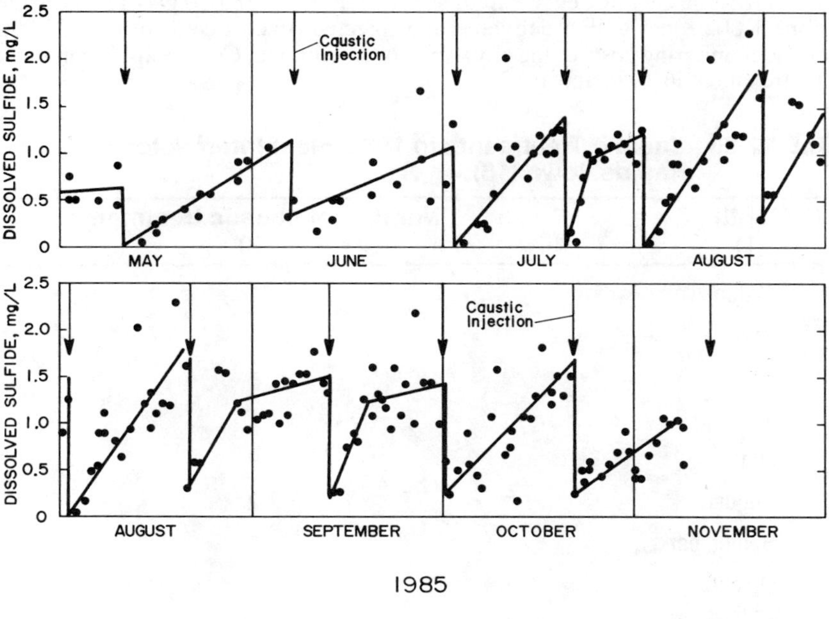

Figure 12-32. Dissolved sulfide levels, main interceptor, Cedar Rapids, Iowa (18)

TABLE 12-24. Gravity Sewer Characteristics, Main Interceptor, Cedar Rapids, Iowa (18)

Parameter (1)	Value (2)
Material	reinforced concrete
Date in service	December 1979
Length, miles	5.9
Diameter, in.	84 and 96
Average Flow, mgd	
daily	30
low	17
peak	36
Average velocity of flow, fps	2.9
Average wastewater 5-day BOD, mg/L	475
Typical wastewater pH	6.5 to 6.9
Typical wastewater temperature, ^{o}F	72 to 79

every two to three weeks maintained dissolved sulfide concentrations near 1 mg/L, with peaks of about 2 mg/L, throughout the summer and early fall.

In 1985, the City of Cedar Rapids spent $22,000 for 38,000 gal of 50% NaOH solution or $0.58 per gallon delivered and injected. Each treatment cost $2,000. Labor and monitoring costs of the city were not determined. Cedar Rapids continued caustic treatment in 1986 and 1987.

TABLE 12-25. Caustic Treatments in 1985, Main Interceptor, Cedar Rapids, Iowa (18)

Month (1)	Number of caustic treatments (2)
March	1
April	0
May	1
June	1
July	2
August	2
September	2
October	1
November	1
Total	11

G. POTASSIUM PERMANGANATE

Potassium permanganate (K_2MnO_4) is a strong oxidizing agent that has many industrial applications, such as chemical synthesis, chemical purification, drinking water purification, odor control, aquaculture, jet engine repair, wire descaling, hazardous waste control, sulfide control, grease removal, and sludge dewatering. It is sold in a crystal, or granule, form that can be dark purple, dark bronze, or gray in color. An additive that is used in the potassium permanganate in high humidity applications, as in water or wastewater, results in a gray color. It is mixed with water as a 6% solution that contains 8.67 lb/gal.

Eye protection equipment, rubber gloves, and respirators should be worn when handling potassium permanganate. When it is kept cool and dry, it is a stable chemical that is not combustible, but it will support combustion. It becomes unstable when contaminated with acids, other oxidizing agents, and organic material.

1. Design Requirements

Potassium permanganate will react chemically with sulfide, forming elemental sulfur under acidic conditions and potassium sulfate under alkaline conditions. Three parts of potassium permanganate will react with each part of sulfide under acid conditions, and 12 parts will be required under alkaline conditions. Several reactions between these extremes may take place, and they are dependent on specific conditions. Most manufacturers recommend using 6 to 7 parts for sulfide control in wastewater. Because potassium permanganate is such a strong oxidizing agent, it is recommended that dose requirements be determined in each case by direct field tests with the sewage to be treated.

Potassium permanganate was used in Saratoga County Sewer District No. 1 in New York (17). It was chosen because a supply was readily available. This, of course, can often be a swaying argument for the use of any specific chemical at a

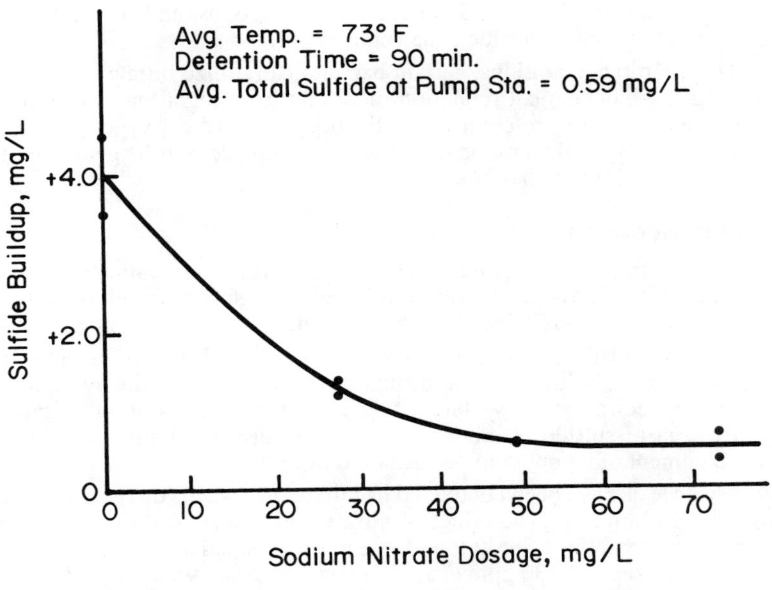

Figure 12-33. Effects of sodium nitrate on sulfide generation in Bluff Cove force main, Los Angeles, California (16)

specific location. In this particular application, permanganate was added at three pump stations, using dry feeders equipped with slurry modules. The potassium permanganate solution was then metered into the influent sewage in the wet well at each pump station. Approximately 125 lb/day of potassium permanganate were used to treat less than 1 mgd of sewage, and were reported successful in controlling odor problems. No data were available for sulfide levels before or after treatment.

2. Equipment

Potassium permanganate crystals can be purchased in 55-lb to 330-lb containers, or in bulk. The containers should be stored on a concrete slab and protected from heat and weather with an enclosure. Solution storage can be in fiberglass or in most plastic tanks. Piping can be PVC. The solution can be batch mixed, or dry feeders can be utilized. Consult the manufacturer for suitable metering pumps. Eye-wash and shower facilities should be provided. Potassium permanganate will stain concrete and metal; suitable coatings are recommended, to make simple clean-up easier.

3. Costs

Most applications of permanganate have been on a small or intermittent scale. At a chemical cost of slightly over $1.00/lb, it is usually uneconomical for high flow or high sulfide wastewaters. For a 1-mgd wastewater flow containing 5 mg/L of dissolved sulfide, and adding permanganate at a 6:1 ratio, a total of 250 lb/day of potassium permanganate would be required. The annual chemical cost would be over $90,000.

H. SODIUM NITRATE

Sodium nitrate has been successfully used for odor control in overloaded oxidation ponds, anaerobic lagoons, sludge storage lagoons, and filters. It has not been widely used to control sulfide in sewers.

Under anaerobic conditions, certain bacteria can utilize nitrate as an oxygen source during the biochemical reduction of nitrate to nitrogen gas. Bacteria will utilize hydrogen acceptors preferentially in the following order: oxygen, nitrate, and sulfate. Theoretically, in the absence of oxygen, no sulfide will be generated until all of the oxygen and nitrate has been reduced.

1. Design Requirements

Laboratory test data reported in 1943 (6) showed that complete protection against odors could be provided by the addition of enough sodium nitrate to furnish the oxygen needed to satisfy 50% of the 5-day BOD.

To treat an oxidation pond, the sodium nitrate is first dissolved in water. Then it is disbursed into the oxidation pond by adding it continually to the influent to the pond or into a recirculation system. It is presumed that sodium nitrate can successfully control sulfide in oxidation ponds because the long residence time allows development of a population of nitrate reducers.

Full-scale tests have been conducted in collection systems on the use of nitrate to control sulfide build-up. The effect of nitrate addition on sulfide build-up was tested in a 12-in. diameter, 1,965 ft long force main, that had a 90-minute detention time at an average flow of 130 gpm (16). The results are shown in Fig. 12-33. Total sulfide was normally 0.6 mg/L at the wet well, and 4.0 to 5.0 mg/L at the end of the force main. Only half of the nitrate added was consumed during the 90-minute period that sewage moved through the force main; the remainder was in the sewage at the end of the force main. The following observations were made: (1) A population of nitrate reducing bacteria had to become established in the slime layer, which required several days of preconditioning; (2) sulfide was not reduced below 1 mg/L; (3) approximately 10 parts of nitrate were required for every part of sulfide eliminated; and (4) nitrate addition may be practical only when sulfide concentrations are high.

2. Equipment

A sodium nitrate addition system consists of a storage tank for the sodium nitrate solution, metering pump, piping, and valves. If a dry form of sodium nitrate is the only available source, then a dry feeder and mixing system will be required. Large installations require bulk storage for the dry form.

3. Costs

Capital costs for a liquid sodium nitrate injection facility do not much differ from those for some alternative sulfide control methods, such as sodium hypochlorite, ferrous chloride, and ferrous sulfate. One source of liquid sodium nitrate contains 2 lb/gal of sodium nitrate, and costs $0.50/gal. The annual chemical cost for partial control of 1 mgd of sewage containing 5 mg/L sulfide, and using a 10:1 dose of sodium nitrate to sulfide, would be $40,000.

I. OZONE

Ozone is a powerful oxidant that will chemically oxidize hydrogen sulfide to elemental sulfur, or possibly sulfite, provided sufficient contact time is allowed. Its principal use has been in the treatment of odorous gas streams and in disinfecting water and wastewater. Ozone is a very unstable gas that requires generation on site, and its effective, economical use for sulfide control in wastewaters has not been demonstrated.

J. BACTERIAL CULTURES AND ENZYMES

Specialized bacterial cultures, or enzymes, are often promoted as economical solutions to the control of sulfide generation. While it may be possible, in theory, to alter the microbiology of either the sulfide generation process or the concrete corrosion process, practical applications of the materials currently marketed have not provided sufficient data for analysis. Further research into the microbiology involved in sulfide generation and concrete corrosion may some day produce a genetic pathway that can provide effective control.

REFERENCES

1. Airco Industrial Gases Inc., *Industrial Gases Data Book*, 6th ed., Murryhill, NJ, 1982.
2. Bowker, R.P.G., Smith, J.M., and Webster, N.A., *Design Manual, Odor and Corrosion Control in Sanitary Sewerage Systems and Treatment Plant*, EPA 625/1-85/018, U.S. Environmental Protection Agency, Oct. 1985.
3. Chambers, C.W., et al, *Chlorination of Wastewater*, Manual of Practice No. 4, Water Pollution Control Federation, Washington, DC, 1976.
4. Condon, R.L., Cooper, R.A., and Englande, A.J., *Instream Aeration to Control Dissolved Sulfides in Sanitary Sewers*, EPA 670/2-73-024, U.S. Environmental Protection Agency, Cincinnati, OH, Sept. 1973.

5. Fong, J.C., "Sulfide Control with Hydrogen Peroxide in Large Diameter Sewers," report to the Chief Engineer and General Manager, County Sanitation Districts of Los Angeles County, July 1986.
6. Heukelekain, H., "Effect of the Addition of Sodium Nitrate to Sewage on Hydrogen Sulfide Production and BOD Reduction," *Sewage Works Journal*, Vol. 15, No. 2, 1943, pp. 255-261.
7. Hoag, L.N., McLaren, F.R., and Hall, G.H. "Control of Odors and Corrosion in the Sacramento Regional Wastewater Conveyance System," a report to the Board of Directors of the Sacramento Regional County Sanitation District by the Sacramento Area Consultants, Sacramento, CA, Sept. 30, 1976.
8. Hollerbach, G.H., "The Use of Molecular Oxygen for Hydrogen Sulfide Control in Sanitary Sewage Systems," presented at the 56th Annual Conference of the Water Pollution Control Federation, Atlanta, GA, Oct. 4, 1983.
9. Jameel, P., and Ross, S., "Solving Sulfide Problems for Sunbelt City," *Brown and Caldwell Quarterly*, Vol. 15, No. 2, April-June 1986, pp. 4-5.
10. Jameel, P., "The Use of Ferrous Chloride to Control Dissolved Sulfides in the Interceptor Sewers at Mesa, Arizona," Brown and Caldwell Report, 1987.
11. Jin, C.G., "Sulfide Control with Sodium Hydroxide in Large Diameter Sewers," report to the Chief Engineer and General Manager, County Sanitation Districts of Los Angeles County, Mar. 1987.
12. Kite, O.A., and Garrett, M.E., "Oxygen Transfer and its Measurement," *Journal of the Institute of Water Pollution Control*, Vol. 82, No. 1, 1983.
13. Mitchell, R.C., "U-Tube Aeration," National Technical Information Service, U.S. Department of Commerce No. PB228127 for the U.S. Environmental Protection Agency, Cincinnati, OH, 1973.
14. Pomeroy, R.D., and Bowlus, F.D., "Progress Report on Sulfide Control Research," *Sewage Works Journal*, Vol. 18, No. 4, 1946, pp. 597-640.
15. Pomeroy, R.D., *Process Design Manual for Sulfide Control In Sanitary Sewerage Systems*, EPA 625/1-74-005, U.S. Environmental Protection Agency, Oct. 1974.
16. Pomeroy, R.D., Parkhurst, J.D. Livingston, J., and Bailey, H.H., *Sulfide Occurrence and Control in Sewage Collection Systems*, EPA 600/X-85-052, U.S. Environmental Protection Agency, Cincinnati, OH, 1985.
17. Rudolf, D.J., "Solution to Odor Problem Gives Unexpected Savings," *Water, Engineering, and Management*, Vol. 129, No. 12, Nov. 1982, pp. 28-29.
18. Schafer, P., and Rockwell, D., "Sulfide Control in Omaha and Cedar Rapids," presented at the July 8-10, 1986 ASCE National Conference on Environmental Engineering, Cincinnati, OH.
19. Schwartz, H.G. Jr., et al, *Design of Wastewater and Stormwater Pumping Stations*, Manual of Practice No. FD-4, Water Pollution Control Federation, Washington, DC, 1981.
20. Sewell, R.J., *Sulfide Control in Sanitary Sewers Using Air and Oxygen*, EPA-670/2-75-060, U.S. Environmental Protection Agency, Cincinnati, OH, June 1975.
21. Speece, R.E., Adams, J.L., and Wooldridge, C.B., "U-Tube Aeration Operating Characteristics," *Journal of the Sanitary Engineering Division*, ASCE, Vol. 95, No. SA3, Proc. Paper 6621, June 1969, pp. 563-574.
22. Speece, R.E., and Orosko, R., "Design of U-Tube Aeration Systems," *Journal of the Sanitary Engineering Division*, ASCE, Vol. 96, No. SA3, Proc. Paper 7348, June 1970, pp. 715-725.
23. Vivona, M.A., and Whalen, G.A., "Controlling Sulfides and Odors in Sewers, Part I," *Public Works*, Vol. 113, No. 3, Mar. 1982, pp. 73-76.
24. Won, D.L., "Sulfide Control with Ferrous Chloride in Large Diameter Sewers," report to the Chief Engineer and General Manager, County Sanitation Districts of Los Angeles County, Nov. 1986.

CHAPTER 13

SAFETY ASPECTS OF HYDROGEN SULFIDE

A. INTRODUCTION

Safety procedures are crucial whenever operations or maintenance personnel enter confined spaces in wastewater collection or treatment facilities. Hydrogen sulfide gas in unsafe concentrations is often found in such facilities. Of increasing importance are the design and operation of foul air ventilation systems, which, under certain conditions, can produce concentrations of hydrogen sulfide gas that are at or near toxic levels.

This chapter provides an overview of the safety aspects of hydrogen sulfide that should be considered when working in wastewater systems: hazards and toxic effects, development of design criteria and safety procedures, safety equipment for confined spaces, training programs, and additional aids.

It should be noted that this chapter does not provide complete coverage of all safety matters dealing with wastewater systems, nor does it provide specific design criteria or standard operating procedures (SOPs). Detailed review of references used in this chapter and other sources is needed to produce adequate criteria and standards for safe operation of wastewater facilities.

B. HAZARDS AND TOXIC EFFECTS

There are no accurate statistics on fatalities or injuries from toxic effects of hydrogen sulfide in the wastewater industry. However, information from various sources indicates that at least several fatalities occur annually in the United States. Assessments of the circumstances of these events show a decided lack of understanding and training of employees entering wastewater structures, lack of necessary safety equipment, and poor staff management and supervision. Accidents often result in multiple fatalities, as would-be rescuers of the first victim unknowingly enter a toxic atmosphere and are also overcome by high H_2S gas concentrations.

Hydrogen sulfide is the most common toxic gas found in wastewater collection and treatment systems. A concentration of 0.1% (1,000 ppm) is lethal. Exposure to reduced levels results in eye and respiratory irritation, headaches, and temporary blindness (see Chap. 2 for more specific details). Hydrogen sulfide concentrations can change rapidly in a confined space where wastewater is present. For instance, a sudden drop in wastewater pH at a location of turbulence, or a malfunction of the ventilation system, can cause a very rapid rise in the H_2S level of the atmosphere. Since the sense of smell is quickly lost at high H_2S concentrations (>100 ppm), the worker subjected to this situation does not immediately detect that there is a problem.

The National Institute for Occupational Safety and Health (NIOSH), and other governmental agencies and private organizations, are attempting to reduce the needless loss of life from these accidents (6, 7). More training and education, better safety procedures, increased attention from supervisors and management staff, and better designs by engineers are needed.

Extremely high concentrations of hydrogen sulfide in air are explosive. The lower explosive limit is 4.3%, and the upper explosive limit is 46% by volume (9). It is almost inconceivable to achieve this range of H_2S concentration in wastewater facilities. The highest concentrations of H_2S typically found in wastewater operations are in digester gas, where concentrations in the 5,000 to 10,000 ppm (v/v) range are common (equal to 0.5 to 1.0%). However, digester gas evolving from the anaerobic digestion process should contain no oxygen. Leaks and vents in digester gas piping often emit very high concentrations of H_2S to the atmosphere, as well as methane (combustible and explosive), carbon dioxide, and other gases. The combined toxic and explosive hazards associated with digester gas require good design practices and careful operating procedures.

C. DEVELOPMENT OF DESIGN CRITERIA AND SAFETY PROCEDURES

Engineers, managers, and operations and maintenance staff need to become more aware of the safety problems associated with H_2S and how others are dealing with these problems. Confined spaces are where the majority of the problems occur, and, therefore, most attention is directed to confined spaces.

1. Definition of Confined Space

A confined space is usually defined as any enclosed space that has limited openings for entry and exit, is not intended for consistent employee occupancy, and does not have sufficient, reliable ventilation to prevent the build-up of a hazardous atmosphere. Some typical confined spaces in the wastewater industry are sewers, manholes, underground chambers and wet wells, sludge thickening and holding tanks, digesters, and foul air handling ducts and structures. Sometimes the structures or rooms containing screening and grit removal facilities qualify as confined spaces, despite being above-ground in many instances.

TABLE 13-1. Workplace Standards for Hydrogen Sulfide Gas [a]

Agency or Organization (1)	Standard (2)	Reference (3)
OSHA	20 ppm average over 15 minutes.	2
	50 ppm average over 10 minutes. Maximum peak during one work shift.	2
NIOSH	10 ppm average over 10 minutes. Maximum peak during one work shift.	5
	50 ppm, instantaneous peak. Evacuation recommended.	5
ACGIH	10 ppm, average over work shift. Maximum allowable average exposure.	1
	15 ppm average over 15 minutes.	1

[a] Limits for typical work exposure in an 8-hr/day, 40-hr/wk job.

2. Safety Standards

Several organizations provide guidance and recommended limits for hydrogen sulfide concentrations in work-place atmospheres (see Table 13-1).

The federal Occupational Safety and Health Administration (OSHA) exposure limits technically apply to nongovernmental operations but are used, nonetheless, as a guide by municipal wastewater authorities. NIOSH and the American Conference on Governmental Industrial Hygienists (ACGIH) guidelines and recommendations apply to any activity, privately or publicly operated.

The recommended H_2S limits of these organizations are established to provide a margin of safety. However, due to the rapid rate at which H_2S gas concentrations can increase in certain situations, these limits are only meaningful if continuous, reliable H_2S monitoring is conducted while employees are located in confined spaces.

There may be only a minute or even a few seconds to react to a rapidly increasing H_2S concentration before serious health effects occur. Therefore, quick detection of H_2S concentration changes can be critical to the lives of workers. Quick detection and rapid response results when a facility or sewerage agency develops a comprehensive confined space entry Standard of Practice (SOP) for personnel. Section 3, which follows, discusses general aspects of confined space entry standards.

3. Development of Confined Space Entry Standard of Practice

The major considerations in developing the SOP are to define the problem, and to develop procedures for employees who will be working within the limits of a confined space. The SOP requires a training program, job planning, a permit or approval system for confined space entry, and a record-keeping program. The confined space entry SOP program must have the full backing and support of upper level management.

Regulations and guidelines. —All safety SOPs should be based on current regulations and guidelines of federal agencies, as supplemented by current state or local safety standards and industry guidelines. Current standards, including ventilation and monitoring aspects, must be reviewed and their requirements understood prior to proceeding with the development of the SOP.

Review and approval. —After developing the confined space entry SOP, it should be reviewed and approved by management, then sent to state or local safety officials for review and comment. Many states have agencies and departments specifically organized to assist in the establishment of safety programs. State agency staff involved in these safety programs are typically not the same staff who are involved in enforcement.

4. Designing to Minimize Problems

In designing wastewater facilities, engineers should consider safety requirements to minimize the chances for hazardous atmospheres to develop, make entry and exit safer, and allow for adequate ventilation and monitoring. In their designs, engineers need to identify the locations of confined spaces and work with the agency or owner to coordinate how entry to these spaces will be controlled. Signs at confined space entrances, which warn employees and others of the dangers and prohibit access except under strict approvals and controls, are becoming more common.

The design of an interceptor system, the specific sulfide control measures used, and any interceptor ventilation systems will profoundly affect the H_2S gas concentration at pumping station wet wells, junction structures, and, particularly at

TABLE 13-2. Typical Equipment Needs for Working in Confined Spaces

1. Atmosphere testing and monitoring equipment to guard against oxygen deficiency, combustible gases, and toxic gases, such as hydrogen sulfide.
2. Equipment to provide for fresh air exchange-this almost always means some type of mechanical ventilation. In some cases it means a fresh air blower and a foul air withdrawal fan.
3. If the confined space has vertical entry and exit, tripod lifting equipment with harness and individual lifelines for each person going into the confined space, and for the standby crew.
4. Protective clothing, such as rubber boots and gloves, rain gear, hard hats, and eye protection.
5. Explosion-proof lights.
6. Communications equipment as required by the scope of the work and for rescue operations.
7. First aid kit to include amyl-nitrite capsules for H_2S exposure.
8. In some cases, self-contained breathing apparatus for each person entering the confined space, and for the standby crew.

NOTE: See Refs. 3, 7, 8, and 9 for more details concerning safety equipment and its proper use.

the treatment plant headworks. Designing the wastewater collection system to minimize sudden changes in flow, pH, and sulfide concentration (if possible) can only help to improve employee safety.

Engineers should be aware of air exchange rates in confined spaces and the different methods of distributing supply air and withdrawing foul air. Ventilation system design and the reliability of the motors, fans, and control systems are more important now that increasing numbers of process units and tankage at treatment plants are being covered for odor control reasons. Hydrogen sulfide concentrations under some of these covers, and in some foul air ventilation and odor scrubbing systems, may be in the range where human safety issues arise.

In short, improved design of new facilities and better coordination between the design engineers and the operations staff should help improve safety.

D. SAFETY EQUIPMENT FOR CONFINED SPACES

There are many types of safety equipment available, and it is important that the equipment chosen fits the job requirements. If possible, the equipment should be user-friendly and durable. All personnel, from supervisors to employees, must have a clear understanding of the equipment's use and maintenance requirements, and should be trained to operate it effectively. As with any program, a records system must be established to document training and maintenance on the equipment. Table 13-2 lists equipment typically associated with confined space entry. Monitoring of hydrogen sulfide gas in confined spaces is often not performed carefully enough to assure the safety of employees.

To assure good performance, the type of equipment is not as important as its use and maintenance. Hydrogen sulfide gas monitoring equipment is normally a back-up or secondary safety system, because ventilation and other controls are typically used to keep H_2S concentrations within safe working limits. In some specialized cases, natural ventilation may be considered adequate to keep H_2S and other gas concentrations low. In these cases, where gas monitoring equipment may be practically the primary safety equipment, it takes on added importance.

Hydrogen sulfide monitoring equipment falls into the following general categories:

1. Personal electronic monitoring devices with relatively fast reaction time (a few seconds), which provide noise or visual alarm, or both, if a certain H_2S concentration (say 10 ppm) is exceeded. More sophisticated devices have a meter or digital read-out of H_2S concentration.
2. Colorimetric devices, typically using lead acetate, which provide a color change when subjected to H_2S. The color is compared to a standard set of colors to indicate the specific concentration.

Gas monitoring equipment suppliers can provide details on the variety of equipment available. Potential purchasers of this equipment may wish to check with several vendors to evaluate the degree of sophistication and accuracy available, and to assess the costs of different models.

E. TRAINING PROGRAM

Training programs are an important way to improve operator safety. The first step in the program should be to obtain strong management support. Management must be informed of the potential problems and must establish proper attitudes about safety extending to all levels of the organization, especially to the field supervisors and crew foreman or other team leaders. Employee and management time and equipment expenses must be budgeted. Specific training periods need to be set aside at regular intervals.

1. Employee Training

After the supervisors and foremen are trained, they should be made responsible for personnel training and operational safety. All personnel working in the vicinity of confined spaces should be made aware of the associated hazards. People who are required to work in a confined space, or who support those working in a confined space, must be trained in every aspect of the SOP.

Training should include:

- Operation, fitting, and maintenance of various types of respirators and their proper application.
- First aid.
- Use and maintenance of safety equipment, including gas monitoring devices.
- Planning all work with safety in mind.
- Permit/approval procedures.
- Rescue procedures.

All of the listed activities should be clearly described in the confined space entry SOP. Training should not be considered complete until the supervisor or other designated official (safety or training officer) has judged that the employee has attained an acceptable degree of proficiency. As part of the procedure, the employee should be interviewed concerning his feeling as to the adequacy of his training. Training is not complete until the training documentation has been completed.

2. Continued Education and Reinforcement

After the confined space entry program has been developed and training completed, supervisors and management need to take a personal interest in the program. They need to evaluate and monitor the program to see that it is understood and is being used in the field by all personnel. Supervisors should communicate effectively with management for continued emphasis and support. Most importantly, supervisors need to take immediate action to see that problems with the program are corrected and retraining conducted as necessary.

F. ADDITIONAL AIDS

In addition to the references listed for this chapter, sewer safety information may be obtained from the following:

- National Association of Sewer Service Companies (NASSCO), 101 Wymore Road, Room 501, Altamonte, Florida 32714. NASSCO has several slide-tape training presentations available on hazard awareness and sewer maintenance safety.
- A videotape on "confined space entry" is available from NIOSH-DTMD C-12, 4676 Columbia Parkway, Cincinnati, Ohio 45226.
- Many regional, state, or local water pollution control associations have training seminars and safety literature available.
- National Fire Prevention Association (NFPA) has slide-tape training presentations on self-contained breathing apparatus.

Table 13-3 lists information on the most common gases found in the wastewater industry. More information on various gases and their effects can be found in industrial hygiene references.

TABLE 13-3 Characteristics of Gases Common to Wastewater Industry (9)

Gas and chemical formula (1)	Specific gravity (2)	Explosive limits LEL (3)	Explosive limits UEL (4)	Max. safe 60-min. exposure (% vol.in air) (5)	Max. safe 8-hour exposure (% vol. in air) (6)	Common properties (7)	Physiological effects (8)	Location of highest concentration (9)	Most common sources (10)	Simplest and safest method of testing (11)
Ammonia NH_3	0.59	16	25	0.03	0.01	colorless, sharp, and pungent	irritates eyes and respiratory tract; toxic at 0.01%	up high	sewer gas	oxygen deficiency indicator; odor
Carbon dioxide CO_2	1.53	non flammable		4.0–6.0	0.5	colorless, odorless, nonflammable; may cause acid taste in large quantities	acts on respiratory nerves; 10% cannot be endured for more than a few minutes	down low but may rise if heated	sludge, sewer gas, combustion carbon and its compounds	oxygen deficiency indicator

Carbon monoxide CO	0.97	12.5 74.2	4.0	0.005	colorless, odorless, tasteless, non-irritating; flammable, explosive, poisonous	combines with hemoglobin of blood causing oxygen starvation; fatal in 1 hr. at 0.1%; unconsciousness in 30 min. at 0.25% and causes headaches in a few hours at 0.02%	up high specifically if in presence of illuminating gas	manufactured fuel gas, flue gas, combustion and fires	CO indicator
Chlorine	2.49	non-flammable	0.0004	0.0001	yellow, green color; irritating, pungent odor; non-flammable and supports combustion	irritates respiratory tract, causes irritation and burning of the skin coughing, and pulmonary edema in small concentrations	down low	chlorine cylinder and feed line leaks	chlorine detector

Ethane **C_2H_6**	1.05	3.1	15	no limit provided oxygen percentage (at least 12%) is sufficient for life		colorless, odorless, tasteless, flammable, explosive, nonpoisonous	acts mechanically to deprive tissues of oxygen; does not support life	down low	natural gas	combustible gas indicator, oxygen deficiency indicator
Hydrogen sulfide **H_2S**	1.19	4.3	46	0.02-0.03	0.001	rotten egg odor in small concentrations; colorless, flammable, and explosive	paralyzes the respiratory system; lessens the sense of smell as concentration increases; rapidly fatal at 0.2%	down low; can be higher if air is hot and humid	coal gas, petroleum, sewer gas and sludge gas	lead acetate paper,lead acetate ampoules, H_2S detecor
Methane **CH_4**	0.55	5	15	no limit providing sufficient oxygen (at least 12%) is present		colorless, odorless, tasteless, explosive, flammable, and nonpoisonous	deprives tissues of oxygen; does not support life	at top, increasing to certain depth	digestion of sludge	combustible gas indicator; oxygen deficiency indicator

Nitrogen N_2	0.97	non flammable			colorless, tasteless, odorless, and nonflammable	in very high concentrations reduces oxygen intake; does not support life	up high and sometimes in low areas	sewer and sludge gas	oxygen deficiency indicator
Oxygen (in air) O_2	1.11	non-flammable			colorless, odorless, tasteless; supports combustion	normal air contains 20.93% O_2. Below 19% considered deficient; 13% dangerous; below 5%-7% fatal	variable at different levels	oxygen deficiency from poor ventilation and chemical combustion of O_2	oxygen deficiency indicator
Sludge gas	varies	5.3	19.3	varies with composition	flammable practically odorless, and colorless	will not support life	up high	digestion of sludge	combustible gas indicator, oxygen deficiency indicator

Gasoline C_5H_{12} - C_9H_{20}	3.0-4.0	1.3	7	0.4-0.7	varies	color, flammable, explosive, odor noticeable at 0.03%	symptoms of intoxication when inhaled, difficult breathing and convulsions; fatal at 2.43%	down low	service stations, storage tanks and dry cleaning operations	combustible gas indicator; oxygen deficiency indicator

REFERENCES

1. American Conference of Governmental Industrial Hygienists, *Documentation of the Threshold Limit Values and Biological Exposure Indices*, 5th ed., 1986, p. 318.
2. Code of Federal Regulations, *OSHA regulations on H_2S gas*, 29 CFR, 1910.1000, Table Z-2
3. Conklin, J.T., "The Practicalities of Manhole Entry", *Water/Engineering and Management*, Scranton Gillette, Chicago, Illinois, Jan. 1986
4. NIOSH, "A Guide to Safety in Confined Spaces," Publication #87-113, U.S. Department of Health and Human Services, Cincinnati, OH, 1987.
5. NIOSH, "Criteria for Recommended Standard: Occupational Exposure to Hydrogen Sulfide," *Publication #77-158*, 1977.
6. NIOSH - "Request for Assistance in Preventing Occupational Fatalities in Confined Spaces," Jan. 1986.
7. Pettit, T.E., "Confined Space: Avoiding the Hazards," *Occupational Health and Safety*, Vol. 52, July 1983, pp. 17- 45.
8. WPCF, "Collected Wisdom," *WPCF Deeds and Data*, Vol. 21, No. 6, Washington, D.C., June, 1984. Includes "Who's to Blame?" by R. D. Pomeroy, "Entering and Working in Digesters," "Wet and Dry Wells," and "Pumping Stations."
9. WPCF, *Safety and Health in Wastewater Systems*, Manual of Practice No. 1. Washington, D.C., 1983.

APPENDIX A
GLOSSARY

Activated sludge. Sludge floc produced in raw or settled wastewater by the growth of zoogleal bacteria and other organisms in the presence of dissolved oxygen, and accumulated in sufficient concentration by returning floc previously formed.

Activated sludge process. A biological wastewater treatment process in which a mixture of wastewater and activated sludge is agitated and aerated. The activated sludge is subsequently separated from the treated wastewater (mixed liquor) by sedimentation, and wasted or returned to the process as needed.

Aeration. (1) The bringing about of intimate contact between air and a liquid by one or more of the following methods:(a) spraying the liquid in the air;(b) bubbling air through the liquid; (c) agitating the liquid to promote surface absorption of air. (2) The supplying of air to confined spaces under nappes, downstream from gates in conduits, etc., to relieve low pressures and to replenish air entrained and removed from such confined spaces by flowing water. (3) Relief of the effects of cavitation by admitting air to the section affected.

Aeration tank. A tank in which sludge, wastewater, or other liquid is aerated.

Aerobic. Requiring, or not destroyed by, the presence of free elemental oxygen.

Aerobic bacteria. Bacteria that require free elemental oxygen for their growth.

Aerobic digestion. Digestion of suspended organic matter by means of aeration. See **digestion**.

Air filter. A filter of cloth or other material used for removing particulates from air.

Alkali. Any of certain soluble salts, principally of sodium, potassium, magnesium, and calcium, that have the property of combining with acids to form neutral salts and may be used in chemical processes such as water or wastewater treatment.

Alkaline. The condition of water, wastewater, or soil which contains a sufficient amount of alkali substances to raise the pH above 7.0.

Amino acid. (1) Organic acid containing one or more amino groups. (2) Any of a class of organic compounds that contain one carboxyl group and one amino group; the alpha-amino acids [$RCH(NH_2)-COOH$] are the building blocks from which proteins are constructed.

Ammonia. A chemical combination of hydrogen (H) and nitrogen (N) occurring extensively in nature. The combination used in water and wastewater engineering is expressed as NH_3.

Anaerobic. Requiring, or not destroyed by, the absence of air or free (elemental) oxygen.

Anaerobic bacteria. Bacteria that grow only in the absence of free elemental oxygen.

Anaerobic digestion. The degradation of organic matter brought about through the action of microorganisms in the absence of elemental oxygen.

Anion. A negatively charged ion in an electrolyte solution, attracted to the anode under the influence of electric potential.

Anode. Positive pole of an electrolytic system; in a vacuum tube, the electrode toward which electrons flow.

Anticorrosion treatment. Treatment to reduce or eliminate corrosion-producing qualities of a water.

Bacteria. A group of universally distributed, rigid, essentially unicellular microscopic organisms lacking chlorophyll. Bacteria usually appear as spheroid. rod-like, or curved entities, but occasionally appear as sheets, chains, or branched filaments. Bacteria are usually regarded as plants.

Bicarbonate alkalinity. Alkalinity caused by bicarbonate ions.

Bioassay. (1) An assay method using a change in biological activity as a qualitative or quantitative means of analyzing a material's response to biological treatment. (2) A method of determining toxic effects of industrial wastes and other wastewaters by using viable organisms or live fish as test organisms.

Biochemical oxygen demand. A standard test used in assessing wastewater strength. See **BOD**.

Biological filtration. The process of passing a liquid through the medium of a biological filter, thus permitting contact with attached zoogleal films that adsorb and absorb fine suspended, colloidal, and dissolved solids and release end products of biochemical action.

Biological slime. The gelatinous film of zoogleal growths covering the medium or spanning the interstices of a biological bed. Also called microbial film.

Biological wastewater treatment. Forms of wastewater treatment in which bacterial or biochemical action is intensified to stabilize, oxidize, and nitrify the unstable organic matter present. Intermittent sand filters, contact beds, trickling filters, and activated sludge processes are examples.

BOD. (1) Abbreviation for biochemical oxygen demand. The quantity of oxygen used in the biochemical oxidation of organic matter in a specified time, at a specified temperature, and under specified conditions. (2) A standard test used in assessing wastewater strength.

Bulking sludge. An activated sludge that settles poorly because of a floc of low density.

Centrifugal dewatering of sludge. The partial removal of water from wastewater sludge by centrifugal action.

Chemical oxygen demand (COD). A measure of the oxygen-consuming capacity of inorganic and organic matter present in water or wastewater. It is expressed as the amount of oxygen consumed from a chemical oxidant in a specific test. It does not differentiate between stable and unstable organic matter and thus does not necessarily correlate with biochemical oxygen demand. Also known as OC and DOC, oxygen consumed and dichromate oxygen consumed, respectively.

Chlorine. An element ordinarily existing as a greenish-yellow gas about 2.5 times as heavy as air. At atmospheric pressure and a temperature of -39.1^{o}, the gas becomes an amber liquid about 1.5 times as heavy as water. The chemical symbol of chlorine is Cl, its atomic weight is 33.457, and its molecular weight is 70.914.

COD. Symbol for chemical oxygen demand. See **chemical oxygen demand**.

Combined sewer. A sewer intended to receive both wastewater and storm or surface water.

Composite wastewater sample. A combination of individual samples of water or wastewater taken at selected intervals, generally hourly for some specified period, to minimize the effect of the variability of the individual sample. Individual samples may have equal volume or may be roughly proportioned to the flow at time of sampling.

Corrosion. The gradual deterioration or destruction of a substance or material by chemical action, frequently induced by electrochemical processes. The action proceeds inward from the surface.

Critical depth. The depth of water flowing in an open channel or partially filled conduit corresponding to one of the recognized critical velocities.

Dechlorination. The partial or complete reduction of residual chlorine in a liquid by any chemical or physical process.

Dehydration. (1) The chemical or physical process whereby water in chemical or physical combination with other matter is removed. (2) The geologic process in the weathering of rocks or soils whereby water is given up and a new mineral compound formed.

Depressed sewer. A section of sewer constructed lower than adjacent sections to pass beneath a valley, watercourse, or other obstruction. It runs full or at pressure greater than atmospheric because its crown is depressed below the hydraulic grade line. Also referred to as inverted siphon.

Diffuser. A porous plate, tube, or other device through which air is forced and divided into minute bubbles for diffusion in liquids. Commonly made of carborundum, alundum, metal, or plastic materials.

Diffusion aerator. An aerator that blows air under low pressure through submerged porous plates, perforated pipes, or other devices so that small air bubbles rise through the water or wastewater continuously.

Digestion. (1) The biological decomposition of organic matter in sludge, resulting in partial gasification, liquefaction, and mineralization. (2) The process carried out in a digester.

Dissolved oxygen. The oxygen dissolved in water, wastewater, or other liquid, usually expressed in milligrams per liter, parts per million, or percent of saturation. Abbreviated DO.

Dissolved sulfide. As used in this MOP, the sum of the species of S^{-2}, HS^{-}, and H_2S found within a water sample. The portion of the dissolved sulfide in each of these three forms is heavily dependent on the pH of the sample.

Diurnal fluctuation. The cyclic rise and fall of the water table or stream-flow during a 24-hr period in response to changes in evapotranspiration draft from groundwater.

Dolomite. An equimolar combination of calcium and magnesium carbonates which occurs in nature as a hard rock. Upon calcination it becomes dolomitic lime, containing 30-50% magnesia and 70-50% percent calcium oxide.

Domestic wastewater. Wastewater derived principally from dwellings, business buildings, institutions, and the like. It may or may not contain groundwater, surface water, or storm water.

Dry-weather flow. (1) The flow of wastewater in a combined sewer during dry weather. Such flow consists mainly of wastewater, with no storm water included. (2) The flow of water in a stream during dry weather, usually contributed entirely by groundwater.

Effluent. (1) A liquid which flows out of a containing space. (2) Wastewater or other liquid, partially or completely treated, or in its natural state, flowing out of a reservoir, basin, treatment plant, or industrial treatment plant, or part thereof. (3) An outflowing branch of a main stream or lake.

Energy gradient. The slope of the energy line of a body of flowing water with reference to a datum plane.

Equivalent per million (epm). A unit of chemical equivalent weight of solute per million unit weights of solution. Concentration in equivalents per million is calculated by dividing concentration in parts per million by the chemical combining weight of the substance or ion.

Filter. A device or structure for removing solid or colloidal material, usually of a type that cannot be removed by sedimentation, from water, wastewater, or other liquid. The liquid is passed through a filtering medium, usually a granular material but sometimes finely woven cloth, unglazed porcelain, or specially prepared paper. There are many types of filters used in water or wastewater treatment. See **trickling filter**.

Filter bed. (1) A type of bank revetment consisting of layers of filtering medium of which the particles gradually increase in size from the bottom upward. Such a filter allows the groundwater to flow freely, but it prevents even the smallest soil particles from being washed out. (2) A tank for water filtration having a false bottom covered with sand, as a rapid sand filter. (3) A pond with sand bedding, as a sand filter or slow sand filter.

Filter loading. Organically, the pounds of biochemical oxygen demand (BOD) in the applied liquid per unit of filter bed area or volume per day. Hydraulically, the quantity of liquid applied per unit of filter bed area or volume per day.

Filtration. The process of passing a liquid through a filtering medium (which may consist of granular material, such as sand, magnetite, or diatomaceous earth, finely woven cloth, unglazed porcelain, or specially prepared paper) for the removal of suspended or colloidal matter.

Final sedimentation. The separation of solids from wastewater in a final settling tank.

Five-day BOD. That part of oxygen demand associated with biochemical oxidation of carbonaceous, as distinct from nitrogenous, material. It is determined by allowing biochemical oxidation to proceed, under conditions specified in Standard Methods, for 5 days.

Flocculation. In water and wastewater treatment, the agglomeration of colloidal and finely divided suspended matter after coagulation by gentle stirring by either mechanical or hydraulic means. In biological wastewater treatment where coagulation is not used, agglomeration may be accomplished biologically.

Flowing-through time. (1) The time required for a volume of liquid to pass through a basin, identified in terms of the characteristic being measured, such as mean time, modal time, minimum time. These various times may be estimated from tracer-recovery curves. Dyes, electrolytes, and radioactive substances are used as tracers. (2) The average time required for a small volume of liquid to pass through a basin from inlet to outlet.

Flow line. The position of the water surface in a flowing stream or conduit for a normal or specified rate of discharge.

Flushing manhole. A manhole provided with a gate so that wastewater may be accumulated and then discharged rapidly for flushing a sewer.

Force main. A pressure pipe joining the pump discharge at a water or wastewater pumping station with a point of gravity flow.

Fouling. A gelatinous, slimy accumulation on the waterway of a conduit, resulting from the activity of organisms in the waters. Fouling is more easily removable than tuberculation. Fouling may be found on concrete, masonry, and metal surfaces, but tuberculation is found only on metal surfaces.

Free available chlorine. The amount of chlorine available as dissolved gas, hypochlorous acid, or hypochlorite ion that is not combined with an amine or other organic compound.

Free residual chlorination. The application of chlorine or chlorine compounds to water or wastewater to produce a free available chlorine residual directly or through the destruction of ammonia or certain organic nitrogenous compounds.

Free surface. The boundary surface of a liquid which is in contact with the atmosphere.

Fresh wastewater. Wastewater of recent origin containing dissolved oxygen.

Gas-solubility coefficient. The quantity of gas absorbed by a unit quantity of water at a given temperature when the water is exposed to a pure atmosphere of the gas under a barometric pressure of 760 mm. Formerly called solubility coefficient.

Grab sample. A single sample of wastewater taken at neither set time nor flow.

Gradient. The rate of change of any characteristic per unit of length or slope. The term is usually applied to such things as elevation, velocity, pressure.

Grease. In wastewater, a group of substances including fats, waxes, free fatty acids, calcium and magnesium soaps, mineral oils, and certain other nonfatty materials. The type of solvent and method used for extraction should be stated for quantitation.

Grease trap. A device for separation of grease from wastewater by flotation so that it can be removed from the surface.

Grit chamber. A detention chamber or an enlargement of a sewer designed to reduce the velocity of flow of the liquid to permit the separation of mineral from organic solids by differential sedimentation.

Grit collector. A device placed in a grit chamber to convey deposited grit to a point of collection.

Grit compartment. The portion of the grit chamber in which grit is collected and stored before removal.

Groundwater. Subsurface water occupying the saturation zone, from which wells and springs are fed. In a strict sense the term applies only to water below the water table. Also called phreatic water, plerotic water.

Headworks. All the structures and devices located at the head or diversion point of a conduit or canal. The term as used is practically synonymous with diversion works; an intake heading.

Heavy metals. Metals that can be precipitated by hydrogen sulfide in acid solution, for example, lead, silver, gold, mercury, bismuth, copper.

High-rate filter. A trickling filter operated at a high average daily dosing rate, usually between 10 and 40 mgd/acre including any recirculation of effluent.

House connection. The pipe carrying the wastewater from the building to a common sewer. Also called building sewer, house sewer.

Household wastes. The water-carried wastes from kitchens, toilets, lavatories, and laundries.

Hydraulic grade line. A hydraulic profile of the piezometric level of water at all points along a line. The term is usually applied to water moving in a conduit, open channel, or stream, but may also be applied to free or confined groundwater. In an open channel, it is the free water surface.

Hydrogen-ion concentration. The weight of hydrogen ion in moles per liter of solution. Commonly expressed as the pH value, which is the logarithm of the reciprocal of the hydrogen-ion concentration.

Hydrogen sulfide. Molecular H_2S, either in the gas phase or within the liquid phase. When H_2S is in the liquid phase, it is in equilibrium with hydrosulfide (HS^-) and sulfide (S^{-2}).

Hydrologic cycle. The circuit of water movement from the atmosphere to the earth and return to the atmosphere through various stages or processes such as precipitation, interception, runoff, infiltration, percolation, storage, evaporation, and transpiration. Also called water cycle.

Imhoff tank. A deep, two-storied wastewater tank originally patented by Karl Imhoff. It consists of an upper continuous-flow sedimentation chamber and a lower sludge-digestion chamber. The floor of the upper chamber slopes steeply to trapped slots through which solids may slide into the lower chamber. The lower chamber receives no fresh wastewater directly, but is provided with gas vents and with means for drawing digested sludge from near the bottom.

Immediate biochemical oxygen demand. (1) The initial quantity of oxygen used by polluted liquid immediately upon being introduced into water containing dissolved oxygen. It may be exercised by end products of prior biochemical action or by chemical substances avid for oxygen. (2) In the standard laboratory procedure, the apparent BOD for 15 min at 20°C.

Industrial wastes. The liquid wastes from industrial processes, as distinct from domestic or sanitary wastes.

Industrial wastewater. Wastewater in which industrial wastes predominate. See **domestic wastewater, industrial wastes**.

Institutional wastewater. Wastewater from institutions such as hospitals, sanitariums, prisons, or charitable institutions.

Intercepting sewer. A sewer that receives dry-weather flow from a number of transverse sewers or outlets and frequently additional predetermined quantities of storm water (if from a combined system), and conducts such waters to a point for treatment or disposal.

Junction chamber A converging section of a conduit, usually large enough for a man to enter, used to facilitate the flow from one or more conduits into a main conduit.

Lateral. A sewer which discharges into a branch or other sewer and has no common sewer tributary to it.

Lift station. A small wastewater pumping station that lifts the wastewater to a higher elevation when the continuance of the sewer at reasonable slopes would involve excessive depths of trench, or that raises wastewater from areas too low to drain into available sewers. These stations may be equipped with pneumatic ejectors or centrifugal pumps.

Lime. Any of a family of chemicals consisting essentially of calcium hydroxide made from limestone (calcite) which is composed almost wholly of calcium carbonate or a mixture of calcium and magnesium carbonate.

Line manhole. A manhole on a sewer at a point where no other sewers connect. It is often located at a point where the sewer changes direction, either in line or grade.

Lining. A protective covering over all, or over a portion, of the perimeter of a conduit or reservoir, intended to prevent seepage losses, to withstand pressure, or to resist erosion. In the case of conduits, lining is also sometimes installed to reduce friction losses.

Liquor. Water, wastewater, or any combination; commonly used to designate liquid phase when other phases are present.

Manhole. An opening in a sewer provided for the purpose of permitting a person to enter or leave the sewer.

Mass spectrometry. A means of sorting out ions by measuring their masses according to their ratios of charge to mass. The instrument used for this measurement is a mass spectrograph. By means of magnetic and electrostatic fields applied at right angles to the positive ray beam, particles having the same value of e/m (energy over mass) are brought to a focus on a photographic plate. From the resulting mass spectrum, the atomic weights of the particles may be deduced.

Mean velocity. The average velocity of a stream flowing in a channel or conduit at a given cross section or in a given reach. It is equal to the discharge divided by the cross-sectional area of the section, or the average cross-sectional area of the reach. Also called average velocity.

Membrane filtration. A method of quantitative or qualitative analysis of bacterial or particulate matter in a water sample by filtration through a membrane capable of retaining bacteria.

Microbial film. A gelatinous film of microbial growth attached to or spanning the interstices of a support medium. Also called biological slime.

Minimum grade. The grade of a sewer, not under pressure, sufficient to maintain the minimum velocity, when the sewer is partly full, that will prevent the deposition of material carried by the water.

Most probable number (MPN). That number of organisms per unit volume that, in accordance with statistical theory, would be more likely than any other number to yield the observed test result or that would yield the observed test result with the greatest frequency. Expressed as density of organisms per 100 mL. Results are computed from the number of positive findings of coliform-group organisms resulting from multiple-portion decimal- dilution plantings.

Nitrification. (1) The conversion of nitrogenous matter into nitrates by bacteria. (2) The treatment of a material with nitric acid.

Nondepositing velocity. The velocity that will maintain suspended sediment in movement.

Nonsettleable solids. Wastewater matter that will stay in suspension for an extended period of time. Such period may be arbitrarily taken for testing purposes as one hour. See **suspended solids.**

Nonuniform flow. A flow in which the velocities vary either in magnitude or direction, or both, from point to point along the stream or conduit, because of variations in discharge, cross section, slope, roughness, alignment. There are two types: steady nonuniform flow and unsteady nonuniform flow.

Odor control. (1) In water treatment, the elimination or reduction of odors in a water supply by aeration, algae elimination, superchlorination, activated carbon treatment, and other methods. (2) In wastewater treatment, the prevention or reduction of objectionable odors by chlorination, aeration, or other processes, or by masking with chemical aerosols.

Odor detection threshold. The point at which, after successive dilutions with odorless air (or water), the odor of an air (or water) sample can just be detected. The threshold odor is expressed quantitatively by the number of times the sample is diluted with odorless air (or water.)

Organic matter. Chemical substances of animal or vegetable origin, or more correctly, of basically carbon structure, comprising compounds consisting of hydrocarbons and their derivatives.

Outfall. (1) The point, location, or structure where wastewater or drainage discharges from a sewer, drain, or other conduit. (2) The conduit leading to the ultimate disposal area. Also see **outfall sewer, wastewater outfall.**

Outfall sewer. A sewer that receives wastewater from a collecting system or from a treatment plant and carries it to a point of final discharge. See **outfall.**

Oxidation. The addition of oxygen to a compound. More generally, any reaction which involves the loss of electrons from an atom.

Oxidation pond. A basin used for retention of wastewater before final disposal, in which biological oxidation of organic material is effected by natural or artificially accelerated transfer of oxygen to the water from air.

Oxidation process. Any method of wastewater treatment for the oxidation of the putrescible organic matter. The usual methods are biological filtration and the activated sludge process.

Oxidation-reduction potential. The potential required to transfer electrons from the oxidant to the reductant; used as a qualitative measure of the state of oxidation in wastewater treatment systems.

Oxygenation capacity. In treatment processes, a measure of the ability of an aerator to supply oxygen to a liquid.

Oxygen balance. (1) The dissolved-oxygen level at any point in a stream resulting from the opposing forces of deoxygenation and reaeration. (2) The relation between the biochemical oxygen demand of a wastewater or treatment plant effluent and the oxygen available in the diluting water.

Oxygen deficiency. (1) The additional quantity of oxygen required to satisfy the oxygen requirement in a given liquid. Usually expressed in milligrams per liter. (2) Lack of oxygen.

Oxygen deficit. The difference between the dissolved oxygen level at saturation and the actual dissolved oxygen concentration in water.

Oxygen demand. The quantity of oxygen utilized in the biochemical oxidation of organic matter in a specified time, at a specified temperature, and under specified conditions. See **BOD**.

Oxygen depletion. Loss of dissolved oxygen from water or wastewater resulting from biochemical or chemical action.

pH. The reciprocal of the logarithm of the hydrogen-ion concentration. The concentration is the weight of hydrogen ions, in grams, per liter of solution. Neutral water, for example, has a pH value of 7 and a hydrogen-ion concentration of 10^{-7}.

Phenolic compounds. Hydroxy derivatives of benzene. The simplest phenolic compound is hydroxy benzene, C_6H_5OH.

Phenolphthalein alkalinity. A measure of the hydroxides plus one-half of the normal carbonates in aqueous suspension. Measured by the amount of sulfuric acid required to bring the water to a pH value of 8.3, as indicated by a change in color of phenolphthalein. It is expressed in parts per million of calcium carbonate.

Population equivalent. A means of expressing the strength of organic material in wastewater. Domestic wastewater consumes, on an average, 0.17 lb of oxygen per capita per day, as measured by the standard BOD test. This figure has been used to measure the strength of organic industrial waste in terms of an equivalent number of persons. For example, if an industry discharges 1,000 pounds of BOD per day, its waste is equivalent to the domestic wastewater from 6,000 persons (1,000÷0.17 = 6,000).

Postchlorination. The application of chlorine to water or wastewater subsequent to any treatment, including prechlorination.

Preaeration. A preparatory treatment of wastewater consisting of aeration to remove gases, add oxygen, promote flotation of grease, and aid coagulation.

Prechlorination. The application of chlorine to water or wastewater prior to any treatment.

Private sewer. A sewer privately owned and used by one or more properties.

Relief sewer. (1) A sewer built to carry the flows in excess of the capacity of an existing sewer. (2) A sewer intended to carry a portion of the flow from a district in which the existing sewers are of insufficient capacity, and thus prevent overtaxing the latter.

Residual oxygen. The dissolved oxygen content of a stream after deoxygenation has begun.

Rigid conduit. A conduit with a cross-sectional shape that cannot be distorted sufficiently to change its vertical dimension more than 0.1% without causing injurious cracks.

Roughness coefficient. A factor, in the Chezy, Darcy-Weisbach, Hazen-Williams, Kutter, Manning, and other formulas for computing the average velocity of flow of water in a conduit or channel, which represents the effect of roughness of the confining material on the energy losses in the flowing water.

Salinity. (1) The relative concentration of salts, usually sodium chloride, in a given water. It is usually expressed in terms of the number of parts per million of chlorine (Cl). (2) A measure of the concentration of dissolved mineral substances in water.

Sanitary sewer. A sewer that carries liquid and water-carried wastes from residences, commercial buildings, industrial plants, and institutions, together with minor quantities of ground, storm, and surface waters that are not admitted intentionally. See **wastewater**.

Sanitary wastewater. (1) Domestic wastewater with storm and surface water excluded. (2) Wastewater discharging from the sanitary conveniences of dwellings (including apartment houses and hotels), office buildings, industrial plants, or institutions. (3) The water supply of a community after it has been used and discharged into a sewer.

Scouring velocity. The minimum velocity necessary to dislodge stranded material from the boundary of a waterway, conduit, or pipeline by a fluid in motion.

Screen. (1) A device with openings, generally of uniform size, used to retain or remove suspended or floating solids in flowing water or wastewater and to prevent them from entering an intake or passing a given point in a conduit. The screening element may consist of parallel bars, rods, wires, grating, wire mesh, or perforated plate, and the openings may be of any shape, although they are usually circular or rectangular. (2) A device used to segregate granular material such as sand, crushed rock, and soil into various sizes.

Screenings. (1) Material removed from liquids by screens. (2) Broken rock, including the dust, of a size that will pass through a given screen, depending on the character of the stone.

Separate sewer. A sewer intended to receive only wastewater and not storm or surface water. Also called sanitary sewer.

Separate sewer system. A sewer system carrying sanitary wastewater and other water-carried wastes from residences, commercial buildings, industrial plants, and institutions, together with minor quantities of ground, storm, and surface waters that are not intentionally admitted. See **wastewater**. Also see **combined sewer**.

Septicity. A condition produced by growth of anaerobic organisms.

Septic wastewater. Wastewater undergoing putrefaction under anaerobic conditions.

Service connection. A pipeline, with its appurtenances, that branches off or connects a water or sewer main with premises.

Service life. The period of time between the date when a physical unit of property was first put into service new as a part of the property and the date when it was retired, or is anticipated to be retired, from service.

Service pipe. The pipeline extending from the water main to the building served.

Settleable solids. (1) That matter in wastewater which will not stay in suspension during a preselected settling period, such as one hour, but either settles to the bottom or floats to the top. (2) In the Imhoff cone test, the volume of matter that settles to the bottom of the cone in one hour.

Settling basin. A basin or tank in which water or wastewater containing settleable solids is retained to remove by gravity a part of the suspended matter. Also called sedimentation basin, sedimentation tank, settling tank.

Settling chamber. (1) A basin or tank in which water or wastewater containing settleable solids is retained to remove by gravity a part of the suspended matter. (2) The second or final element of the so-called biolytic tank, which is a combination of a flocculating tank and a settling tank. (3) Sometimes, the sedimentation compartment of a two-story tank, as in the case of an Imhoff tank.

Settling tank. A basin or tank in which water or wastewater containing settleable solids is retained to remove by gravity a part of the suspended matter. Also called sedimentation basin, sedimentation tank, settling basin.

Sewage. The spent water of a community. Term now being replaced in technical usage by preferable term wastewater. See **wastewater**.

Sewage gas. (1) Gas resulting from the decomposition of organic matter in wastewater. (2) Gas produced during the digestion of sludge.

Sewage works. All-inclusive term for wastewater collection, pumping, treatment, and disposal facilities. Term declining in use. See **sewer system, wastewater facilities**.

Sewer. A pipe or conduit that carries wastewater or drainage water.

Sewerage. System of piping, with appurtenances, for collecting and conveying wastewater from source to discharge. Term declining in use. See **sewer system, wastewater facilities**.

Sewer appurtenances. Structures, devices, and appliances, other than pipe or conduit, that are integral parts of a sewer system.

Sewer gas. (1) Gas evolved in sewers that results from the decomposition of the organic matter in the wastewater. (2) Any gas present in the wastewater system, even though it is from such sources as gas mains, gasoline, cleaning fluid.

Sewer manhole. A shaft or chamber providing access from the surface of the ground to a sewer.

Sewer outfall. The outlet or structure through which wastewater is finally discharged.

Sewer outlet. The point of final discharge of wastewater or treatment plant effluent.

Sewer system. Collectively, all of the property involved in the operation of a sewer utility. It includes land, wastewater lines and appurtenances, pumping stations, treatment works, and general property. Occasionally referred to as a sewerage system.

Sludge. See **activated sludge**.

Storm sewer. A sewer that carries storm water and surface water, street wash and other wash waters, or drainage, but excludes domestic wastewater and industrial wastes. Also called storm drain.

Substrate. (1) The substances used by organisms in liquid suspension. (2) The liquor in which activated sludge or other matter is kept in suspension.

Sulfate-reducing bacteria. Bacteria capable of assimilating oxygen from sulfate compounds, thereby reducing them to sulfides. See **sulfur bacteria**.

Sulfide. Commonly refers to a variety of forms of sulfur, usually within the liquid phase, including soluble hydrosulfide (HS^-), molecular H_2S, organic sulfide complexes, and inorganic metal sulfides (FeS, ZnS, etc.). Technically, sulfide should refer to the sulfide ion (S^{-2}).

Sulfur bacteria. Bacteria capable of using dissolved sulfur compounds in their growth; bacteria deriving energy from sulfur or sulfur compounds.

Sulfur cycle. A graphical presentation of the conservation of sulfur in nature, living animal matter through dead organic matter, various stages of decomposition, plant life, and the return of living animal matter, showing the changes that occur in this element in course of the cycle. It is used to illustrate biological action as well as aerobic and anaerobic acceleration of the transformation of this element by wastewater and sludge treatment.

Superchlorination. Chlorination wherein the doses are deliberately selected to produce free or combined residuals so large as to require dechlorination.

Supernatant liquor. (1) The liquor overlying deposited solids. (2) The liquid in a sludge-digestion tank that lies between sludge at the bottom and floating scum at the top.

Surface aeration. The absorption of air through the surface of a liquid.

Suspended load. The sediment in suspension in a stream, that is transported at essentially the velocity of the water.

Suspended matter. (1) Solids in suspension in water, wastewater, or effluent. (2) Solids in suspension that can be removed readily by standard filtering procedures in a laboratory. See **suspended solids**.

Suspended solids. (1) Solids that either float on the surface of, or are in suspension in, water, wastewater, or other liquids, and which are largely removable by laboratory filtering. See **Suspended matter**. (2) The quantity of material removed from wastewater in a laboratory test, as prescribed in "Standard Methods for the Examination of Water and Wastewater" and referred to as nonfilterable residue.

Threshold odor. See **Odor detection threshold**

Trade wastes. The liquid wastes from industrial processes, as distinct from domestic or sanitary wastes. Called industrial wastes.

Trash rack. A grid or screen placed across a waterway to catch floating debris.

Treated sewage. Wastewater that has received partial or complete treatment.

Trickling filter. A filter consisting of an artificial bed of coarse material, such as broken stone, clinkers, slate, slats, brush, or plastic materials, over which wastewater is distributed or applied in drops, films, or spray from troughs, droppers, moving distributors, or fixed nozzles, and through which it trickles to the underdrains, giving opportunity for the formation of zoogleal slimes which clarify and oxidize the wastewater. See **filter**.

Trunk main. A large pipe serving as a supply main or feeder main in a water distribution system.

Trunk sewer. A sewer that receives many tributary branches and serves a large territory.

Trunk system. A system of major sewers serving as transporting lines and not as local or lateral sewers.

Turbidimeter. An instrument for measurement of turbidity, in which a standard suspension usually is used for reference.

Turbidity. (1) A condition in water or wastewater caused by the presence of suspended matter, resulting in the scattering and absorption of light rays. (2) A measure of fine suspended matter in liquids. (3) An analytical quantity usually reported in arbitrary turbidity units determined by measurements of light diffraction.

Turbulence. (1) The fluid property that is characterized by irregular variation in the speed and direction of movement of individual particles or elements of the flow. (2) A state of flow of water wherein the water is agitated by cross currents and eddies, as opposed to laminar, streamline, or viscous flow. See **turbulent flow**.

Turbulent flow. (1) The flow of a liquid past an object such that the velocity at any fixed point in the fluid varies irregularly. (2) A type of fluid flow in which there is an unsteady motion of the particles and the motion at a fixed point varies in no definite manner. Sometimes called eddy flow, sinous flow.

Turbulent velocity. The velocity of water flowing in a conduit above which the flow will always be turbulent, and below which the flow may be either turbulent or laminar, depending upon circumstances.

Ultimate biochemical oxygen demand. (1) Commonly, the total quantity of oxygen required to satisfy completely the first-stage biochemical oxygen demand. (2) More strictly, the quantity of oxygen required to satisfy completely both the first-stage and the second-stage biochemical oxygen demands.

Undigested sludge. Settled sludge promptly removed from sedimentation tanks before decomposition has much advanced. Also called raw sludge.

Vacuum filter. A filter consisting of a cylindrical drum mounted on a horizontal axis, covered with a filter cloth, and revolving with a partial submergence in liquid. A vacuum is maintained under the cloth for the larger part of a revolution to extract moisture. The cake is scraped off continuously.

Valence. An integer representing the number of hydrogen atoms with which one atom of an element (or one radical) can combine (negative valence) or the number of hydrogen atoms the atom or radical can displace (positive valence).

Valve. (1) A device installed in a pipeline to control the magnitude and direction of the flow. It consists essentially of a shell and a disk or plug fitted to the shell. (2) In a pump, a waterway, passage through which is controlled by a mechanism.

Vapor pressure. (1) The pressure exerted by a vapor in a confined space. It is a function of the temperature. (2) The partial pressure of water vapor in the atmosphere. (3) Partial pressure of any liquid.

Velocity. See following terms modifying velocity: **nondepositing, scouring, turbulent**.

Velocity head. (1) The vertical distance or height through which a body must fall freely, under the force of gravity, to acquire the velocity that it possesses. It is equal to the square of the velocity divided by twice the acceleration of gravity. (2) The theoretical vertical height through which a liquid body may be raised by its kinetic energy. It is equal to the square of the velocity divided by twice the acceleration due to gravity.

Vent. In plumbing, a pipeline installed to provide a flow of air to or from a drainage system, or to provide a circulation of air within such system to protect trap seals from siphonage and back pressure. See **vent pipe.**

Vent pipe. A pipeline attached to an air vent. It is usually constructed in a vertical position and, under normal conditions, is filled with water to the elevation of the hydraulic gradient. The upper end is open. When pressure in the penstock is reduced below atmospheric, the water passes out and air rushes in through the vent pipe and through the air vent into the penstock, preventing its collapse due to differences in inside and outside pressure. See **vent.**

Vent stack. In plumbing, a vertical vent pipe installed primarily to provide circulation of air to and from any part of the drainage system.

Vent system. In plumbing, a pipe or pipes installed to provide a flow of air to or from a drainage system or to provide circulation of air within such a system to protect trap seals from siphonage or back pressure.

Ventilated manhole cover. A manhole cover with openings in it for ventilation.

Wastewater. The spent water of a community. From the standpoint of source, it may be a combination of the liquid and water- carried wastes from residences, commercial buildings, industrial plants, and institutions, together with any groundwater, surface water, and storm water that may be present. In recent years, the word wastewater has taken precedence over the word sewage.

Wastewater analysis. The determination of chemical composition, concentration, and biological condition of wastewater and treatment plant effluents.

Wastewater composition. (1) The relative quantities of the various solid, liquid, and gaseous constituents of wastewater. (2) The chemical and physical characteristics of the solid and liquid constituents of wastewater, and their relationships apart from the degree of concentration.

Wastewater disposal. The act of disposing of wastewater by any method. Not synonymous with wastewater treatment. Common methods of disposal are: dispersion, dilution, broad irrigation, privy, cesspool.

Wastewater facilities. The structures, equipment, and processes required to collect, carry away, and treat domestic and industrial wastes, and dispose of the effluent.

Wastewater influent. Wastewater as it enters a wastewater treatment plant or pumping station.

Wastewater irrigation. Irrigation of crops with wastewater.

Wastewater outfall. The outlet or structure through which wastewater is finally discharged.

Wastewater oxidation. The process whereby, through the agency of living organisms in the presence of oxygen, the organic matter contained in wastewater is converted into a more stable or a mineral form.

Wastewater treatment. Any process to which wastewater is subjected in order to remove or alter its objectionable constituents, and thus render it less offensive or dangerous.

Wastewater treatment works. (1) An arrangement of devices and structures for treating wastewater, industrial wastes, and sludge. Sometimes used as synonymous with waste treatment plant or wastewater treatment plant. (2) A water pollution control plant.

Water. (1) A transparent, odorless, tasteless liquid, a compound of hydrogen and oxygen, H_2O, freezing at 32°F or 0°C and boiling at 212°F or 100°C, which, in more or less impure state, constitutes rain, oceans, lakes, rivers, and other bodies; it contains 11.188% hydrogen and 88.812% oxygen, by weight. It may exist as a solid, liquid, or gas and, as normally found in the lithosphere, hydrosphere, and atmosphere, may have other solid, gaseous, or liquid materials in solution or suspension. (2) To wet, supply, or irrigate with water.

Water pollution control plant. An arrangement of devices and structures for the control of waterborne pollution of waterways. Also referred to as treatment plant, with appropriate adjective describing source of wastewater.

Water reclamation. The recovery of wastewaters for useful purposes by treatment processes and subsequent return to either a surface or groundwater source.

Watershed planning. (1) Planning the use and treatment of land and water that will most nearly meet all the objectives of soil and water conservation from the standpoint of both the individual farms and all the residents of the watershed. (2) Considering the interrelated uses of water from several watersheds to meet regional water supply requirements as well as local watershed objectives; may be part of state or national water resource planning.

APPENDIX B
NOTATION

A = Concrete alkalinity expressed as calcium carbonate equivalent decimal fraction (Eq. 6-9)

A = Alkalinity, meq/L (Eq. 3-9)

A = Area = $3.14r^2$ (Eq. 6-14)

beta = Factor combining hydraulic variables

b = Surface width of the stream, feet

b/P' = Ratio of width of wastewater stream at surface to exposed perimeter of the pipe wall above the water surface

$[BOD]_5$ = (Five-day) biochemical oxygen demand concentration, mg/L

$[BOD]_c$ = Climactic BOD, mg/L

BOD_r = BOD remaining at time t, mg/L

BOD_t = $BOD_u\ (1-e^{-k1t})$

BOD_u = Ultimate carbonaceous BOD, mg/L

C = Rate of concrete corrosion by sulfuric acid at the pipe crown, in./yr (Eq. 10-1)

C = Average exposed pipe perimeter corrosion rate, in./yr (Eq. 6-9)

C = Average corrosion rate, in./yr (Eq. 6-10)

CO_3^{-2} = Concentration of carbonate ion, mol/L

CCF = Crown corrosion factor

D = Pipe diameter, ft

d_m = Mean hydraulic depth (area of flow divided by surface width),

$[DS]$ = Dissolved sulfide, mg/L

$\frac{d[S]}{dt}$ = Rate of change of sulfide concentration, mg/L-hr

EBOD = Effective BOD, mg/L

$[EBOD]_c$ = Climactic EBOD, mg/L

H^+ = Hydrogen ion concentration, mol/L

HCO_3^- = Concentration of bicarbonate ion, mol/L

H_2S = Molar concentration of hydrogen sulfide

(HS^-) = Molar concentration of the hydrosulfide ion

J = Fraction of dissolved sulfide present as H_2S as a function of pH

K_T = Rate constant at temperature T

k = Coefficient of efficiencyfor acid reaction considering the estimated fraction of acid remaining on the wall (Eq. 6-9)

L = Length of reach in meters/1,000 or feet/1,000 (Eq. 6-7)

L = Service life, years (Eq. 6-10)

M = Effective sulfide flux coefficient in meters per hour

M.W. = Molecular weight of substances, g

N = Empirical factor accounting for sulfides losses by oxidation and escape to atmosphere

OH^-= Concentration of hydroxide ion, mol/L

p = Wetted perimeters, ft

p'= Perimeter of pipe exposed to atmosphere, ft

pH = Negative logarithm of the activity of the hydrogen ion

pK^1 = Negative logarithm of the practical ionization constant in water

P^* = Pressure at which TO is given, atm

Q = Discharge volume in cubic feet per second

r = Hydraulic radius, feet

S = Invert slope (Eq. 9-1)

S = Slope (hydraulic gradient), ft/ft (Eq 6-12)

S = Total sulfide concentration mg/L (Eq. 6-1)

s = Energy gradient of the stream (Eq. 6-1)

s = Slope of energy line of the wastewater stream m/m or ft/ft (Eq. 6-8)

S_1 = Sulfide concentration at the upstream end of the reach, mg/L

S_2 = Sulfide concentration at downstream end of the reach, mg/L

Se = Energy gradeline or energy slope, ft/ft

$S_{\lim}$ or S_∞ = Limiting total sulfide concentration, mg/L (determined from Figs. 6-2 through 6-4)

S_C = Total sulfide mg/L, leaving the structure at point "C"

T = Wastewater temperature, °C

T_c = Climactic wastewater temperature, °C

t = Flow time, hours

$t_2 - t_1$ = t = Flow time in given reach with constant slope, diameter and flow, hours

T* = Temperature at which TO is given, °C

TCF = Turbulence Corrosion Factor, equal to 1.0 for uniform flow conditions well removed from points of turbulence

TO = Threshold odor

v = Stream velocity, ft/s (Eq. 6-8)

v = Velocity in meters per second (Eq. 6-1)

w = Specific weight of water, pounds per cubic foot

Y_o = Depth of flow

Z = Pomeroy Z factor (Eq. 6-12)

z = Thickness of allowable concrete (thickness covering the reinforcing steel) loss, inches

β = Factor combining hydraulic variables

θ = Temperature coefficient

ϕ_{sw} = Flux of hydrogen sulfide to the pipe wall, g/m^2-hr.

APPENDIX C
METRIC FORMS OF EQUATIONS

For metric units (r and D in meters):

$$\frac{d[S]}{dt} = 1 \times 10^{-3}\,[\text{EBOD}]\; r^{-1}\,(1 + 0.74r) \tag{6-13m}$$

$$[S]_{\text{lim}} = \frac{0.33 \times 10^{-3}\,[\text{EBOD}]}{(sv)^{0.375}} \times \frac{d_m}{r} \tag{6-4m}$$

Note: For N = 0.64 rather than 0.96, change constant 0.33 to 0.50.

For metric units:

$$S_2 = [S]_{\text{lim}} - \left[\frac{[S]_{\text{lim}} - S_1}{\log^{-1} \dfrac{(sv)^{0.375}\Delta t}{2.40_{d_m}}}\right] \tag{6-6m}$$

Note: For N = 0.64 rather than 0.96, change constant 2.40 to 3.60.

For metric units:

$$C_{\text{avg}} = \frac{11.5\,k\,\phi_{sw}}{A}\ \text{mm/yr} \tag{6-9m}$$

$$\phi_{sw} = 0.69\,J\,[DS]\left(\frac{b}{P'}\right)(sv)^{0.375}\ (\text{gms/m}^2\text{–hr}) \tag{6-8m}$$

APPENDIX D
SAMPLING AND ANALYTICAL TECHNIQUES

A. INTRODUCTION

Sulfide is often present in groundwater —especially in hot springs —and it is common in wastewaters, coming in part from the decomposition of organic matter, sometimes from industrial wastes, but mostly from the bacterial reduction of sulfate. Hydrogen sulfide escaping into the air from sulfide-containing wastewater causes odor nuisances. The threshold odor concentration of H_2S in clean water is between 0.01 and 0.1 mg/L. H_2S is very toxic and has claimed the lives of numerous workers in sewers. It attacks metals directly, and indirectly has caused serious corrosion of concrete sewers, because it is oxidized biologically to sulfuric acid on the pipe wall.

From an analytical standpoint, three categories of sulfide in water and wastewater are distinguished:

1. **Total sulfide** includes dissolved H_2S and HS^-, as well as acid-soluble metallic sulfides present in the suspended matter. The S^{2-} is negligible, amounting to less than 0.5% of the dissolved sulfide at pH 12, less than 0.05% at pH 11, etc. Copper and silver sulfides are so insoluble that they do not respond in the ordinary sulfide determinations; they can be ignored for practical purposes.
2. **Dissolved sulfide** is that remaining after the suspended solids have been removed by flocculation and settling.
3. **Un-ionized hydrogen sulfide** may be calculated from the concentration of dissolved sulfide, the pH of the sample, and the practical ionization constant of hydrogen sulfide.

B. SAMPLING AND STORAGE

Take samples with a minimum of aeration. Preserve a sample for a total sulfide determination by putting four drops of 2N zinc acetate into a 100-mL bottle, filling completely with the sample, and stoppering.

C. QUALITATIVE TESTS

A qualitative test for sulfide is often useful and is advisable in the examination of certain industrial wastes containing interfering substances which may give a false negative result in the methylene blue procedure.

Antimony test. Add 0.5 mL saturated solution of potassium antimony tartrate to about 200 mL sample in a bottle, and follow with 0.5 mL 6N HCl in excess of phenolphthalein alkalinity. The yellow Sb_2S_3 is discernible at a sulfide concentration of 0.5 mg/L. Comparisons with samples of known sulfide concentrations make the technique roughly quantitative. The only known interferences are metallic ions such as lead, which hold the sulfide so firmly that it does not produce antimony sulfide, and dithonite, which decomposes in acid solution to produce sulfide.

Silver sulfide-silver electrode test. The potential of a silver sulfide-silver electrode assembly relative to a reference electrode varies with the activity of the sulfide ion in solution. By correcting for the ion activity coefficient and pH, this potential allows an estimate of the sulfide concentration. Standardize the electrode frequently against a sulfide solution of known strength. An electrode of this type can be used as an endpoint indicator for titration of dissolved sulfide with a standard solution of a silver or lead salt, but slow response is always a problem.

Lead acetate paper and silver foil tests. Confirm odors attributed to H_2S with lead acetate paper. On exposure to the vapor of a slightly acidified sample, the paper becomes blackened by the formation of PbS. A strip of silver foil is more sensititve than lead acetate paper. Clean the silver by dipping in a sodium cyanide solution, and rinse. Silver is particularly suitable for long-time exposure in the vicinity of possible H_2S sources because the black Ag_2S is permanent, whereas PbS slowly oxidizes.

D. SELECTION OF QUANTITATIVE METHODS

Iodine reacts with sulfide in acid solution, oxidizing it to sulfur. A titration based on this reaction is an accurate method for determining sulfide at concentrations above 1 mg/L, if interferences are absent and if loss of H_2S is avoided. The unmodified iodine method (D) is useful for standardizing the colorimetric method, and is suitable for analyzing samples freshly taken from wells or springs. The method can be used for wastewater and partly oxidized water from sulfur springs if interfering substances are separated first.

The colorimetric method (C) is based on the reaction of sulfide, ferric chloride, and dimethyl-p-phenylenediamine (p-aminodimethylaniline) under conditions that produce methylene blue. Ammonium phosphate is added after color development to remove the ferric chloride color. The procedure is applicable at sulfide concentrations only up to 20 mg/L.

E. SEPARATION OF SOLUBLE AND INSOLUBLE SULFIDES

Unless the sample is entirely free from suspended solids (dissolved sulfide equals total sulfide), to measure dissolved sulfide first remove insoluble matter. Accomplish the separation by producing an aluminum hydroxide floc that is settled, leaving a clear supernatant for analysis.

1. Apparatus. Glass bottles with stoppers. Use 100 mL if sulfide will be determined by the methylene method, and 500 to 1,000 mL if by the titrimetric method.
2. Reagents: (1) Sodium hydroxide solution, NaOH, 6N, or (2) aluminum chloride solution, 6N —because of the hygroscopic and caking tendencies of this chemical, purchase 100-g (or 1/4-lb) bottles of the hexahydrate, $AICl_3{\cdot}6H_2O$. Dissolve the contents of a previously unopened 100-g bottle in 144 mL distilled water (or the contents of a 1/4-lb bottle in 164 mL water).
3. Procedure. To a 100-mL glass bottle add 0.2 mL (4 drops) 6N NaOH. Fill the bottle with sample and add 0.2 mL (4 drops) 6N $AICl_3$. Stopper the bottle with no air under the stopper. Rotate back and forth about a transverse axis vigorously for one minute or longer to flocculate the contents. Vary the volumes of these chemicals to get good clarification without using excessively large amounts and to produce a pH of 6 to 9. If a 500- or 1,000-mL bottle is used, add proportionally larger amounts of reagents. Let settle until reasonably clear supernatant can be drawn off. With proper flocculation, this may take 5 to 15 minutes. Do not wait longer than necessary.

F. SAMPLE PRETREATMENT TO REMOVE INTERFERING SUBSTANCES OR TO CONCENTRATE THE SULFIDE

The iodine titration method suffers interference by any reducing substances that react with iodine, including thiosulfate, sulfite, and various organic compounds, both solid and dissolved.

Strong reducing agents also interfere in the colorimetric test by preventing the formation of the blue color. Thiosulfate at concentrations above 10 mg/L may retard color formation or completely prevent it. Sulfide itself prevents the reaction if its concentration is very high, in the range of several hundred milligrams per liter. Test qualitatively industrial wastes likely to contain these substances but showing no color by the antimony method to avoid the possibility of a falsely negative report. Iodide, which is likely to be present in oil-field wastewaters, may diminish color formation if its concentration exceeds 2 mg/L. Ferrocyanide produces a blue color.

Eliminate interferences due to sulfite, thiosulfate, iodide, and many other soluble substances, but not ferrocyanide, by first precipitating the sulfide as ZnS, then removing all or most of the solution and replacing with distilled water. Use the same procedure, even when not needed for removal of interferences, to concentrate the sulfide content into a smaller volume for greater analytical sensitivity.

1. Apparatus. Glass bottles with stoppers.
2. Reagents: (1) Zinc acetate, 2N: dissolve 220 g $Zn(C_2H_3O_2)_2 \cdot 2H_2O$ in 870 mL water; this makes 1 L solution. (2) sodium hydroxide solution, NaOH,6N.
3. Procedure. Put 0.15 mL (3 drops) 2N zinc acetate solution into a 100-mL glass bottle, fill the bottle with sample, and add 0.10 mL (2 drops) 6N NaOH solution. Stopper with no air bubbles under the stopper and mix by rotating back and forth about a transverse axis. For the titrimetric procedure, use a 500-mL bottle or other convenient size, with proportionally larger volumes of reagents. The volume of reagents added may be varied, depending on the nature of the sample, since the objective is to obtain a precipitate that is not excessively bulky but that settles readily. Add enough NaOH to produce a pH above 9. Let the precipitate settle for 30 minutes. The treated sample is relatively stable and can be held for several hours. However, if much iron is present, oxidation may be fairly rapid.

   If the titrimetric method is to be used, filter the precipitate through glass fiber filter paper and continue at once with titration. If the methylene blue method is used, let the precipitate settle for 30 minutes and decant off as much water as possible without loss of precipitate. Refill the bottle with distilled water, re-suspend the precipitate, and withdraw a sample. If interfering substances are present in high concentration, settle, decant, and refill a second time. If the sulfide concentration is known to be low, add only enough water to bring the volume to one-half or one-fifth of the original volume. Use this technique for analyzing samples of very low sulfide concentrations. After determining the sulfide concentration colorimetrically, multiply the result by the ratio of the final to the initial volume.

G. METHYLENE BLUE METHOD

1. Apparatus

- Matched test tubes, approximately 125 mm long and 15 mm OD.
- Droppers, delivering 20 drops/mL methylene blue solution. To obtain uniform drops it is essential to hold the dropper in a vertical position and to allow the drops to form slowly.

- If photometric rather than visual color determination will be used, either: (1) Spectrophotometer, for use at a wavelength of 625 nm with cells providing light paths of 1 cm and 1 mm, or (2) filter photometer, with a filter providing maximum transmittance near 600 nm.

2. Reagents

- Amine-sulfuric acid stock solution: dissolve 27 g N,N-dimethyl-p-phenylenediamine oxalate (also called p-aminodimethylaniline oxalate) in a cold mixture of 50 mL conc H_2SO_4 and 20 mL distilled water. Cool and dilute to 100 mL with distilled water. The amine oxalate should be fresh; an old supply may be oxidized and discolored to a degree that results in interfering colors in the test. Store in a dark glass bottle. When this stock is diluted and used in the procedure with a sulfide-free sample, it must yield a colorless solution.
- Amine-sulfuric acid reagent: dilute 25 mL amine-sulfuric acid stock solution with 975 mL 1+1 H_2SO_4. Store in a dark glass bottle.
- Ferric chloride solution: dissolve 110 g $FeCl_3 \cdot H_2O$ in 40 mL water.
- Sulfuric acid solution, H_2SO_4, 1+1.
- Diammonium hydrogen phosphate solution: dissolve 400 g $(NH_4)_2HPO_4$ in 800 mL distilled water.
- Methylene blue solution I: use USP grade dye or one certified by the Biological Stain Commission. The dye content should be reported on the label and should be 84% or more. Dissolve 1.0 g in distilled water and make up to 1 L. This solution will be approximately the correct strength, but because of variation between different lots of dye, standardize against sulfide solutions of known strength and adjust its concentration so that 0.05 mL (1 drop) = 1.0 mg/L sulfide.

Standardization: put several grams of clean, washed crystals of sodium sulfide, $Na_2S \cdot 9H_2O$, into a small beaker. Add somewhat less than enough water to cover the crystals. Stir occasionally for a few minutes, then pour the solution into another vessel. This solution reacts slowly with oxygen, but the change is unimportant in a period of a few hours. Make the solution daily. To 1 L distilled water, add 1 drop of solution and mix. Immediately determine the sulfide concentration by the methylene blue procedure and by the titrimetric procedure. Repeat the procedures, using more than 1 drop of H_2S solution or smaller volumes of water, until at least five tests have been made, with a range of sulfide concentrations between 1 and 8 mg/L. Calculate the average percent error of the methylene blue result as compared to the titrimetric result. If the average error is negative, that is, the methylene blue results are lower than the titrimetric results, dilute the methylene blue solution by the same percentage, so that a greater volume will be used in matching colors. If the methylene blue results are high, increase the strength of the solution by adding more dye.

- Methylene blue solution II: dilute 10.00 mL of the adjusted methylene blue solution 1 to 100 mL.

3. Procedure

Color development. Transfer 7.5 mL sample to each of two matched test tubes, using a special wide-tip pipet or filling to the marks on the test tubes. Add to Tube A 0.5 mL amine-sulfuric acid reagent and 0.15 mL (3 drops) $FeCl_3$ solution. Mix immediately by inverting the tube slowly, only once. To Tube B add 0.5 mL 1+1 H_2SO_4 and 0.15 mL (3 drops) $FeCl_3$ solution and mix. The presence of sulfide ion will be indicated by the appearance of blue color in Tube A. Color development is usually complete in about 1 minute, but a longer time is often required for the fading out of the initial pink color. Wait 3 to 5 minutes, then add 1.6 mL $(NH_4)_2HPO_4$ solution to each tube. Wait 3 to 15 minutes and make color comparisons. If zinc acetate was used, wait at least 10 minutes before making a visual color comparison.

Color determination.

1. Visual color estimation. Add methylene blue solution I or II, depending on the sulfide concentration and the desired accuracy of the test, dropwise, to the second tube, until the color matches that developed in the first tube. If the concentration exceeds 20 mg/L, repeat the test with a portion of the sample diluted to one tenth. With methylene blue solution I, adjusted so that 0.056 mL (1 drop)=1.0 mg/L sulfide when 7.5 mL of sample are used:

   mg/L sulfide = No. drops solution I + 0.1 (No. drops solution II)
2. Photometric color measurement. A cell with a light path of 1 cm is suitable for measuring sulfide concentrations from 0.1 to 2.0 mg/L. Use shorter or longer light paths for higher or lower concentrations. The upper limit of the method is 20 mg/L. Zero the instrument with a portion of the treated sample from Tube B. Prepare calibration curves on the basis of the colorimetric tests made on Na_2S solutions simultaneously analyzed by the titrimetric method, plotting concentration versus absorbance. A straight-line relationship between concentration and absorbance can be assumed from 0 to 1.0 mg/L. Read the sulfide concentration from the calibration curve.
3. Precision and accuracy. The accuracy is about ± 10%. The standard deviation has not been determined.

H. TITRIMETRIC (IODINE) METHOD

1. Reagents

- Hydrochloric acid, HCl, 6N.
- Standard iodine solution, 0.0250N: dissolve 20 to 25 g potassium iodide, KI, in a little water and add 3.2 g iodine. After the iodine has dissolved, dilute to 1,000 mL and standardize against 0.02050N sodium thiosulfate, using starch solution as indicator.
- Standard sodium thiosulfate solution, 0.0250N.
- Starch solution.

2. Procedure.

If the sulfide was precipitated with zinc and water was decanted, make the titration in the original bottle. If the precipitate was dewatered on a glass filter paper, return the filter with precipitate to the original bottle.

Measure from a buret into a 500-mL flask an amount of iodine solution estimated to be an excess over the amount of sulfide present. Add distilled water, if necessary, to bring the volume to about 20 mL. Add 2 mL 6N HCl. Pipet 200 mL sample into the flask, discharging the sample under the surface of the solution. If the iodine color disappears, add more iodine so that the color remains. Back-titrate with sodium thiosulfate solution, adding a few drops of starch solution as the end point is approached, and continuing until the blue color disappears.

3. Calculation. One mL of 0.025N iodine solution reacts with 0.4 mg of sulfide:

$$\text{mg/L S} = \frac{400\,(a - b)}{\text{mL sample}}$$

in which a = mL 0.025N iodine, and b = mL 0.025N thiosulfate solution.

4. Precision. The precision of the end point varies with the nature of the sample. In clean waters, it should be determinable within 1 drop, which is equivalent to 0.1 mg/L in a 200-mL sample.

I. CALCULATION OF UN-IONIZED HYDROGEN SULFIDE

Hydrogen sulfide and HS^-, which together constitute dissolved sulfide, are in equilibrium with hydrogen ions:

$$H_2S \rightleftarrows H^+ + HS^-$$

The ionization constant of H_2S is used to calculate the distribution of dissolved sulfide between the two forms. The practical constant written in the logarithmic form pK′ is used. The constant varies with temperature and ionic strength of the solution. The ionic strength effect can be estimated most easily from the conductivity. Because the effect of ionic strength is not large, values that are sufficiently dependable generally can be assumed if the nature of the sample is known. Table D-1 gives values of pK′ for various temperatures and ionic strengths. The temperature effect is practically linear over the range from 15°C to 35°C; interpolations or extrapolations can be used. The last line of Table D-1 corresponds approximately to seawater.

From the pH of the sample and the appropriate value of pK′ calculate pH-pK′. From Fig. 6-14 read the proportions of dissolved sulfide present as H_2S (left-side scale of Fig. 6-14). Let this proportion equal J and

$$J \times (\text{dissolved sulfide}) = \text{un-ionized } H_2S \text{ expressed as S}$$

J. SAMPLE COLLECTION, TRANSPORTATION, AND PRESERVATION

It is obvious that great care in analysis is of little value if the sampling step is inadequate. It is important that a number of factors be carefully considered in any sampling procedure.

The largest errors in analytical work arise from sampling, sample preservation, and compositing. Usually, a large phenomenon involving millions of gallons is being characterized by looking at four to eight liters of sample taken over 24 hours. Laboratory equipment and analytical techniques are generally quite precise, but sampling at best is an approximation. As much attention as possible must therefore be given to this step.

1. Representative Sampling

The following factors need to be carefully considered in sampling:

TABLE D-1. Values of pK′, Logarithmic Practical Ionization Constants for Hydrogen Sulfide

Conductivity at 25° C	pK′ at Given Temperature		
	20°C	25° C	30°
(1)	(2)	(3)	(4)
0		7.03*	
100	7.08	7.01	6.94
200	7.07	4.00	6.93
400	7.06	6.99	6.92
700	7.05	6.98	6.91
1,200	7.04	6.97	6.90
2,000	7.03	6.96	6.89
3,000	7.02	6.95	6.88
4,000	7.01	6.94	6.87
5,200	7.00	6.93	6.86
7,200	6.99	6.92	6.85
10,000	6.98	6.91	6.84
14,000	6.97	6.90	6.83
22,000	6.96	6.89	6.82
50,000	6.95	6.88	6.81

* Theoretical.

- Frequency
- Location
- Quantity
- Containers
- Storage and transport
- Contamination

2. Location

Location of the sampling points is obviously of great importance. Poor location can, to a degree, be compensated for by increasing the locations and the number of samples, but this in turn increases the difficulty of analysis both in the laboratory and of the data so obtained. Correct location is of course dependent upon a thorough knowledge of the process or receiving waters. Simply requiring a shotgun grouping of sampling points may confuse the analysis rather than clarify it.

The sampling location must be one of representative flow. At sampling boxes or manholes, care should be taken in collecting the sample and removing it from the box or manhole to preclude scraping the walls.

It is important that exact sampling locations be designated and used for each repetitive sample.

- Use flow diagrams for treatment plant or industrial discharge.

- Use USGS maps or similar accurate charts for ocean discharges or discharges to streams, lakes, or ground.
- Use in-plant and aerial photographs for other discharges to provide accurate location.

Carefully mark sampling stations and provide a number designation. Some uniform numbering system is necessary so that sample location data may be easily retrieved. For plants, a similar type of in-plant process numbering might be used. If possible, such standardization should be designed with the possibility of ultimate computer data storage and retrieval.

3. Sampling Methods

Sampling methods must be selected with the component to be tested in mind. Composite samples are used to assess treatment results and to test for:

- Suspended solids
- Settleable solids
- Ether solubles
- BOD and COD
- Floatable solids
- Turbidity
- Toxicity (however, dissolved oxygen must be added for the test to be run)
- Radioactivity
- Total phosphate
- Total nitrogen
- Other nutrients.

Grab samples are used for:

- pH
- Cyanide
- Phenolic compounds
- Ammonia nitrogen
- Total chlorine residual
- Total identifiable chlorinated hydrocarbons
- Other exotic organics
- Dissolved oxygen
- Sulfide
- Other materials likely to break down or change under the conditions of sampling or sample storage

- Bacteriological or biological determinations.

4. Sample Storage and Transport

Although the preservatives mentioned previously are usable, it is always better to start analyses as soon as possible. Storage at temperatures no greater than 4°C is also advisable, whether preservatives are used or not, to minimize change. It is therefore important that transport to the analyst doing the work and a scheduling for immediate analysis be prearranged prior to any sampling.

As already noted, certain materials must be taken as grab samples and analyzed immediately without storage or transport. These include:

- The nitrogen series: ammonia, nitrate, nitrite, organic-N
- Dissolved sulfide
- Phosphorous (recommended, although may be reasonably stable)
- pH
- DO
- Cyanide
- Phenols
- Chlorine, chloramines, chlorine residual
- Cr^{+6}
- Turbidity
- Color

REFERENCES

1. Camp, T.R., *Water and Its Impurities*, Reinhold Publishing Corp., New York, New York, 1963, p. 63.
2. Lawrence, A.W., McCarty, P.L., and Guerin, F.J.A., "The Effects of Sulfides on Anaerobic Treatment," *Int. J. Air Water Pollut.*, Vol. 10, 1966, p. 207.
3. Nusbaum, I., "Determination of Phenols by the p-nitrosodimethylaniline Method: A Discussion," *Sewage Ind. Wastes*, Vol. 25, 1953, p. 312.
4. Nusbaum, I., "Determining Sulfides in Water and Wastewater," *Water Sewage Works*, Vol. 112, 1965, p. 113.
5. Platford, R.F., "The Activity Coefficient of Sodium Chloride in Seawater," *J. Mar. Res.*, Vol. 23, 1965, p. 55.
6. Pomeroy, R.D., "The Determination of Sulfides in Sewage," *Sewage Works J.*, Vol. 8, 1936, p. 472.
7. Pomeroy, R.D., "Hydrogen Sulfide in Sewage," *Sewage Works J.*, Vol. 13, 1941, p. 498.
8. WPCF, *Simplified Laboratory Procedure for Wastewater Examination*, 3rd Edition, Water Pollution Control Federation, Washington, D.C., 1985.

APPENDIX E
EXAMPLE PROBLEMS

Where To Find Example Problems

A. Evaluating Sulfide Build-up Potential

1. EPA flow-slope relationship, Fig. 4-8

a. Gravity sewer on uniform slope

1. Given Data: Trunk Sewer with no intermediate junction BOD_c = 254 mg/L, T_c = 30°C, Q = 8 cfs (6-hr average peak flow), slope = 0.0006, L = 3000 ft.
2. Solution: Using Eq. 6-11:

$$EBOD_c = (BOD_c)(1.07)^{(T_c-20)} = (254)(1.07)^{(30-20)}$$
$$= 499.7 \text{ mg/L} = 500 \text{ mg/L}$$

Q = 8 cfs S = 0.0006

The slope-flow relationship is below Curve B; therefore, sulfide build-up potential is likely.

b. Gravity sewer with drops

1. Given Data: Same trunk sewer as above, except drops 0.60 ft twice through manholes at 1,000-ft intervals.
2. Solution: Q = 8 cfs

$$S = \left(0.0006 + \frac{1.2 \text{ ft}}{3000 \text{ ft}}\right) = 0.001$$

Thus, with average slope = 0.001 and EBOD = 500 mg/L, the slope-flow relationship is on Curve A, and the sulfide build-up potential is minimal.

c. Gravity sewer with uniform slope

1. Given data: Same sewer as A.1.b. with EBOD = 350 mg/L, slope = 0.001, Q = 8 cfs.
2. Solution: Fig. 4-8 shows that for a Curve A condition, a flow of 8 cfs requires an effective slope of 0.001. With an EBOD = 350 mg/L, the effective slope for no build-up would be:

$$(0.001) \times \left(\frac{350}{500}\right)^{1/2} = 0.0008 = \text{effective slope}$$

Therefore, for the given slope of 0.001, the flow-slope relationship plots below Curve A; hence, sulfide build-up potential is anticipated.

2. Pomeroy Z Formula Criterion

a. Gravity sewer steep slope

1. Given data: Gravity trunk sewer with no junctions, EBOD = 500 mg/L, S = 0.06, d/D = 0.50, Q = 8 cfs.
2. Solution: Looking at Table 6-4, Chap. 6, we see that if d/D = 0.50, then P/b = 1.57, and using Eq. 6-12

$$Z = \frac{(\text{EBOD})^{(P/b)}}{(Q)^{1/3}(S)^{1/2}} = \frac{(500)(1.57)}{(8)^{1/3}(0.06)^{1/2}} = 1{,}602\ Z$$

From Table 6-3, we see that this value indicates sulfide build-up will rarely occur.

b. Gravity sewer shallow slope

1. Given data: Gravity trunk sewer as in A.2. a.1, except that the slope = $S = 0.0008$.
2. Solution: Calculate Pomeroy Z factor using Eq. 6-12:

$$Z = \frac{(500)(1.57)}{(8)^{1/3}(0.0008)^{1/2}} = 15{,}114$$

According to Table 6-3, this value indicates sulfide generation is common.

B. SULFIDE BUILD-UP PREDICTION

1.Gravity sewers using the Pomeroy-Parkhurst Equations

a. Single reach example.

1. Given data: $(\text{BOD})_c$ = 300 mg/L, T_c = 28°C, length= 5,300 ft, design life = 50 years, k factor = 0.7, sewage pH = 7.4, slope = 0.00085 ft/ft, climactic ratio c = 0.5, upstream sulfide = 1.0 mg/L, pipe diameter = 24 in., insoluble sulfide = 0.2 mg/L, design flow for pipe half full (d/D = 0.50), Q = 3.3 cfs, and Manning n = 0.013.

2. Solution: Use Eq. 6-11 to solve for climactic EBOD. Then use Eqs. 6-4 and 6-6 to solve for sulfide limiting value and downstream sulfide, respectively.

$$(\text{EBOD})_c = (\text{BOD})_c \times (1.07)^{(T_c-20)}$$
$$= (300\ \text{mg/L})(1.07)^{28-20} = 515\ \text{mg/L}$$

Since $R = \frac{A}{P}$, and $d_m = \frac{A}{b}$ then $\frac{d_m}{R} = \frac{P}{b}$

From Table 6-4, $d/D \cong 0.5$ gives:

$$\frac{A}{d^2} \times d^2 = 0.393 \times (2)^2 = 1.572 = A$$

and $\frac{P}{b}$ =1.57, substituting,

$$V = \frac{Q}{A} = \frac{3.3\ \text{cfs}}{1.572\ \text{ft}^2} = 2.10\ \text{fps}$$

Now calculate S_{lim} using Eq. 6-4 as modified above:

$$S_{lim} = \frac{(0.00052)\ (EBOD)_c}{(sv)^{0.375}} \times \frac{P}{b}$$

$$= \frac{(0.00052)\,(515)\,(1.57)}{(0.00085 \times 2.10)^{0.375}}$$

$$= 4.51\ mg/L$$

Flow through time in hours

$$\Delta t = \frac{L}{V(3{,}600)} = \frac{5{,}300}{(2.10)\,(3{,}600)}$$

$$\Delta t = 0.70\ hr$$

Now calculate downstream sulfide concentration, S_2, using Eq. 6-6:

$$S_2 = S_{lim} - \frac{S_{lim} - S_1}{\log^{-1}\left[\frac{(sv)^{0.375}\,t}{1.15\,d_m}\right]}$$

in which upstream sulfide $S_1 = 1.0$ mg/L,

$$d_m \text{ (Table 6–4)} = d_m/D = 0.393\ (2.0)$$
$$= d_m = 0.786\ ft.$$

$$S_2 = 4.51 - \frac{4.51 - 1.0}{\log^{-1} \frac{(0.00085 \times 2.1)^{0.375} \times 0.7}{1.15\,d_m}}$$

$$S_2 = 4.51 - \frac{3.5}{\log^{-1} \frac{(0.0652)}{0.9039}} = 4.51 - \frac{3.5}{1.808}$$

$$S_2 = 1.54\ mg/L$$

b. Multiple reach example

1. Given data: We will use same basic data as before for the first reach.

Reach 2 data: pipe diameter = 30 in., slope = 0.0008 ft/ft, and L = 10,000 ft.
Reach 3 data: pipe diameter = 21 in., slope = 0.001 ft/ft, and L = 5,000 ft.

2. Solution for Reach 2 (from Kings):

$$Q_F = \frac{0.463}{n}\, d^{8/3}\ S^{1/2}$$

gives

$$Q_{full} = \frac{0.463}{0.013} (2.5)^{8/3} (0.0008)^{1/2}$$

$$= 11.60 \text{ cfs}$$

therefore,

$$\frac{Q}{Q_F} = \frac{3.3}{11.60} = 0.284$$

From Fig. 6-13 we see $d/D = 0.36$ for $Q/Q_F = 0.284$. Interpolating from Table 6-4:

d/D	A/D^2	P/b	d_m/D
0.36	0.255	1.34	0.265

From this

$$\left(\frac{A}{D^2}\right) D^2 = 0.255 \, (2.5)^2$$

$$= 1.59 \text{ ft}^2 = A$$

Substituting,

$$v = Q/A = \frac{3.3 \text{ cfs}}{1.59 \text{ ft}^2} = v = 2.08 \text{ fps}$$

Now calculate the sulfide limit S_{lim} from Eq. 6-4:

$$S_{\text{lim}} = \frac{(0.00052)\ (515)\ (1.34)}{(0.0008 \times 2.08)^{0.375}}$$

$$S_{\text{lim}} = 3.95 \text{ mg/L}$$

Flow through time

$$t = \frac{10{,}000}{2.08\ (3{,}600)} = 1.34 \text{ hr}$$

and

$$\frac{d_m}{D} D = d_m = 0.265(2.5) = d_m = 0.66 \text{ ft}$$

Now calculate the downstream sulfides S_2 using the downstream sulfides of Reach 1 as the upstream sulfides S_1 = 1.54 mg/L for Reach 2 (from Eq. 6-6):

$$S_2 = 3.95 - \frac{3.95 - 1.54}{\log^{-1} \dfrac{(0.0008 \times 2.08)^{0.375}\ (1.34)}{1.15 \times 0.66}}$$

$$S_2 = 3.95 - \frac{2.41}{1.45}$$

$$S_2 = 2.28\ \text{mg/L}$$

and this value now becomes the upstream sulfide S_1 for Reach 3.

Solution for Reach 3 (from Kings):

$$Q_f = \frac{0.463}{0.013}\ (1.75)^{8/3}\ (0.001)^{1/2} = 5.01\ \text{cfs}$$

therefore

$$\frac{Q}{Q_{full}} = \frac{3.3\ \text{cfs}}{5.01\ \text{cfs}} = 0.66$$

From Fig. 6-13, we see d/D = 0.59. Interpolating from Table 6-4:

d/D	A/D²	A/P	d_m/D
0.59	0.482	1.78	0.491

From this,

$$\left(\frac{A}{D^2}\right) D^2 = 0.482\ (1.75)^2 = 1.48\ \text{ft}^2$$

Substituting:

$$v = \frac{Q}{A} = \frac{3.30\ \text{cfs}}{1.48\ \text{ft}^2} = v = 2.23\ \text{fps}$$

Now calculate the sulfide limit S_{lim} from Eq. 6-4:

$$S_{\text{lim}} = \frac{(0.00052)\ (515)\ (1.78)}{(0.001 \times 2.23)^{0.375}} = \frac{0.48}{0.10}$$

$$S_{\text{lim}} = 4.72\ \text{mg/L}$$

Flow through time

$$\Delta t = \frac{5{,}000}{2.23\ (3{,}600)}$$
$$= 0.62\ \text{hr}$$

and

$$\frac{(d_m)}{D} D = d_m = 0.491\ (1.75) = 0.86\ \text{ft}$$

TABLE E-1. Results for Multiple Reach Example

Reach (1)	Rel. Depth (2)	Flow, cfs (3)	Flow, fps (4)	Sulfide Limit, mg/L (5)	S_1, mg/L (6)	S_2, mg/L (7)
1	0.50	3.30	2.10	4.51	1.00	1.54
2	0.36	3.30	2.08	3.95	1.54	2.28
3	0.59	3.30	2.23	4.71	2.28	2.61

Note: Effective biochemical oxygen demand = 515 mg/L.

Now calculate the downstream sulfide S_2 where $S_2 = 2.28$ mg/L of Reach 2 becomes S_1 of Reach 3 (from Eq. 6-6):

$$S_2 = 4.71 - \frac{4.71 - 2.28}{\log^{-1} \dfrac{(0.001 \times 2.23)^{0.375}\ (0.62)}{1.15 \times 0.86}}$$

$$= 4.71 - \frac{2.43}{1.16} = 2.61 \text{ mg/L}.$$

Summary: Results are shown in Table E-1.

c. Multiple reach example with variable flow depth

1. Given data: A three-reach, approximately four-mile-long interceptor sewer is to be designed. The sewage characteristics are expected to be as follows:Climactic BOD = 300 mg/L; climactic temperature = 28°C; acid reaction factor k = 0.80; pH = 7.4; climactic ratio c = 0.35;upstream sulfide S_1 at the beginning of Reach 1 is expected to be 0.5 mg/L during the climactic period; and insoluble sulfide is 0.20 mg/L. Design life required is 50 years. The relative depth of flow d will be 0.2 for the first 10 years and 0.45 for the next 40 years.The interceptor reach data are as follows:

Reach	Diam, in.	Slope	Length, ft
1	24	0.00080	5,000
2	36	0.00200	10,000
3	42	0.00088	5,000

2. Required: Determine the probable total sulfide levels in mg/L which may be expected during the climactic period for each increment of the design life of this sewer.
3. Solution for Reach 1: Use Eq. 6-11 for solution of EBOD.

$$\text{EBOD} = (300)(1.07)^{28-20} = 515 \text{ mg/L}.$$

Design depth $d/D = 0.20$. So, from Table 6-4, $v/v_f = 0.617$, $P/b = 1.16$, $d_m/D = 0.140$. Using Manning equation for a pipe flowing full (from Kings):

$$v_f = \frac{0.590}{n}(d)^{2/3}(S)^{1/2}$$

$$v_f = \left(\frac{0.590}{0.013}\right)(2.0)^{2/3} \times (0.0008)^{1/2}$$

$$= 2.04\ \text{ft/s}$$

$$v = 0.617$$

$$v_f = 1.26\ \text{ft/s}$$

Calculating S_{lim} from Eq. 6-4:

$$S_{\text{lim}} = \frac{(0.00052)(515)(1.16)}{[(0.0008)(1.26)]^{0.375}} = 4.13\ \text{mg/L}$$

$$S_{\text{lim}} = \frac{(0.00052)(515)(1.48)}{[(0.0008)(1.95)]^{0.375}} = 4.47\ \text{mg/L}$$

Flow through time

$$\Delta t = \frac{5{,}000}{1.26}(3{,}600)$$

$$\Delta t = 1.10\ \text{hr}$$

Calculating S_2 from Eq. 6-6:

$$S_2 = 4.13 - \frac{4.13 - 0.5}{\log^{-1}\dfrac{[(0.0008)\ (1.26)]^{0.375}\ (1.10)}{1.15 \times 0.28}}$$

$$S_2 = 4.13 - \frac{3.63}{\log^{-1}(0.26)} = 2.12$$

So $S_2 = 2.12$ mg/L for Reach 1, first increment (ten years). For the second (40-yr) increment, $d/D = 0.45$, $v_f = 2.04$ ft/s, $v/v_f = 0.954$, $P/b = 1.48$, $dm/D = 0.345$, and $v = (0.954)(2.04) = 1.95$ ft/s = 0.59 m/s.

$$\Delta t = \frac{5{,}000}{(1.95)(3{,}600)} = 0.71\ \text{hr}$$

$$d_m = (0.345)(2.0) = 0.69\ \text{ft}$$

From Eq. 6-4:

$$S_{\text{lim}} = \frac{(0.00052)(515)(1.48)}{[(0.0008)(1.95)]^{0.375}} = 04.47\ \text{mg/L}$$

And from Eq. 6-6:

$$S_2 = 4.47 - \frac{4.47 - 0.5}{\log^{-1} \frac{[(0.0008)\ (1.95)]^{0.375}\ (0.71)}{(1.15)\ (0.69)}}$$

$$S_2 = 4.47 - \frac{3.97}{1.20} = 1.16\ ^{mg}\!/\!_L\ .$$

So S_2 = 1.16 mg/L for Reach 1, second increment, of 40 years.

4. Solution for Reach 2: d/D = 0.20, S_1 = 2.12 mg/L (from Reach 1), v/v_f = 0.617, P/b = 1.16, $d_{m/D}$ = 0.140, and

$$v_f = \left(\frac{0.590}{0.013}\right) (3.0)^{2/3}\, (0.002)^{1/2} = 4.24\ ^{ft}\!/\!_s$$

$$v = (4.24)\,(0.617) = 2.61\ ^{ft}\!/\!_s$$

$$\Delta t = \frac{10{,}000}{(2.61)\,(3{,}600)} = 1.06 \text{ hr}$$

$$d_m = (0.140)\,(3.0) = 0.42 \text{ ft}$$

From Eq. 6-4:

$$S_{\text{lim}} = \frac{(0.00052)\,(515)\,(1.16)}{[(0.0002)\,(2.61)]^{0.375}} = 2.23\ ^{mg}\!/\!_L$$

From Eq. 6-6:

$$S_2 = 2.23 - \frac{2.23 - 2.12}{\log^{-1} \frac{[(0.0002)\ (2.61)]^{0.375}\ (1.06)}{(1.15)\ (0.42)}}$$

$$S_2 = 2.23 - \frac{0.11}{2.02} = 2.18\ ^{mg}\!/\!_L$$

S_2 = 2.18 $^{mg}\!/\!_L$ for Reach 2, first increment, 10 years.

For the second increment (40 years), d/D = 0.45, S_1 = 1.16 mg/L (from Reach 1), v/v_f = 0.954, P/b = 1.48, d_m/D = 0.345, and

$$\Delta t = \frac{10{,}000}{(4.05)(3{,}600)} = 0.69 \text{ hr}$$

$$d_m = (0.345)\,(3.0) = 1.04 \text{ ft}$$

FromEq. 6-4:

$$S_{\text{lim}} = \frac{(0.00052)\,(515)\,(1.48)}{[(0.002)\,(4.05)]^{0.375}} = 2.41\ ^{mg}\!/\!_L$$

From Eq. 6-6:

$$S_2 = 2.41 - \frac{2.41 - 1.16}{\log^{-1} \frac{[(0.002)\,(4.05)]^{0.375}\ (0.69)}{(1.15)\,(1.04)}}$$

$$S_2 = 2.41 - \frac{1.25}{1.24}$$

So S_2 = 1.41 mg/L for Reach 2, second increment, 40 years.

5. Solution for Reach 3: d/D = 0.20, S_1 = 2.18 mg/L (from Reach 2),

$$v_f = \left(\frac{0.590}{0.013}\right)(3.5)^{2/3}(0.00088)^{1/2} = 3.12 \text{ ft/s}$$

$$v = (0.617)\, v_f = 1.95 \text{ ft/s}$$

$$\Delta t = \frac{5{,}000}{(1.92)(3{,}600)} = 0.72 \text{ hr}$$

$$d_m = 0.140 \times 3.5 = 0.49 \text{ ft}$$

From Eq. 6-4:

$$S_{\text{lim}} = \frac{(0.00052)(515)(1.16)}{[(0.00088)(1.92)]^{0.375}} = 3.40 \text{ mg/L}$$

From Eq. 6-6:

$$S_2 = 3.40 - \frac{3.40 - 2.18}{\log^{-1} \dfrac{[(0.00088)\ (1.92)]^{0.375}\ (0.72)}{(1.15)\ (0.49)}}$$

$$S_2 = 3.40 - \frac{1.22}{1.31} = 2.47 \text{ mg/L}$$

So S_2 = 2.47 mg/L for Reach 3, first increment, 10 years. For the second increment (40 years), d/D = 0.45, S_1 = 1.41 mg/L (from Reach 2), v = (0.954)(3.12) = 2.98 ft/s

$$\Delta t = \frac{5{,}000}{(2.98)(3{,}600)} = 0.47 \text{ hr}$$

$$d_m = (0.345)(3.5) = 1.21 \text{ ft}$$

From Eq. 6-4:

$$S_{\text{lim}} = \frac{(0.00052)(515)(1.48)}{[(0.00088)(2.98)]^{0.375}} = 3.68 \text{ mg/L}$$

From Eq. 6-6:

$$S_2 = 3.68 - \frac{3.68 - 1.41}{\log^{-1} \dfrac{[(0.00088)\ (2.98)]^{0.375}\ (0.47)}{(1.15)\ (1.21)}}$$

$$S_2 = 3.68 - \frac{2.27}{1.09} = 1.59 \text{ mg/L}$$

So S_2 = 1.59 mg/L for Reach 3, second increment, 40 years.

6. Results are shown in Table E-2 and Fig. E-1.

TABLE E-2. Results for Multiple Reach Example with Variable Flow Depth

Reach (1)	Rel. depth (2)	Flow, cfs (3)	Flow, fps (4)	Sulfide limit, mg/L (5)	S_1, mg/L (6)	S_2, mg/L (7)
(a) Pipe Diam 1 = 24 in., slope = 0.0008 ft/ft, length = 5,000 ft						
1 , 1	0.20	0.56	1.26	4.13	0.50	2.12
1 , 2	0.45	2.67	1.95	4.47	0.50	1.16
(b) Pipe Diam 2 = 36 in., slope = 0.002 ft/ft,length = 10,000 ft						
2 , 1	0.20	2.62	2.61	2.23	2.12	2.18
2 , 2	0.45	12.46	4.05	2.41	1.16	1.41
(c) Pipe Diam 3 = 42 in., slope = 0.00088 ft/ft, length = 5,000 ft						
3 , 1	0.20	2.62	1.92	3.40	2.18	2.47
3 , 2	0.45	12.47	2.98	3.68	1.41	1.59

2. Force Mains and Inverted Siphons

a. Single reach force main.

1. Given data: $(BOD)_c = 400$ mg/L, $T_c = 25°C$, length = 3,080 ft, pipe diameter = 15 in., flow Q = 14.2 cfs, and, upstream sulfides S_1 = 2.0 mg/L.
2. Solution: (EBOD) = 400 x $(1.07)^{25-20}$ = 561 mg/L. Force main flows full, therefore $V = Q/A = Q/\ (D/2)^2$

$$V = \frac{14.2}{(0.39)} \quad 11.57 \text{ ft/ s}$$

$$\Delta t = \frac{3{,}080}{(11.57)\,(3{,}600)} = 0.07 \text{ hr}$$

$$r = d/4 = 1.25/4 = 0.31 \text{ ft}$$

Using Eq. 6-13 with $M = 1.0 \times 10^{-3}$ m/hr

$$\frac{dS}{dt} = (3.28)\ M(\text{EBOD})\,(1 + 0.48r)\ r^{-1}$$

$$\frac{dS}{dt} = \frac{(3.28)\,(1.0 \times 10^{-3})\,(561)\,[1 + (0.48)\,(0.31)]}{0.31}$$

$$= 6.8 \text{ mg/L} - \text{hr}$$

$$\left(\frac{dS}{dt}\right)(\Delta t) = \text{sulfide build-up} = (6.8)(0.07)$$

$$= 0.5 \text{ mg/L}$$

$$S_2 = S_1 + (dS) = 2 + 0.5 = 2.5 \text{ mg/L}$$

b. Two-barrel inverted siphon.

1. Given data: $(BOD)_c$ = 289 mg/L, T_c = 27°C, Q_{15} = 4.15 cfs, Q_{24} = 10.68 cfs, D_1 = 24 in. and L_2 = 365 ft. Upstream sulfides for both barrels = 0.50 mg/L.
2. Solution: $(EBOD)_c = (289)(1.07)^{27\text{-}20}$ = 464 mg/L

$$A_{24} = \left(\frac{2.0}{2}\right)^2 = 3.14 \text{ ft}^2,\ A_{15} = \left(\frac{1.25}{2}\right)^2$$

$$= 1.23 \text{ ft/s}$$

$$\frac{Q_{24}}{A_{24}} = V_{24} = \frac{10.68}{3.14} = 3.40 \text{ ft/s}$$

$$\frac{Q_{15}}{A_{15}} = V_{15} = \frac{4.15}{1.23} = 3.4 \text{ ft/s}$$

$$(\Delta t) = \frac{365}{(3.40)(3{,}600)} = 0.03 \text{ hr},\ r_{24}$$

$$= 0.50 \text{ ft},\ r_{15} = 0.31 \text{ ft}$$

$$\frac{dS_{24}}{dt} = \frac{(3.28)(1.0 x 10^{-3})(464)[1 + (0.48)(0.50)]}{0.50}$$

$$= 3.77 \text{ mg/L} - \text{hr}$$

$$\frac{dS_{15}}{dt} = \frac{(3.28)(1.0 x 10^{-3})(464)[1 + (0.48)(0.31)]}{0.31}$$

$$= 5.64 \text{ mg/L} - \text{hr}$$

So sulfide build-up (24 in.) = (3.77)(0.03) = 0.11 mg/L
and sulfide build-up (15 in.) = (5.64)(0.03) = 0.17 mg/L.

Determine two-barrel build-up by mass balance (Eq. 6-23):

$$S_{build\text{-}up} = \frac{(0.11)(10.68) + (0.17)(4.15)}{10.68 + 4.15}$$

$$= 0.13 \text{ mg/L}$$

Sulfide in flow exiting siphon = S_1 + build-up = 0.50 + 0.13 = 0.63 mg/L.

C. CALCULATION OF CORROSION RATE AND CORROSION PROTECTION

1. For the single reach example, paragraph B.1.a., calculate the maximum or climactic pipe crown corrosion rate in inches per year, assuming asbestos cement pipe with an alkalinity of 0.50.

Solution:

$$DS = \text{total sulfide - insoluble sulfide} \quad DS = 1.5 - 0.20 = 1.30 \text{ mg/L}$$

during climactic period.

Peak corrosion rate during climactic period: from Table 6-4, for $d/D = 0.50$, $b/P^l = 0.637$, slope = 0.00085, $v = 2.10$ ft/s, and pH = 7.4. From Table 6-5, for pH = 7.4, $J = 0.28$, and from Eq. 6-8:

$$\begin{aligned}\phi_{sw} &= 0.45\,(sv)^{0.375}\,J\,DS\,(b/P') \\ &= 0.45\,(0.00085 \times 2.10)^{0.375}\,(0.28)\,(1.30)\,(.637) \\ &= 0.0097 \text{ g/m-hr}\end{aligned}$$

From Eq. 6-9:

$$C_{ave} = \frac{0.45\,k\,\phi_{sw}}{A}$$

Assume $k = 0.70$ (given in B.1.a.), so

$$C_{ave} = \frac{0.45\,(0.7)\,(0.0097)}{0.5} = .0061 \text{ in/yr}$$

Crown or springline corrosion rate during climactic period (Eq. 6-25) assume $CCF = 1.50$:

$$\begin{aligned}C_{\max} &= C_{ave} \times CCF = 0.0061 \times 1.50 \\ &= 0.0092 \text{ in/yr}\end{aligned}$$

2. What is the annual average pipe crown corrosion rate? The climactic ratio c of annual average to peak (climactic period) sulfide was given in B.1.a. as 0.50.

Solution: Calculate the maximum crown corrosion as in the previous example, except use annual average [DS] rather than climactic [DS]:

Annual average [DS] = 0.5 x climactic [DS]

= 0.5 x 1.30 x 1.50 = 0.065 mg/L

From Eq. 6-8:

$$\phi_{sw} = 0.45\,(0.00085 \times 2.10)^{3/8}\,(0.28)\,(0.65)\,(0.367)$$

$$= 0.0049 \text{ g/m}^2 - \text{hr}$$

From Eq. 6-9:

$$C_{ave} = \frac{0.45\,k\,\phi_{sw}}{A} = 0.45\,(0.7)\,\frac{(0.0049)}{0.50} = 0.0031 \text{ in/yr}$$

3(a) Calculate the corrosion protection Az required for a 50-yr life, both for the mainline pipe and for a section downstream of a junction structure, using $CCF = 2.0$, and $TCF = 2.0$, and $k = 0.70$.

Solution: From (2) above, the annual average ϕ_{sw} is 0.0049 g/m^2-hr.

Using Eq. 6-27, for mainline pipe (uniform flow conditions):

$$\begin{aligned} Az &= 0.45\, k\, \phi_{sw}\, L \times CCF \times 1.0 \\ &= 0.45\,(0.70)\,(0.0049)\,(50) \times 2.0 \times 1.0 = 0.154 \end{aligned}$$

3(b) For the pipe downstream of the junction structure, 10 diameters = 10 x 2 = 20 ft, use 50 ft minimum. From Eq. 6-28, Az required is:

$$\begin{aligned} Az &= 0.45\, k\, \phi_{sw}\, L \times CCF \times TCF \\ &= 0.45\,(0.70)\,(0.0049)\,(50)\;(2)\;(2) = 0.309 \end{aligned}$$

So, for the mainline pipe (uniform flow conditions), Az= 0.15. For the downstream pipe, Az =0.31, for flow at least 50 ft downstream, in the 24-in. pipe @ 0.00085 slope. The example problem shows additional corrosion protection is required downstream of intercept structure.

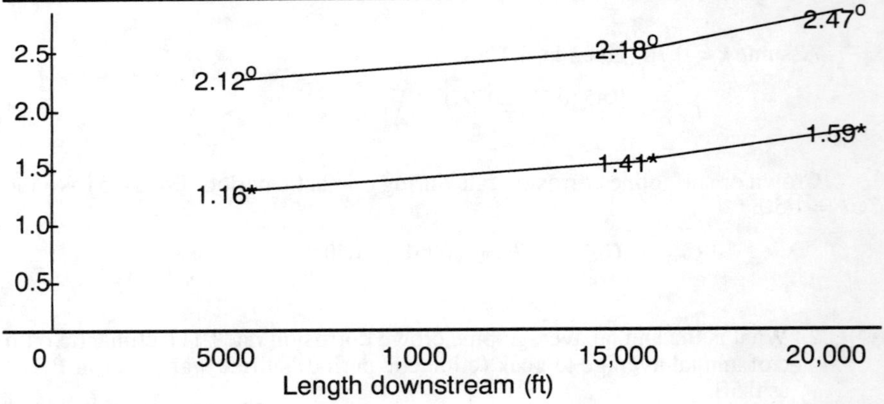

Legend: 0 year 1 to 10
x year 11 to 50

FIG. E-1. Plot of Sulfide Build-up (Example Problem B.1.C.)

APPENDIX F
MANHOLE INSPECTION FORM
PHYSICAL CONDITION SURVEY

UNIT_____________________________DATE____________________HOUR__________

LOCATION____

DWG. NO.______ YR. BUILD ( ) STATION ________MAP GUIDE #___________

PIPE SIZE AND TYPE   UPSTREAM_____DOWNSTREAM_________

SEWAGE SURVEY   H2S____T.S.____D.S.____TEMP.____pH____

RIM TO INVERT_______________

CONDITION OF FLOW UPSTREAM: (SMOOTH   MODERATE   TURBULENT)
CONDITION OF FLOW DOWNSTREAM: (SMOOTH   MODERATE   TURBULENT)
VELOCITY OF FLOW: (SLOW   MODERATE   FAST) DEPTH___

PIPE MEASUREMENTS: UPSTREAM: HORIZONTALLY_____VERTICALLY___
DOWNSTREAM   HORIZONTALLY_____VERTICALLY___

pH MEASUREMENTS:

UPSTREAM   6' IN @ CROWN_________3' IN____3' IN & 2' ABOVE W.S.____
DOWNSTREAM 6' IN @ CROWN_________3' IN____3' IN & 2' ABOVE W.S.____

CONDITION OF M.H.   CIRCLE TYPE OF M.H.   (A)   (B)   (C)   (D)
(E CASE I CASE IICASE III)

BRICKS OR PRECAST:____
____

MORTAR:____
____

STEPS:____
____

OTHER:____
____

CONDITION OF STRUCTURE:

NORTHSIDE:____
____

EASTSIDE:____
____

SOUTHSIDE:____
____

WESTSIDE:____
____

ROOF: pH ( )____
____

BOTTOM: (DEBRIS)____

M.H. MEASUREMENTS: NORTH TO SOUTH______EAST TO WEST_________

M.H. pH 1 FOOT UP   N/S_____E/S_____S/S_____W/S____

CAMERA LOG

Roll #________________Date__________Film Type______________ASA

Manhole #_____________Trunk__________Location ____________________

1________Manhole Label_______________________OPERATOR:____________

2____

3____

4____

5____

6____

7____

8____

9____

10____

11____

12____

13____

14____

15____

16____

17____

18____

19____

20____

21____

22____

23____

24____

25____

Date Labeled__________Comments:

MAP GUIDE #______

CONDITION OF UPSTREAM LINE

CROWN:____

____

____

____

SIDES:____

____

____

____

SOFFIT:____

CONDITION OF DOWNSTREAM LINE

CROWN:____

____

____

____

SIDES:____

____

____

____

SOFFIT:____

ADDITIONAL COMMENTS:______________________________

____

____

____

OBSERVER____________________

CREW____

SANITATION DISTRICTS OF LOS ANGELES COUNTY
CONFINED SPACE ENTRY RECORD SHEET

Date:____________________ Confined Space Location__________________

Crew Members:

Supervisor or Lead Person:

Others 1._______________

2._______________

3._______________

4._______________

5._______________

Job Assignments:

Person to Enter Confined Space_______________

Person to Hold Life Line _______________

Standby Employees_______________

____

____

Confined Space Pre-Entry Safety Check List		
Item	Check if Satisfactory	Initial
1. Location of nearby working telephone by all crew members		
2. Traffic control equipment in place?		
4. Air blower operating OK?		
3. Adequate protective clothing & equipment in place		
5. O_2/Explosimeter Meter Working OK		
6. Life Line (Rope) Condition OK?		
7. Safety Harness OK?		
8. End of life line adequately secured?		
9. Emergency air supply available for rescue?		
10. Hydrogen sulfide detector available and OK?		

SANITATION DISTRICTS OF LOS ANGELES COUNTY
CONFINED SPACE ENTRY RECORD SHEET

Person Assigned to Fill Out This Log Sheet:_________

Confined Space Operations Log Sheet						
Event	Time	Percent Explosimeter Reading[1]	Percent Oxygen Reading[2]	Hydrogen Sulfide Reading[3]	Detectable Odors	
					Ob-	Describe
		% of L.E.L.	$\%O_2$	ppm H_2S	Yes/No	If Yes
Manhole opened						
Ventilation started						
Entry Into Manhole, T_e						
T_e + 15 min. T_e + 30 min. T_e + 45 min. T_e + 60 min. T_e + 75 min. T_e + 90 min.						
Exit from manhole						

Notes:
1. Abort entry if explosimeter reading exceeds 20% of L.E.L
2. Abort entry if oxygen concentration reading is less than 19.5%.
3. Abort entry if hydrogen sulfide reading exceeds 10 ppm.

APPENDIX G
SULF.BAS PROGRAM LISTING*

```
10 FOR X=1 TO 10  :PRINT  :NEXT X
20 PRINT TAB(10) *************************************************************
30 PRINT TAB(10)                                                            ***
40 PRINT TAB(10)                                                            ***
50 PRINT TAB(10)                        PROGRAM 'SULF.BAS'                  ***
60 PRINT TAB(10)                                                            ***
70 PRINT TAB(10)                                                            ***
80 PRINT TAB(10) *************************************************************
110 REM-- This program is a sub-program of "SULF.BAS"
              Its five sub-programs are:
                              "SULF.BAS" (or "S.BAS")
                              "PI.BAS"
                              "RD.BAS"
                              "CLC.BAS"
                              "OUT.BAS"
330 PRINT TAB(27) "Project Name?  "; :INPUT"", PRNAM$
          :PRINT
          :PRINT TAB(23) "Project Engineer's Name? "; :INPUT"", ENAM$
          :PRINT
          :PRINT TAB(33) "Date" "; :INPUT"", DAT$
340        IF PRNAM$="" THEN PRNAM$="No Project Name"
350        IF ENAM$="" THEN EMAN$="Project Engineer:  No Name"
             ELSE
             ENAM$="Project Engineer:  "+ENAM$
360        IF DAT$="" THEN DAT$="No Date"
370 PRINT:PRINT
380 PRINT TAB(20) "Do you want to have information printed out?"
          :PRINT : PRINT TAB(22) "(Make sure the printer is on if you do.)"
          :PRINT : PRINT TAB(34)"(Y/RETURN)  ";  : INPUT"", P$
          :IF P$="Y" THEN P$="Y"  :REM-- Cap 'Y' to operate printer
390 PRINT:PRINT:PRINT
400 PRINT TAB(27)         " ***MENU***"
410 PRINT
420 PRINT TAB(27)"1. INPUT PRELIMINARY DATA"
430 PRINT TAB(27)"2. INPUT REACH DATA"
440 PRINT TAB(27)"3. PROCESS DATA (AND OUTPUT)"
450 PRINT TAB(27)"4. OUTPUT (OR RE-OUTPUT) RESULTS"
460 PRINT TAB(27)"5. END"
470 PRINT
480 PRINTTAB(27)"  WHICH WOULD YOU LIKE? ";
490 INPUT "", MC
          :IF MC<1 OR MC>5 THEN
          PRINT : PRINT TAB(27)"Input number from 1 to 5: ";
          :GOTO 490 :REM--Input Menu Choice again
500 ON MC GOTO 510, 520, 530, 540, 550
510 CHAIN "PI.BAS", 110, ALL
520 CHAIN "RD.BAS", 110, ALL
530 CHAIN "CLC.BAS", 120, ALL
540 CHAIN "OUT.BAS", 110, ALL
590 END
```

* An IBM PC-compatible 5 1/4-inch 360K floppy disk containing the public domain SULF.BAS program listed above is available for $30 including postage and handling from:

Special Professional Education
Box 9, Harvill Building, University of Arizona
Tucson, Arizona 85721
(602) 621-3054

```
10 REM  ***************************************************
20 REM                                                  ***
30 REM                   PROGRAM 'PI.BAS'               ***
40 REM                                                  ***
50 REM                   (PRIMARY INPUT)                ***
60 REM                                                  ***
70REM *****************************************************
100 REM-- This program is a sub-program of "SULF.BAS"
           Its five sub-programs are:
                         "SULF.BAS" (or "S.BAS")
                         "PI.BAS"
                         "RD.BAS"
                         "CLC.BAS"
                         "OUT.BAS"
110 PRINT:PRINT:PRINT  :REM-- Other program's starting point
120 IF P$="Y" THEN LPRINT TAB(28)"> PRIMARY DATA INPUT <"
          :LPRINT
130 PRINT TAB(20)
          "Do you want an incremental life analysis?"
          :PRINT
          :PRINT TAB(34)"(Y/RETURN)  ";:INPUT"", IL$
          :IF IL$="Y" OR IL$="y" THEN IL$="Y"
140        IF IL$<>"Y" THEN NI=1
          :GOTO 180
          :REM-- If true, this is NOT an incremental analysis, so skip input of number of
          increments.
150 PRINT
          :PRINT TAB(29) "Number of increments? ";
160        INPUT"", NI
          :IF NI=0 THEN PRINT
          :PRINT TAB(26) "Must input number:  Input = ";
          :GOTO 160  :REM--Ask for NI again
170        IFP$="Y" THEN LPRINT TAB(18)
          "(Incremental Life Analysis with"; NI; "Increments)"
          :LPRINT
          :LPRINT TAB(10)STRING$ (61,45)
180 IF DM1FLG THEN ERASE LIFE
          ELSE
          DM1FLG=1  :REM-- If array has already been DIMensioned,
          it must first be ERASEd before re-DIMensioning
190 DIM LIFE (NI)
200 PRINT:PRINT
210 PRINT TAB(10)"*** RELATIVE DEPTH and FLOW QUANTITY ***"
220 PRINT
230 PRINT TAB(10) "There are several ways in which you may input concerning Rel."
240 PRINT TAB(10) "Depth and Flow Quantity:"
250 PRINT
260 PRINT TAB(10)"1. Input Design Rel. Depth and computer will compute a Design"
270 PRINT TAB(13)"Flow Qnty and Flows for further reaches."
280 PRINT
290 PRINT TAB(10)"2. Input variable Rel. Depths (one for each reach) and compu-"
300 PRINT TAB(13)"ter will compute Qnty's for each reach."
310 PRINT
320 PRINT TAB(10)"3. Input a Design Flow Qnty and computer will determine a De-"
330 PRINT TAB(13)"sign Rel. Depth.  From there, the system will be as in 1."
340 PRINT
350 PRINT TAB(10)"4. Input variable Flow Qnty's (one for each reach) and compu-"
360 PRINT TAB(13)"ter will determine Rel. Depth's for each reach."
370 PRINT :IF P$="Y" THEN LPRINT
380 PRINT TAB(33)
          "Your choice: ";
          :IF P$="Y" THEN LPRINT TAB(28)
```

```
            "Your System Choice: ";
390          INPUT "", SC
            "IF SC<>INT (SC) OR SC<1 OR SC>4 THEN
            PRINT : PRINT TAB(20)"Input number from 1 to 4: ";
            :GOTO 390  :REM-- Ask for SC again
400          IF P$="Y" THEN LPRINT SC
410 PRINT :IF P$="Y" THEM LPRINT
420 IF SC=1 THEN A$=SPACE$(10)+"------- CONSTANT REL. DEPTH ------"
            :GOSUB 770
430 IF SC=2 THEN A$=SPACE$(9)+"-------- VARIABLE REL. DEPTH's -------"
            :GOSUB 770
440 IF SC=3 THEN A$=SPACE$(10)+"---------- CONSTANT FLOW QNTY --------"
            :GOSUB 770
450 IF SC=4 THEN A$=SPACE$(9)+" -------- VARIABLE FLOW QNTY's --------"
            :GOSUB 770
460 PRINT :IF P$="Y" THEN LPRINT
470 A$="**** START PRIMARY INPUT ****"
            :L$=""
            :GOSUB 770
480 PRINT :IF P$="Y" THEN LPRINT
490 A$="Bio-chemical Oxygen Demand?"
            :L$="= (mg/1)"
            :GOSUB 770
            :INPUT"", BOD :IF P$="Y" THEN LPRINT "=";BOD; "mg/1"
500 PRINT :IF P$="Y" THEN LPRINT
510 A$="Climactic Temperature?"
            :L$="= (deg C)"
            :GOSUB 770
            :INPUT"", TEMP :IF P$="Y" THEN LPRINT "="; TEMP; "deg C"
520 PRINT :IF P$="Y" THEN LPRINT
530 REM                        [[[[[[[[[[[ ]]]]]]]]]]]
540 FOR I=1 TO NI
            :A$="Design Life"
            :IF IL$="Y" THEN
            A$=A$+", increment"= +STR$ (I)
550          A$=A$+"?"
            :L$="= (yrs)"
            :GOSUB 770
            :INPUT"", LIFE(I) :IF P$="Y" THEN LPRINT "="; LIFE(I); "yrs"
560 NEXT I
570 REM                        [[[[[[[[[[[ ]]]]]]]]]]]
580 PRINT :IF P$="Y" THEN LPRINT
590 A$="Acid-Reaction Factor K?"
            :L$="= (0-1)"
            :GOSUB 770
            :INPUT"", K :IF P$="Y" THEN LPRINT "="; K
600 PRINT : IF P$="Y" THEN LPRINT
610 A$="Sewage pH?"
611 L$="= (7-8)"
612 GOSUB 770
613 INPUT"", PH :IF P$="Y" THEN LPRINT "="; PH
614 J=1/(1+(10^(PH-7)))
620 PRINT : IF P$="Y" THEN LPRINT
630 A$="Climactic Ratio C?"
            :L$="= (0-1)"
            :GOSUB 770
            :INPUT"", C :IF P$="Y" THEN LPRINT "="; C
640 PRINT : IF P$="Y" THEN LPRINT
650 A$="Upstream Sulfide Level?"
            :L$="= (mg/1)"
            :GOSUB 770
            :INPUT"", USL :IF P$="Y" THEN LPRINT "="; USL; "mg/1"
```

```
660 PRINT : IF P$="Y" THEN LPRINT
670 A$="Insoluble Sulfides?"
        :L$="= "mg/1"
        :GOSUB 770
        :INPUT"", IS :IF P$="Y" THEN LPRINT "="; IS; "mg/1"
680 PRINT :IF P$="Y" THEN LPRINT
690 A$="**** PRIMARY INPUT COMPLETED ****"
        :L$=""
        :GOSUB 770
700 PRINT:PRINT : IF P$="Y" THEN LPRINT:LPRINT
710 PRINT TAB(21) "Do you want to go on to Reach Data Input?"
720 PRINT : PRINT TAB(19)"(RETURN to go on, M for Menu, R to Re-input:)  ";
        :INPUT"", C$
730 IF C$="" THEN CHAIN "RD.BAS", 110, ALL
740 IF C$="M" OR C$="m" THEN CHAIN "SULF.BAS", 390, ALL
750 IF C$="R" OR C$="r" THEN GOTO 460 :REM-- Start input over
760 PRINT : PRINT TAB(29) "(Input must be as specified)"
        :GOTO 720  :REM--Ask for Choice$ again
770 REM ===== PRINT & LPRINT SUBROUTINE ======
780   PRINT TAB(10) AS; TAB(56) L$; TAB(67);
        :IF P$="Y" THEN LPRINT TAB(10) A$; TAB(56);
790 RETURN   :REM ===== END SUBROUTINE ======
800 END

10 REM  **********************************************************
20 REM                                                         ***
30 REM                  PROGRAM "RD.BAS"                       ***
40 REM                                                         ***
50 REM                  (REACH DATA INPUT                      ***
60 REM                                                         ***
70 REM  **********************************************************
100 REM-- This program is a sub-program of "SULF.BAS"
          Its five sub-programs are:
                    "SULF.BAS" (OR "S.BAS")
                    "PI.BAS"
                    "RD.BAS"
                    "CLC.BAS"
                    "OUT.BAS"
110 PRINT:PRINT:PRINT  :REM--CLEAR SCREEN HERE
120 IF P$="Y" THEN LPRINT TAB(29)">>> REACH DATA INPUT <<<"
        :LPRINT
        :LPRINT TAB(32) "No. of Reaches =";
130 PRINT TAB(18)
        "How many reaches would you like to analyze?" ";
140      INPUT"", NR
        :IF NR=0 THEN
        PRINT TAB(25) "Must input number: Input = ";
        :GOTO 140 :REM--Ask for NR again
150      IF P$="Y" THEN LPRINT NR
160 IF DM2FLG THEN ERASE D, S, L, AS, AQ, RD, Q,V, Z, SLIM, S1, S2, AZ
        ELSE
        DM2FLG=1
        :REM-- If arrayed variables have already been DIMensioned,
        they must first be ERASEd before re-DIMensioning
170 DIM D(NR), S(NR), L(NR), AS(NR), AQ(NR),
        RD(NR,NI), Q(NR,NI),
        V(NR,NI), Z(NR,NI), SLIM(NR,NI), S1(NR,NI), S2(NR,NI), AZ(NR,NI)
180 PRINT :IF P$="Y" THEN LPRINT
190 IF SC=2 OR SC=4 GOTO 300  :REM-- If true, skip input of non-variable RD or Qnty
200 A$=STRING$(61,45) :GOSUB 750
210 REM                 [[[[[[[[[[ ]]]]]]]]]]
220 FOR I=1 TO NI
```

```
230           IF SC=1 THEN
                 :A$="Design Rel. Depth"
                 :IF IL$="Y" THEN
                 A$=A$+", increment"=STR$(I)+"?"
                 ELSE
                 A$=A$+"?"
240           IF SC=1 THEN
                 L$="= (0-1)"
                 :GOSUB 750
                 :INPUT"", RD(1.I) :IF P$="Y" THEN LPRINT "="; RD(1.I)
250           IF SC=3 THEN
                 :A$="Design Flow Qnty"
                 :IF IL$="Y" THEN
                 A$=A$+", increment"+STR$(I)+"?"
                 ELSE
                 A$=A$+"?"
260           IF SC=3 THEN
                 L$="= (cfs)"
                 :GOSUB 750
                 :INPUT"", Q(1,I) :IF P$="Y" THEN LPRINT "="; Q(1,I); "cfs"
270 NEXT I
280 REM                    [[[[[[[[ ]]]]]]]]]]
290 PRINT :IF P$="Y" THEN LPRINT
300 A$="******* START REACH DATA INPUT ******"
           :L$=""
           :GOSUB 750
310 FOR R=1 TO NR
320           IF R>1 THEN
                 A$=STRING$(61,45) :L$="" :GOSUB 750
                 :REM-- Print dotted line if NOT 1st Reach
330              PRINT :IF P$="Y" THEN LPRINT
340           A$="Pipe Diameter"+STR$(R)+"?"
                 :L$="= (inches)"
                 :GOSUB 750
                 :INPUT"", D(R) :IF P$="Y" THEN LPRINT "="; D(R); "inches"
350              D(R)=D(R)/12 :REM-- Pipe Diam must be in ft for calculations
360 A$="Slope of Reach"+STRS(R)+"?"
                 :L$="= (ft/ft)"
                 :GOSUB 750
                 :INPUT"", S(R) :IF P$="Y" THEN LPRINT "="; S(R); "ft/ft"
370           A$="Length of Reach"+STRS(R)+"?"
                 :L$="= (ft)"
                 :GOSUB 750
                 :INPUT"", L(R) :IF P$="Y" THEN LPRINT "="; L(R); "ft"
380           PRINT :IF P$="Y" THEN LPRINT
390           IF SC=1 OR SC=3 GOTO 490 :REM-- If true, skip input of variable Rel. Depth or Qnty
400           REM                [[[[[[[[[[ ]]]]]]]]]]]
410           FOR I=1 TO NI
420              IF SC=2 THEN
                     A$="Design Rel. Depth for Reach"+STR$(R)
                     :IF IL$="Y" THEN
                     A$=A$+", increment"+STRS(I)+"?"
                     ELSE
                     A$=A$+"?"
430           IF SC=2 THEN
                     :L$="= (0-1)"
                     :GOSUB 750
                     :INPUT"", RD(R,I) :IF P$="Y" THEN LPRINT "="; RD(R,I)
440              IF SC=4 THEN
                     A$="Design Flow Qnty for Reach "+STR$(R)
                     :IF IL$="Y" THEN
                     A$=A$+", increment"+STR$(I)+"?"
```

```
                    ELSE
                    A$=A$+"?"
450               IF SC=4 THEN
                    :L$="= (cfs)"
                    :GOSUB 750
                    :INPUT"", Q(R,I) :IF P$="Y" THEN LPRINT "="; Q(R,I); "cfs"
460            NEXT I
470            PRINT :IF P$="Y" THEN LPRINT
480 REM                      [[[[[[[[[[ ]]]]]]]]]
490            IF R=1 GOTO 570
                 :REM-- No need to ask for Intercepted Sulfide if 1st Reach
500            A$="Intercepted Sulfide at start of Reach "+STR$(R)+"?"
510            PRINT TAB (10) A$
                 :PRINT TAB (34)"(RETURN if none)"; TAB(56)"= (mg/1); TAB(67);
                 :INPUT"", AS(R)
520                IF AS(R)=0 THEN
                 PRINT
                 :PRINT TAB(21)"(No Intercepted Sulfide at this Reach)"
                 :GOTO 560
                 ELSE
                 IF P$="Y" THEN LPRINT TAB(10) A$; TAB(56) "="; AS(R); "mg/1"
530            A$="Intercepted Flow Qnty at start of Reach"+STR$(R)+"?"
                 :L$="= (cfs)"
                 :GOSUB 750
540                INPUT"",AQ(R)
                 :IF AQ(R)=0 THEN PRINT TAB(10)
                 "(Flow must be input if Sulfide value was input)"
                 :PRINT TAB (56)"Input = ";
                 :GOTO 540 :REM--Ask for Intercepted Flow Qnty again
550                IFP$="Y" THEN LPRINT "="; AQ(R); "cfs"
                 :LPRINT
560                PRINT
570            IF ER$="N" GOTO 590
580 NEXT R
590 A$=STRING$(61,45) :PRINT TAB(10)A$
600 PRINT TAB(22) "Is Reach-Data correct? (N/RETURN) "; :INPUT"", ER$
          :IF ER$="N" OR ER$="n" THEN  ER$="N" ELSE GOTO 670
610 PRINT TAB(16) "Which Reach is error in" (One reach at a time) ";
          :INPUT"", R
620 PRINT
          :PRINT TAB(28) "Re-Input data for Reach"; R; ":"
630 A$=STRING$(61,45) :L$="" :GOSUB 750 :REM-- Print dotted line
640 A$=SPACE$(14)+"--- CORRECTED DATA FOR REACH"+STR$(R)+" ---" :GOSUB 750
650  PRINT :IF P$="Y" THEN LPRINT
660 GOTO 330
670 PRINT
680 A$="**** REACH DATA INPUT COMPLETED ****"
          :L$=""
          :GOSUB 750
690  PRINT:PRINT
700 PRINT TAB (17)"Do you want to go on to process and printout?"
710 PRINT TAB (17)"(RETURN to go on, M for Menu:) "; :INPUT"", C$
720 IF C$="" THEN CHAIN "CLC.BAS", 120, ALL
730 IF C$="M" OR C$="m" THEN CHAIN "SULF.BAS", 390, ALL
740 PRINT TAB(10) "(Input must be as specified)"
          :GOTO 710 :REM--Ask for Choice$ again
750 REM ======= PRINT & LPRINT SUBROUTINE ======
760               PRINT TAB (10); A$; TAB(56); L$; TAB(67);
                    :IF P$="Y" THEN LPRINT TAB(10); A$; TAB(56);
770 RETURN: REM =========== END SUBROUTINE ==========
780 END
```

```
10 REM ***********************************************************
20 REM                                                          ***
30 REM                      PROGRAM "CLC.BAS"                   ***
40 REM                                                          ***
50 REM                      (PROCESSING DATA)                   ***
60 REM                                                          ***
70 REM ***********************************************************
100 REM-- This program is a sub-program of "SULF.BAS"
           Its five programs are:
                        "SULF.BAS" (or "S.BAS")
                        "PI.BAS"
                        "RD.BAS"
                        "CLC.BAS"
                        "OUT.BAS"
120 PRINT:PRINT:PRINT :REM--CLEAR SCREEN HERE
130 PRINT TAB(32)"(Processing Data)
140 FOR I=1 TO NI
           :S1(1,I)=USL
        :NEXT I
        :REM-- Initialize all incremental S1's
150 IF SC=2 OR SC=4 GOTO 200 :REM-- If true, skip initialization of RD's or Q's
                             - Variable values were input
160 FOR I=1 TO NI
        :FOR R=1 TO NR
170          IF SC=1 THEN RD(R,I)=RD(1,I)
180          IF SC=3 THEN Q(R,I)=(1,I)
190         NEXT R
   :NEXT I
   :REM-- Set all variable RD's or Q's equal to initial value for each reach
200 REM ==== START CALCULATIONS ==========
210 EBOD=BOD*1.07^(TEMP-20)
220 R=1 TO NR
        :FOR I=1 TO NI
230          IF SC=1 OR SC=2 THEN
                X=1-2*RD(R,I)
                :THETA=-ATN(X/SQR(-X*X+1))+1.5708
                :GOSUB 590 :REM-- Calculate the other THETA-based variables
240          IF SC=3 OR SC=4 THEN GOSUB 550
                :REM-- Determine THETA from Q(R,I) ANDIes
250         REM--THETA, A, P, PP, B, DM, & V(R,I) are now defined
260         Q(R,I)=QT :REM-- Set Qnty equal to calculated value
270 REM FORCE MAIN SULFIDE CALCULATION
280 IF RD(R,I)< .999 THEN 380
290 Q(R,I)=Q(R-1,I)
300 V(R,I)=Q(R,I)/A
310 TIM=L(R)/(V(R,I)*3600)
320 SLIM(R,I)=0
340 DSDT=.01312*EBOD*(D(R)+.12)
350 S2(R,I)=S1(R,I)+(DSDT*TIM)
360 AZ(R,I)=0
370 GOTO 500
380          TIM=L(R)/V(R,I)/3600 :REM-- Time in hours
390           SV= (S(R)*V(R,I))^(3/8)
400          SLIM(R,I)=.00052*EBOD*P/B/SV
410          IF SLIM(R,I)1(R,I)THEN
                GOTO 440: REM If sulfide is decreasing, then no need to check z
                or slope -- use calculated S2 value
420          Z(R.I)=EBOD*P/B/SQR(S(R))/Q(R,I)^(1/3)
430          IF Z(R,I)<5000 OR S(R)>.006 THEN
                S2 (R,I)=S1(R,I)
                :GOTO 460 :REM--Under circumstances Sulfide buildup will
                not occur, so skip definition of calculated S2
```

```
440            ANTI=10^(TIM*SV/1.15/DM)
450            S2(R,I)=SLIM(R,I)-(SLIM(R,I)-S1(R,I))/ANTI
460            REM--S2 is now defined
470            IF S1(R,I)>S2(R,I) THEN
                   GS2=S1(R,I)
                   ELSE
                   GS2=S2(R,I) :REM-- The greater of S1 or S2, for AZ design
480            PHI=.45*SV*J*C*B/PP*(GS2-IS)
490            AZ(R,I)=.45*K*PHI*LIFE(I)*2
500            IF R <NR THEN
                   S1(R+1,I)=(S2(R,I)*Q(R,I)+AS(R+1)*AQ(R+1))/(Q(R,I)+AQ(R+1))
510            REM-- If NOT final Reach, set S1 for the next Reach
                   (same inc) equal to current (mass-balanced) S2.
520          NEXT I
   :NEXT R
530 ============== CALCULATIONS COMPLETED ==============
540 CHAIN "OUT.BAS", 110 ALL
550  REM   REL.DEPTH FROM QNTY SUBROUTINE
560          TMP=0 :REM--Initialize TeMPorary storage
570          LO=0 : HI=3.14159 :REM--Initialize limiters
580          THETA=(HI-LO)/2+LO
590          REM---------- START HERE IF SC = 1 OR 2 ----------------
600          A+D(R)^2/4*(THETA-SIN(THETA)*COS(THETA))
610          P=D(R)*THETA
620          PP=D(R)*(3.14159-THETA)
630          B=D(R)*SIN(THETA)
640          DM=A/B
650          HR=A/P
660          V(R,I)=1.49/.013*SQR(S(R))*HR^(2/3)
670          QT=A*V(R,I)
680          IF SC=1 OR SC=2 GOTO 750
                :REM--If true, then variables needed are now defined, so RETURN
690          IF QT=TMP GOTO 770 : REM-- If true, then QT is maxed out
700          TMP=QT :REM-- Set TEMP equal to old value of QT for next run thru
710          IF INT(1000*QT)<INT(1000*Q(R,I)) THEN LO=THETA :GOTO 580
720           IF INT(1000*QT)>INT(1000*Q(R,I)) THEN HI=THETA :GOTO 580
730          REM--Values now defined: THETA,A,P,PP,B,DM,R,V(R,I) & QT(~=Q(R,I))
740          IF SC=3 OR SC=4 THEN RD(R,I)=(1-COS(THETA))/2
                :REM--If true, calculate Rel. Depth
750 RETURN
760 REM END SUBROUTINE
770 REM ****ERROR >>> QNTY TOO LARGE *****
780          PRINT:PRINT:PRINT
790          PRINT TAB (10)
           "******************************************
800          PRINT
810          IF SC=3 AND NI > 1THEN PRINT TAB(13)
                "(Value for Design Flow Qnty, increment"; I; "is too large)"
                ELSE
                IF SC=3 THEN PRINT TAB(19)
                "(Value for Design Flow Qnty is too large)"
820          IF SC=4 AND NI > 1THEN PRINT TAB(12)
                "(Design Flow Qnty for Reach"; R; ",  increment";I;  "is too large)"
                ELSE
                IF SC=4 THEN PRINT TAB(19)
                "(Design Flow Qnty for Reach"; R;  "is too large)"
830          PRINT
                :PRINT TAB(26)"Current value ="; Q(R,I); "cfs"
                :PRINT
                :PRINT TAB(27)"RE-INPUT NEW VALUE: ";
                :INPUT"", Q(R,I)
840          IF SC=3 THEN Q(1,1)=Q(R,I)
```

```
                :REM--If SC=3, reset Design Flow Qnty
850           PRINT
                :PRINT TAB(10)
                "*********"
860           PRINT
                :PRINT TAB(32)"(Re-Processing Now)
870           IF SC+1 GOTO 550 :REM-- Re-determine THETA and continue where left off
880           IF SC=3 GOTO 160 :REM-- Start all over
890 REM ********** QNTY ERROR ROUTINE END *************
900 END

10 REM *****************************************************
20 REM                                                   ***
30 REM              PROGRAM "OUT.BAS"                    ***
40 REM                                                   ***
50 REM              (PRINTING OUTPUT)                    ***
60 REM                                                   ***
70 REM *****************************************************
100 REM-- This program is a sub-program of "SULF.BAS"
                Its five sub-programs are:
                              "SULF.BAS" (or "S.BAS")
                              "PI.BAS"
                              "RD.BAS"
                              "CLC.BAS"
                              "OUT.BAS"
110  REM--CLEAR SCREEN HERE
120 LIFSUM=0
            :FOR I=1 TO NI
            :LIFSUM=LIFSUM+LIFE(I)
            :NEXT I
            :REM-- Sum of incremental Life periods (Needed for AZ Summary
                  Subroutine, but calculated here because it remains constant)
130 REM              [[[[[[[[[ ]]]]]]]]]
140 PRINT:PRINT: :REM-- Last program comes in here
145 REM              [[[[[[[[[ ]]]]]]]]]
150 PG=1 :PAG$="Page"+STR$(PG) :REM-- Initialize PaGe number
160 FS=INT((76-LEN(ENAM$)-LEN(DAT$)-LEN(PAG$))/6)
            :REM-- Number of spaces per each Space Field area
165 REM              [[[[[[[[[ ]]]]]]]]]
168 L=O :REM-- Initialize Line Number
170 REM              [[[[[[[[[ ]]]]]]]]]
180 GOSUB 1100 :L=L+5
            :REM-- Print Project Name, Engineer Name,
            Date, & Page # (This takes 5 lines)
190 REM              [[[[[[[[[ ]]]]]]]]]
200 GOSUB 1210 :L=L+5
            :REM-- Print variable caption heading (5 lines for this)
210 REM              [[[[[[[[[ ]]]]]]]]]
220 FOR R=1 TO NR
            :FOR I=1 TO NI
230             X=2 :GOSUB 1380 :REM-- Vertical bar
240             IF IL$<>"Y" THEN
                    A$=STR$(R) :LP$=A$ :X5 :GOSUB 1390
                    :GOTO 290 :REM-- Skip RCH,INC print
250             A$=STR$(R) :LP$=A$ :X=3 :GOSUB 1390
                    :A$="," :LP$=A$ :X=6 :GOSUB 1390
                    :REM-- Print R Subscript with comma
260             I$=STR$(I) :I$=RIGHT$(I$,LEN(I$)-1)
                    :REM-- For I subscript-- must have no leading blanks,
                    because if I =10 it will print INC in wrong column
                    if it has a leading blank
270             IF I<10 THEN X=8 ELSE X=7
```

```
                 :REM-- Put I in correct column
280          A$=I$ :LP$=A$ :GOSUB 1390 :REM-- Print I subscript
290          X=10 :GOSUB 1380 :REM-- Vertical bar
300          PRINT TAB(14) USING "#.##"; RD(R,I);
                 :IF P$="Y" THEN LPRINT TAB(14) USING "#.##; RD(R,I);
310          X=20 "GOSUB 1380 :REM-- Vertical bar
320          PRINT TAB(22) USING "###.##"; Q(R,I);
                 :IF P$="Y" THEN LPRINT TAB(22) USING "###.##"; Q(R,I);
330          X=30 :GOSUB 1380 :REM-- Vertical bar
340          PRINT TAB(32) USING "###.##"; V(R,I);
                 :IF P$="Y" THEN LPRINT TAB(32) USING "###.##"; V(R,I);
350          X=40 :GOSUB 1380 :REM-- Vertical bar
360          PRINT TAB(42) USING "###.##"; SLIM(R,I);
                 :IF P$="Y" THEN LPRINT TAB(42 USING "###.##"; SLIM(R,I);
370          X=50 :GOSUB 1380 :REM-- Vertical bar
380          PRINT TAB(52) USING "###.##"; S1(R,I);
                 :IF P$="Y" THEN LPRINT TAB TAB(52) USING "###.##"; S1(R,I);
390          X=60 :GOSUB 1380 :REM- Vertical bar
400          PRINT TAB(62) USING "###.##"; S2(R,I);
                 :IF P$="Y" THEN LPRINT TAB(62) USING "###.##; S2(R,I);
410          X=70 :GOSUB 1380 :REM-- Vertical bar
420          PRINT TAB(72) USING "##.##"; AZ(R,I);
                 :IF P$="Y" THEN LPRINT TAB(72) USING "##.##"; AZ(R,I);
430          X=79 :GOSUB 1380 :REM-- Vertical bar
440          REM       [[[[[[[[[ ]]]]]]]]
450          REM-- If incremental analysis AND it is NOT the
                 last increment, then print inc-separator line:
460          IF I<NI THEN A$=
"---------|-----------|-----------|-----------|---------|---------|----------|"
:LP$=
"---------|-----------|-----------|-----------|---------|---------|----------|"
                 :X=2 :GOSUB 1390
470          REM     [[[[[[[[[ ]]]]]]]]]]
480          IF IL$="Y" AND I+NI THEN GOSUB 990
                 :REM-- If incremental analysis AND it is
                 the last Inc, print AZ summary
490          REM     [ [[[[[[[ ]]]]]]]]]
500          IF IL$<>"Y" AND R<NR  OR IL$="Y" AND I=NI AND R <NR THEN
                 T$=SRRING$(76,45)
                 :A$="|"+T$+"|" :LP$="!"+T$+"!"
                 :X=2 :GOSUB 1390
                 :REM-- If true, print Reach end line
510         NEXT I
    :NEXT R
520 IF IL$<>"Y" GOTO 700 :REM-- If true, no need to print Life Inc's
530 REM              [[[[[[[[[ ]]]]]]]]]
PRINT LIFE INCREMENTS
                     [[[[[[[[[ ]]]]]]]]]
540 T$=STRING$(76,61)
          :A$="|"+T$+"|" :LP$="!"+T$+"!"
          :X=2 :GOSUB 1390
          :REM-- Double dotted line
550 T$=SPACE$(28)+"LIFE INCREMENTS (yrs)"+SPACE(27)
          :A$="|"+T$"|" :LP$="!"+T$+"!"
          :GOSUB 1390
560 T$=STRING$(76,45)
          :A$="|"+T$+"|" :LP$="!"+T$+"!"
          :GOSUB 1390
          :REM-- Dotted line
570 IF NI/5=INT(NI/5) THEN
          NLI=NI/5
          ELSE
```

```
          NLI=INT(NI/5)+1
          :REM-- Define Number of Lines of Increment print
          580 IF NI/NLI=INT(NI/NLI) THEN
          NIL=NI/NLI :NILL=NIL
          ELSE
          NIL=INT(NI/NLI) :NILL=NI-NIL*(NLI-1)
          :REM-- If top part is true, then the Number of Inc's per the
          last Line will be the same as Number of Inc's in previous Lines
590 ON NIL GOSUB 1310, 1320, 1330, 1340, 1350
          :REM-- Set Tab parameters according to number of Inc's per line
600 I=1 :REM-- Initialize I for 1st Increment
610 FOR LI=1 TO NLI
620        X=2 :GOSUB 1380 :REM-- Vertical bar
630        IF LI<NLI THEN ILIM=I+NIL-1
             ELSE
             ILIM=I+NILL-1
             :ON NILL GOSUB 1310, 1320, 1330, 1340, 1350
             :REM-- Set range (ILIM) for Inc, and if last lineof Inc print, reset
              tab parameters for last line
640        TB=0 :REM-- Re-Initialize Tab parameter subscript
650        FOR I=I TO ILIM
             :A$="INC"+STR$(I)+" ="STR$(LIFE(I)) :LP$=A$
             :TB=TB+1 :X=X(T) :GOSUB 1390
          :NEXT I
          :REM-- Print Life Increment
660        X=79 :GOSUB 1380 :REM-- Vertical bar
670        IF LI THEN
             T$=STRING$(76,45)
             :A$="|"+T$+"|" :LP$="!"+T$+"!"
             :X=2 :GOSUB 1390
             :REM-- If NOT final line of Inc print, then print dotted line
680 NEXT LI
690 REM              [[[[[[[[[[[ ]]]]]]]]]]]
                          PRINT EBOD
                     [[[[[[[[[[[ ]]]]]]]]]]]
700 T$=STRING$(76,61)
          :A$="|"+T$"|" :LP$="!"+T$+"!"
          :X=2 :GOSUB 1390
          :REM-- Double dotted line
710 T$=SPACE$(15)+"Effective Bio-Chemical Oxygen Demand + "
          :A$="|"+T$ :LP$="!"+T$
          :X=2 :GOSUB 1390
          :PRINT USING "###"; EBOD;
          :IF P$="Y" THEN LPRINT USING "###"; EBOD;
720        A$="mg/1" :LP$=A$
          :X=61 :GOSUB1390
          :REM-- Print EBOD label
730 X=70 :GOSUB 1380 :REM-- Vertical bar
740 REM              [[[[[[[[[[[ ]]]]]]]]]]
750 FOR R=1 TO NR
          :FOR I=1 TO NI
760          IF SLIM(R,I)<S1(R,I) OR Z(R,I)=5000 AND S(R)< =.006 THEN GOTO 890
               :REM-- Next Reach
770          IF ZSFLG GOTO 780
               ELSE
               T$=STRING$(76,61)
               :A$+"|"+T$+"|" :LP$="!"+T$+"!"
               :X=2 :GOSUB 1390
               :REM-- If any S2's were equal to S1's, print separation line
780          ZSFLG=1
               :REM-- Set ZSFLG NOT equal to zero so that, if there are any more
                          S2's set equal to S1's, a separation line is NOT printed
```

```
790            X=2 :GOSUB 1380 :REM-- Vertical bar
800            R$=STR(R) :R$=RIGHT$(R$, LEN(R$) -1)
                  :I$=STR$(I) :I$=RIGHT$(I$, LEN(I$)-1)
                  :REM-- R$ is the string representation of R with no leading
                  blank spaces; I$  is the same, similarly with I
810            A$="Note: S2("+R$
                  :IF NI1 THEN
                  A$=A$+","+I$+")"
                  ELSE
                  A$=A$+")"           :REM-- S2 with subscript(s)
820            A$=A$+" = S1("+R$
                  :IF NI>1 THEN
                  A$=A$+","+I$+")"
                  ELSE
                  A$=A$+")"           :REM-- S1 with subscript(s) added
830            A$=A$+" because "
840            IF Z(R,I) <5000 THEN
                  A$=A$+"Z ="+STR$(INT(Z(R,I)))+  5000"
                  :REM-- Z message added, IF  5000
850            IF Z(R,I)<5000 AND S(R).006 THEN
                  A$=A$+ " and "
860            IF S(R) >.006 THEN
                  A$=A$+"S ="+STR$(S(R)0+ > .006"
                  :REM-- S message added, IF  >.006
870            LP$=A$ :X=6 :GOSUB 1390 :REM-- print S2=S1 message
880            X=79 :GOSUB 180 :REM-- Vertical bar
890           NEXT I
             :NEXT R
900 REM [[[[[[[[[[[[[ ]]]]]]]]]]]]]
910 T$=STRING$(76,45)
              :A$="'"+T$+"'" :LP$="'"+T$"'"
              :X=2 :GOSUB 1390
              :REM-- Output End Line
920 ZSFLG=0 :REM-- Set ZSFLG back to zero in case of re-output
930 REM ************ OUTPUT COMPLETED ***************
940 PRINT:PRINT
945 IF P$="Y" THEN ;PRINT ""
950 PRINT TAB (10)"BACK to MENU?  (M if so, R to Re-print, RETURN to END:)";
              :INPUT C$
960 IF C$="M" OR C$="m" THEN CHAIN "SULF.BAS", 390, ALL
970 IF C$="R" OR C$="r" THEN GOTO 110 :REM-- Reprint
980 IF C$="" THEN CHAIN "SULF.BAS", 590
990 REM ================== AZ SUMMARY SUBROUTINE ================
 1000          AZSUM=0
                :FOR I=1 TO NI
                :AZSUM= AZSUM+AZ(R,I)
                :NEXT I
                :REM-- Sum of incremental AZ's for each reach
 1010           T$=STRING$(76,45)
                   :A$+"|"+T$+"|" :LP$="!"+T$+"!"
                   :X=2 :GOSUB 1390
                   :REM-- Dotted line
 1020           X=2 :GOSUB 1380 :REM-- Vertical line
 1030           A="Pipe Diam"+STR$(R)+" ="+STR$(D(R)*12)+" inches" :LP$=A$
                   :X=4 :GOSUB 1390
                   :REM-- Print Tab(4) Pipe Diameter
 1040           A$="Slope +"+STR$(S(R))+" ft/ft" :LP$=A$
                   :X=32 :GOSUB 1390
                   :REM-- Print Tab(31) Slope of Pipe
 1050           IF LIFSUM<100 THEN
                   PRINT TAB(55) USING "##"; LIFSUM;
                :LPRINT TAB(55) USING "##"' LIFSUM
```

```
                 ELSE
                 PRINT TAB(54) USING "###"' LIFSUM
              :LPRINT TAB (54) USING "###"; LIFSUM
1060          A$=" Year Life AZ + " :LP$=A$
                 :X=57 :GOSUB 1390
                 :PRINT TAB(73) USING "#.###"; AZSUM;
                 :IF P$="Y" THEN LPRILNT TAB(73) USING "#.###"; AZSUM;
                 :REM-- Print Sum of AZ increments
1070          X=79 :GOSUB 1380 :REM-- Vertical line
1080          IF R<NR THEN
                 T$=STRING$(76,43)
                 :A$="|+T$+"|" :LP$="!"+T$+"!"
                 :X=2 :GOSUB 1390
                 :REM-- Print Increment End Line IF NOT last Reach
1090 RETURN :REM ================ END SUBROUTINE ====================
1100 ==========PROJECT & ENGINEER's NAME, DATE, & PAGE # SUBROUTINE=======
1110         A$="."+STRING$(76,45)+"." :LP$=A$
                :X=2 :GOSUB 1390
                :REM-- Top dotted line
1120         X=2 :GOSUB 1380 :REM-- Vertical bar
1130            A$=SPACE$((76-LEN(PRNAM$))/2)+PRNAM$ :LP$=A$
                :X=3 :GOSUB1390
                :REM-- Project Name
1140         X-79 :GOSUB 1380 :REM-- Vertical bar
1150         T$=STRING$(76,45)
                :A$="|"+T$"|" :LP"!"+T$+"!"
                :X=2 :GOSUB 1390
                :REM-- Title Divider Line
1160            N$=SPACE$(FS)+ENAM$+SPACE$(FS)
            :D$=SPACE$(FS)+DAT$+SPACE$(FS)
            :PG$=SPACE$(FS)+PAG$
1170         A$="|"+N$"|"+D$+"|"+PG$ :LP$="!"+N$+"!"+D$+"!"+PG$
                :X=2 :GOSUB 1390
                :REM== Engineer's Name, Date, & Page #
1180         X=79 :GOSUB 1380 :REM-- Vertical bar
1190         T$=STRING$(76,45)
                :A$="'"+T$+"'" :LP$="'"+T$+"'"
                :X=2 :GOSUB 1390
                :REM-- Bottom of heading line
1200 RETURN :REM ============END SUBROUTINE =================
1210 REM ======= PRINTING HEADING SUBROUTINE =================
1220 A$"."+STRING$(76,45)+"." :LP$=A$ :X=2 :GOSUB 1390
:REM-- Top dotted line
1230                          A$=
"|       |REL   |FLOW    |FLOW     |SULFIDE    |UP.STRM    |DN.STRM      |AZ       |"
              :LP$=
"!       !REL   !FLOW    !FLOW     ! SULFIDE   ! UP.STRM   ! DN.STRM     ! AZ      !"
              :GOSUB 1390
              :REM-- 2nd line of heading
1240         TI$="RCH,INC" :T2= " RCH "
              :A$="|" :LP$="!"
              :IF NI >1 THEN A$=A$+T1$ :LP$+T1$
              ELSE
              A$=A$+T2$ :LP$=LP$+T2$
1250         A$=A$+
   "|DEPTH|QNITY   |VLCTY     |LIMIT      |SULF S1     |SULF S2     |FACTOR    |"
              :LP$=LP$+
   "! DEPTH !QNITY  !VLCTY     !LIMIT      !SULF S1     !SULF S2     !FACTOR    !"
              :GOSUB 1390
              :REM-- 3rd line
1260         A$=
"|            |(d/DIA)  |(cfs)     |(fps)      |(mg/1)      |(mg/1)     |(mg/1)|   |"
```

```
             :LP$=
"!           ! (d/DIA)   !(cfs)     !(fps)     !(mg/1)     !(mg/1)     !(mg/1)     !"
             :GOSUB 1390
             :REM— 4th line
1270       T$=STRING(76,61)
             :A$=" | "+T$" | " :LP$="!"+T$+"!"
             :GOSUB 1390
             :REM— Bottom dotted line
1280 RETURN :REM =============END SUBROUTINE==================
1290 REM ========= SUBROUTINE FOR CENTERING LIFE INCREMENTS =======
1300       REM— X will be DIMensioned automatically to default value of 10,
             but maximum usage will only be X(5). (Because the maximum
             number of LIFE INC's per line is 5.)
1310       X(1)=36 :RETURN
             :REM— 1 Increment per line
1320       X(1)=22 :X(2)=50 :RETURN
             :REM— 2 Increments per line
1330       X(1)=15 :X(2)=36 :X(3)=57 :RETURN
             :REM— 3 Increments per line
1340       X(1)=10 :X(2) =27 :X(3)=45 X(4)=62 :RETURN
             :REM— 4 Increments per line
1350       X(1)=6 :X(2)=21 :X(3)=36 :X(4)=51 :X(5)=66 :RETURN
             :REM— 5 Increments per line
1360 ===========END LIFE INCREMENT SUBROUTINES =======
1370 REM ========= PRINT & LPRINT SUBROUTINE =============
1380       A$=" | " :LP$="!"
             :REM— GOSUB here to print vertical bar
1390       PRINT TAB(X) A$;
             :IF P$="Y" THEN LPRINT TAB(X) LP$;
             :REM— GOSUB here to print anything else
1400 RETURN :REM ========= END SUBROUTINE ===============
1410 END
```

THE FOLLOWING NOTES APPLY TO SUBPROGRAM "CLC.BAS":

To use a coefficient other than 0.001 in the Force Main sulfide equation, change the constant .01312 in statement 340.

To use the "conservative" coefficients in the Pomeroy-Parkhurst gravity sulfide prediction model, change .00052 to .00078 in statement number 400 and the constant 1.15 to 1.73 in statement number 440. Coefficients SHOULD BE VERIFIED by in- line tests on local operating sewer lines, when a significant fraction of the sewage is from other than domestic sources.

To change the MINIMUM value of Pomeroy's Z criteria (5000) or MAXIMUM slope (.006) beyond which sulfide build-up will not occur (will not be recognized by the program even if the model predicts it), change the limits in statement number 430.

To apply a CCF (Crown corrosion factor) other than 2.0, change statement # 490.

See discussion of program operation and example problems in Chap. 6.

Index